Advantages and Pitfalls of Pattern Recognition

Computational Geophysics Series

Volume 3

Advantages and Pitfalls of Pattern Recognition

Selected Cases in Geophysics

Horst Langer

Susanna Falsaperla

Conny Hammer

Series Editor

Viacheslav Spichak

ELSEVIER

Elsevier
Radarweg 29, PO Box 211, 1000 AE Amsterdam, Netherlands
The Boulevard, Langford Lane, Kidlington, Oxford OX5 1GB, United Kingdom
50 Hampshire Street, 5th Floor, Cambridge, MA 02139, United States

Notices
Knowledge and best practice in this field are constantly changing. As new research and experience broaden our understanding, changes in research methods, professional practices, or medical treatment may become necessary.

Practitioners and researchers must always rely on their own experience and knowledge in evaluating and using any information, methods, compounds, or experiments described herein. In using such information or methods they should be mindful of their own safety and the safety of others, including parties for whom they have a professional responsibility.

To the fullest extent of the law, neither the Publisher nor the authors, contributors, or editors, assume any liability for any injury and/or damage to persons or property as a matter of products liability, negligence or otherwise, or from any use or operation of any methods, products, instructions, or ideas contained in the material herein.

Library of Congress Cataloging-in-Publication Data
A catalog record for this book is available from the Library of Congress

British Library Cataloguing-in-Publication Data
A catalogue record for this book is available from the British Library

ISBN: 978-0-12-811842-9

For information on all Elsevier publications visit our website at
https://www.elsevier.com/books-and-journals

Publisher: Candice Janco
Acquisition Editor: Amy Shapiro
Editorial Project Manager: Samantha Allard
Production Project Manager: Prem Kumar Kaliamoorthi
Cover Designer: Mark Rogers

Typeset by TNQ Technologies

Contents

Preface .... *xi*
Acknowledgments .... *xvii*

Part I: From data to methods

Chapter 1: Patterns, objects, and features *3*
1.1 Objects and patterns 3
1.2 Features 3
1.2.1 Types 3
1.2.2 Feature vectors 4
1.2.3 Feature extraction 5
1.2.3.1 Delineating segments 5
1.2.3.2 Delineating regions 8
1.2.4 Transformations 10
1.2.4.1 Karhunen–Loève transformation (Principal Component Analysis) 10
1.2.4.2 Independent Component Analysis 12
1.2.4.3 Fourier transform 13
1.2.4.4 Short-time Fourier transform and spectrograms 14
1.2.4.5 Discrete wavelet transforms 15
1.2.5 Standardization, normalization, and other preprocessing steps 21
1.2.5.1 Comments 21
1.2.5.2 Outlier removal 22
1.2.5.3 Missing data 22
1.2.6 Curse of dimensionality 23
1.2.7 Feature selection 24
Appendix 1 Basic notions on statistics 27
A1.1 Statistical parameters of an ensemble 27
A1.2 Distinction of ensembles 31

Chapter 2: Supervised learning *33*
2.1 Introduction 33
2.2 Discriminant analysis 35
2.2.1 Test ban treaty—some history 35
2.2.2 The M_S–m_b criterion for nuclear test identification 35

2.2.3 Linear Discriminant Analysis 36
2.3 The linear perceptron 40
2.4 Solving the XOR problem: classification using multilayer perceptrons (MLPs) 46
2.4.1 Nonlinear perceptrons 48
2.5 Support vector machines (SVMs) 53
2.5.1 Linear SVM 53
2.5.2 Nonlinear SVM, kernels 56
2.6 Hidden Markov Models (HMMs)/sequential data 59
2.6.1 Background—from patterns and classes to sequences and processes 59
2.6.2 The three problems of HMMs 62
2.6.3 Including prior knowledge/model dimensions and topology 67
2.6.4 Extension to conditional random fields 68
2.7 Bayesian networks 69
Appendix 2 73
Appendix 2.1 Fisher's linear discriminant analysis 73
Appendix 2.2 The perceptron 76
Appendix 2.3 SVM optimization of the margins 78
Appendix 2.4 Hidden Markov models 80
Appendix 2.4.1 Evaluation 80
Appendix 2.4.2 Decoding—the Viterbi algorithm 82
Appendix 2.4.3 Training—the expectation−maximization /Baum−Welch algorithm 83

Chapter 3: Unsupervised learning 87

3.1 Introduction 87
3.1.1 Metrics of (dis)similarity 88
3.1.2 Clustering 90
3.1.2.1 Partitioning clustering 91
3.1.2.2 Hierarchical clustering 104
3.1.2.3 Density-based clustering 109
3.2 Self-Organizing Maps 114
3.2.1 Training of an SOM 117
Appendix 3 119
Appendix 3.1 Analysis of variance (ANOVA) 119
Appendix 3.2 Minimum distance property for the determinant criterion 120
Appendix 3.3 SOM quality 121

Part II: Example applications

Chapter 4: Applications of supervised learning 127

4.1 Introduction 127
4.2 Classification of seismic waveforms recorded on volcanoes 128
4.2.1 Signal classification of explosion quakes at Stromboli 132

4.2.2 Cross-validation issues 137
4.3 Infrasound classification 139
4.3.1 Infrasound monitoring at Mt Etna—classification with SVM 141
4.4 SVM classification of rocks 144
4.5 Inversion with MLP 151
4.5.1 Identification of parameters governing seismic waveforms 151
4.5.2 Integrated inversion of geophysical data 152
4.6 MLP in regression and interpolation 158
4.7 Regression with SVM 163
4.7.1 Background 163
4.7.2 Brief considerations on pros and cons of SVM and MLP in regression problems 167
4.8 Classification by hidden Markov models and dynamic Bayesian networks: application to seismic waveforms of tectonic, volcanic and lunar origin 168
4.8.1 Background 168
4.8.2 Signals related to volcanic and tectonic activity 169
4.8.3 Classification of icequake and nonterrestrial seismic waveforms as base for further research —HMM 174
4.8.3.1 Icequakes 174
4.8.3.2 Moon quakes 176
4.8.3.3 Classification of seismic waveforms using dynamic Bayesian networks 179
4.9 Natural hazard analyses—HMMs and BNs 180
4.9.1 Estimating volcanic unrest 180
4.9.2 Reasoning under uncertainty—tsunami early warning tasks 181
Appendix 4.1 Normalization issues 183
Appendix 4.2 SVM Regression 184
Appendix 4.3 Bias—Variance Trade-off in Curve Fitting 185

Chapter 5: Applications with unsupervised learning 189
5.1 Introduction 189
5.2 Cluster analysis of volcanic tremor data 191
5.3 Density based clustering 197
5.4 Climate zones 203
5.5 Monitoring spectral characteristics of seismic signals and volcano alert 213
5.6 Directional features 222
Appendix 5 230
Appendix 5.1 Davies-Bouldin index 230
Appendix 5.2 Dunn index 231
Appendix 5.3 Silhouette index 232
Appendix 5.4 Gap index 232
Appendix 5.5 Variation of information 233

Part III: A posteriori analysis

Chapter 6: A posteriori analyses—advantages and pitfalls of pattern recognition techniques 237
6.1 Introduction 237
6.2 Testing issues 238
6.3 Measuring error 239
6.4 Targets 246
6.5 Objects 251
6.6 Features and metrics 253
6.7 Concluding remarks 257
6.7.1 Multilayer perceptrons 257
6.7.2 Support Vector Machines 258
6.7.3 MLP and SVM in regression analysis 258
6.7.4 Hidden Markov models and Bayesian networks 258
6.7.5 Supervised and unsupervised learning 259

Chapter 7: Software manuals 261
7.1 Example scripts related to Chapter 2 261
7.1.1 Linear discrimination, principal components, and marginal distributions 261
7.1.2 The perceptron 267
7.1.3 Support Vector Machines 268
7.1.4 HMM example routines 272
7.2 Example scripts and programs related to Chapter 3 (unsupervised learning) 275
7.2.1 K-means clustering 275
7.2.2 Mixed models 275
7.2.3 Expectation maximization clusters 278
7.2.4 Fuzzy clustering 279
7.2.5 Hierarchical clustering 280
7.2.6 Density-based clustering 280
7.2.7 Unsupervised learning toolbox: KKAnalysis 282
7.2.7.1 Preliminaries 282
7.2.7.2 Installation 283
7.2.7.3 Files 283
7.2.7.4 Getting started 285
7.2.7.5 Configuring KKAnalysis—the "settings" 296
7.3 Programs related to applications (Chapter 4) 300
7.3.1 Back propagation neural network (BPNN) 300
7.3.2 SVM library 302

7.4 Miscellaneous 307
7.4.1 DMGA—generating ground deformation, magnetic and gravity data 307
7.4.2 Treating fault plane solution data........ 311

Bibliography........ 315

Index........ 327

For additional information on the topics covered in the book, visit the companion site:
https://www.elsevier.com/books-and-journals/book-companion/9780128118429

Preface

The digital era has caused an outstanding change in the acquisition of information concerning our planet. We are accustomed to an uninterrupted monitoring by means of satellite imagery, measurements of ground deformation in the context of geodynamical studies, seismic and geochemical data acquisition, etc. Continuous data acquisition in Earth sciences, in general, and geophysics, in particular, leads to the accumulation of a huge amount of information. Terabytes and Terabytes of data pile up in digital archives over short times. Often, we are left without a key to these archives, which turn them into "data graves," containing precious information difficult to unearth. In addition, many geological processes are slow phenomena, the study of which comes along with the need to cover time spans as long as possible. Therefore, the necessity to "unearth" old archives becomes of paramount importance.

Data collection usually brings along considerable technological effort and cost, which must be balanced by the profit gained from the information acquired. In some way, data are the background of actions and decisions, once we have understood the relations between our observations and the processes we are interested in. However, how can we establish these relations? Looking at a long sequence of numbers is useless unless we extract parameters useful for our task of drawing conclusions and take action, if necessary. Sometimes decisions have to be fast, as in the case of an impending natural threat, such as a volcanic eruption or the development of a thunderstorm, for which efficient data handling and interpretation are mandatory. Near real-time processing and the application of automatic procedures to support decision-making are strongly desired. Apart from that, an important aspect, especially in Earth sciences, is the reanalysis of archives. Old data broaden our knowledge concerning the characteristics of natural phenomena and their development with time. The use of long time spans helps us improve the significance of our conclusions and decisions. No need to mention that postprocessing of huge data masses accumulated over long time spans requires patience and/or highly automatized processing schemes for the extraction of essential information, avoiding that a user feels overwhelmed and gives up the task.

A propaedeutic step in pattern recognition is the definition of "objects" related to the phenomena we deal with. A single value in a time series does not constitute an object or pattern; a seismic phase is indeed a pattern. Weather is another example for an object. Observed in a given time span, it can be understood as a pattern described by various values, which refer to temperature, humidity, cloud coverage, etc. In meteorology, we may distinguish patterns characterized by strong cloud coverage, high temperature, and humidity, being possible precursors of a thunderstorm. Similarly, the susceptibility of an area prone to slope failure can be characterized by different parameters, such as precipitation, geographic relief, and subsurface structure. In geology, we can study the object "rock" and describe it by a number of components, for instance, key minerals. In the famous "Streckeisen" diagram, we use the minerals quartz, feldspar (Orthoclase, Plagioclase), and feldspathoids. The "TAS" scheme is another example for the description of the object "rock." We then aim at the identification of structures within our population of patterns and refer to these structures as classes. For example, the rock "granite" is identified on the base of its position in the "Streckeisen" diagram; it forms during collision and subduction of tectonic plates.

When handling a multivariate dataset, we face specific problems in statistical treatment and graphical representation of results. Conventional 2D graphs, where one component is plotted versus another, can be nicely displayed on a sheet of paper, but we have to choose among a large number of possible combinations. In theory, having n components, we should design $n*(n-1)/2$ bidimensional graphs. Even plotting all possible graphs may be insufficient, as the components cannot be supposed to be independent from each other. Thus, the identification of homogeneous data groups and heterogeneities between them may not be possible in any of the plots. We must also be aware of the possibility that a component may provide key information only for a limited number of patterns rather than for the whole ensemble. The conventional 2D graphs represent so-called "marginal distributions," and the problems aforementioned with this kind of representation are well known in multivariate analysis. A solution can be found in techniques of pattern classification.

Pattern recognition can be understood as an element of classification, that is, the process of assigning objects to a category or class. Objects are characterized by a number of features, which can be metrical, ordinal, or nominal data. The set of—in our case numerical—features forms a feature vector, the so-called pattern. The choice of the features essentially depends on practical considerations and is governed by two rules: (i) the features should allow us to identify objects uniquely, and (ii) the smaller the feature vector, the better. Sometimes features can be gained directly from the description of the object, but frequently the preprocessing of data, such as the normalization of numerical values, is necessary to meet these goals properly.

The task of classification is tackled following various strategies. Perhaps, the oldest one is the so-called genetic classification, in which the origin (or the cause) of an object is considered. In climatology, three types of genetic classification may be distinguished based on (1) geographic determinants of climate, (2) surface energy budget, and (3) air mass analysis. In geology, we distinguish between sedimentary and igneous rocks based on their genesis. Igneous rocks can be further divided into volcanic and plutonic rocks, again considering the process at the base of their formation; meanwhile, the mineralogical composition is irrelevant in this distinction. In seismology, we often use the source of a seismic signal as a criterion of classification, for example distinguishing seismic noise from the seismic radiation generated by magma dynamics inside a volcano (so-called volcanic tremor) or the record of a nuclear explosion from a trace recorded in case of an earthquake.

Supervised classification can be understood as the inverse process of genetic classification. First, we see an object and wonder where it comes from. From the 1960s on, intense research has been carried out on the distinction between nuclear explosions and other events, such as earthquakes, chemical explosions, and quarry blasts. The object we have in hand may be a seismic record (or a number of seismic records) and we ask to infer the origin from its characteristics. In geology, we may ask whether we can trace back rock characteristics to an origin—for instance, the provenience of a volcanic product from an eruptive center. Supervised classification uses a priori information inferred from example objects, supposing to know which class they belong to. In modern techniques of supervised classification, this can be achieved without (or with very limited) a priori definition of similarity.

Supervised classifiers typically use an iterative procedure, which tries to find a mathematical formalism to reproduce the expert's way of assigning a class membership to a pattern. The iterative process is often called as training or learning phase of the classifier. Besides, parameters governing operational characteristics of the classifiers have to be identified either by trial and error or by optimization procedures, such as genetic algorithms. With advanced supervised pattern classification methods, for instance support vector machines (SVM) or the multilayer perceptron (MLP), the mathematical structure of the classifier can be, in principle, arbitrarily complex. This brings the huge advantage of generality, in the sense that there is no limitation in the typology of the discrimination function delimiting the classes from each other. Important in this context is that there are enough examples to learn from.

On the contrary, unsupervised classification is based on a suitable definition of similarity between patterns rather than on a priori knowledge of their class membership. The task of unsupervised classification can be formulated as finding groups with a minimum degree of heterogeneity, being most distant from each other. The degree of heterogeneity is defined as a distance measure, or metric, for example, the Euclidean distance, the Mahalanobis distance, Manhattan (or city block distance), etc. Unsupervised classification is preferred when the definition of targets is difficult. Of course, the targets may not be known, and perhaps are only identifiable after a thorough study of structures within the dataset. In other cases, the relation between pattern characteristics and target undergoes changes throughout the dataset. Unsupervised classification is often addressed to as "clustering," that is, the identification of data groups with similar characteristics—where similarity is defined on the base of an a priori defined metrics. The shape of clusters can be spherical, elliptical, or hyperbolic, but also clusters with very irregular shapes can be found. Unsupervised techniques are able to "detect" structures in data, even though their description is not straightforward at first glance. All these characteristics resemble to principles of human cognition, such as learning a language, identifying a character even being hand-written, recognizing similarities among objects, and analyzing interrelations between patterns.

Therefore, there is a strong link with artificial intelligence—the science where we make automates do similar things as human brains do.

Beside grouping single objects, we may also be interested in their interrelation. This aspect is important not only in pattern recognition, but also in forecast. In geophysics, a sequence of objects may be related to the development of some phenomenon, being it a typhoon, a flood, or a volcanic unrest. This implies that a specific pattern is meaningful not only for the components making up its feature vectors, but also for the context defined by other patterns. This is also a typical problem in speech and text analysis, where the meaning of a word or a number not only depends on single characters or digits, but also on their order. Techniques regarding this aspect of related objects can be addressed to as context-dependent methods. Among them, we shall discuss hidden Markov models and (dynamic) Bayesian Networks in more detail.

Pattern recognition is strongly related to data mining, revealing structures within datasets and facilitating the set-up of rules for decisions and actions. The techniques described in this book are based on mathematically formulated procedures; their results are therefore reproducible. Their implementation on modern computers allows us the automatic processing of large amounts of information, which implies a strong relation with the field of machine learning.

The book presented here comprises both the theoretical background of pattern recognition methods as well as practical suggestions for their application. An important aspect is the proper formulation of the classification problem, which implies an appropriate definition of the objects and their description by features. Chapter 1 deals with objects, features, and metrics. Chapter 2 presents the theoretical background of supervised learning. It begins with a simple discrimination problem—the distinction of earthquakes and nuclear explosions on the base of surface and body wave magnitudes. Principles of more advanced techniques, such as the MLP and SVM are demonstrated for that simple case. Hidden Markov models and (dynamic) Bayesian networks are context-based methods, where both features of single objects as well as their interrelation are of interest. Chapter 3 outlines the concepts of unsupervised learning, among these various approaches of clustering as well as Self-Organizing Maps, which are a popular technique of vector quantization. Chapter 4 and 5 present applications of supervised and unsupervised learning partly taken from the literature, partly collected during research projects of the authors themselves. A number of examples regards Mt Etna volcano (Italy), which is often addressed to as a "volcano laboratory" for its persistent activity, favorable logistic conditions that allow the deployment of cutting-edge equipment for multidisciplinary measurements, and long tradition of monitoring also for surveillance purposes. Beside already existing and published material, the authors also present new applications to underscore the potential use of pattern classification techniques in geophysics as well as in a wide field of disciplines. Chapter 6 deals with a critical a posteriori analysis of pattern recognition results, which goes beyond the simple enumeration of some error. This chapter offers keys to answer questions such as "what can be expected from the pattern recognition techniques? Is a somewhat unsatisfying result a failure of the method or there are lessons to learn regarding the problem?." A further, crucial point addressed to is how clustering quality can be measured.

Finally, the book comes along with example programs and datasets. Most of programs are MATLAB™ scripts, which allow the user the reproduction of some of the figures in the book. Besides, there are ready-to-use packages regarding MLP, SVM, and unsupervised learning, in particular clustering and Self-Organizing Maps. The computer codes should allow the reader to perform experiments both using the delivered datasets as well as their own data.

Acknowledgments

This book would not have been written without the advice and help of many colleagues. The authors wish to thank Prof. Giuseppe Nunnari (University of Catania), Prof. Luigi Occhipinti (now University of Cambridge), and Prof. Giovanni Muscato (University of Catania). They guided the first steps of Susanna Falsaperla and Horst Langer in the realm of artificial neural networks, applied to supervised learning. Conny Hammer profited from the advice of supervisors and colleagues, in particular Dr. Matthias Ohrnberger, Dr. Kristin Vogel (University of Potsdam), Dr. Moritz Beyreuther (University of Munich), and Prof. Donat Faeh (Swiss Seismological Service, SED, Zurich). Alfio Messina (INGV—Sezione Roma 2) did much of the "dirty work" preparing the computer codes. In particular, he did most of the coding of the KKAnalysis package for unsupervised learning that comes along with this book. We are grateful to Prof. Chih-Jen (National Taiwan University) for the permission to exploit the LIBSVM library regarding the Support Vector Machines. Authors wish to thank Marcello D'Agostino and Dr. Danilo Reitano (INGV-Osservatorio Etneo, Sezione di Catania) for the technical assistance during the implementation of the online processing of volcanic tremor recorded at Mt. Etna, and Alfio Amantia (INGV—Osservatorio Etneo, Sezione di Catania) for the beautiful photographs of Stromboli volcano. Dr. Rosa Anna Corsaro (INGV —Osservatorio Etneo, Sezione di Catania) provided us the data regarding the geochemical composition of rock samples collected on Mt. Etna. Hypocenter locations of earthquakes on Mt. Etna were passed to us by Dr. Tiziana Tuvè (INGV—Osservatorio Etneo, Sezione di Catania). Besides, we wish to thank Gilian Foulger (Durham University), Jakob Zscheischler (University of Bern), and Hans Chen (Lund University) for their permission of using figures of their work, which were very helpful for the presentation of the concepts discussed in this book.

We sincerely wish to thank the Elsevier staff, in particular Mrs. Marisa La Fleur, Katerina Zaliva, and Samantha Allard who assisted us in all technical aspects during the preparation of the book. Last but not least, we thank the Editor of the Elsevier book series on "Computational Geophysics," Prof. Viacheslav Spichak (Russian Academy of Science) who carefully went through the manuscript. His suggestions and comments were of primary importance regarding the quality of our work.

PART I

From data to methods

CHAPTER 1

Patterns, objects, and features

1.1 Objects and patterns

Before entering into the details of pattern recognition, we should define basic terms. The first question is "What is a pattern?" After careful reasoning, we become aware that this term encompasses a variety of phenomena. Patterns are described by a variety of characteristics, which can be of various types. For instance, how do we describe a fine summer day? Certainly, we expect a high temperature, a low degree of humidity, sunshine, no or little clouds, and no rain. That is, we describe the pattern "fine day" by a series of parameters. In a similar way, we characterize a rock sample by its chemical or mineralogical composition, along with other parameters (e.g., density, strength, etc.). Defining a pattern that way, we can also use the term "object," which is described by "features" and "feature vectors." In this context, we carry out pattern recognition or classification considering the single objects independent of each other. In other words, an object K will be assigned to class M only on the basis of its proper characteristics described in the feature vectors.

However, the definition of patterns may go beyond this. Suppose we succeed to identify a few objects as pattern '9', '7', '0', and '6'. Now consider the pattern 9706. The role of each cipher critically depends on its context—the position where it is found—and the pattern depends on the sequence of objects. In applications like speech recognition or text analysis, this type of context-dependent classification is a critical issue. Instead of dealing with objects described by a single feature vector, we face with frames of those vectors, and the class a pattern belongs to critically depends both on the single feature vectors—for instance, the feature vector for '9'—and the order of the vectors.

1.2 Features

1.2.1 Types

The description of objects uses data of different types. For example, we can describe a person by height, weight, and age—numerical data. Then we may add other information, such as fat, normal, and slim—which are ordinal data. We may also add categorical information, for instance, blond, brown, or black haired. All these different data types may entail specific steps of processing in order to bring them into a form appropriate for our purposes.

Advantages and Pitfalls of Pattern Recognition. https://doi.org/10.1016/B978-0-12-811842-9.00001-7

Ordinal data can be ranked, that is we can treat them in some way as "low", "middle", or "high". The "Beaufort scale" is a ranked measure of wind speed, based on the effects observed. For instance, a hurricane can disrupt trees, whereas a smooth breeze just creates small waves at the sea surface. In earthquake seismology, the intensity of an earthquake in the famous "Mercalli scale" is specified considering its effects on buildings or the behavior of the populations. Both Beaufort and Mercalli scales are typical examples of ranked values, and their relation to numerical parameters—especially with regard to the Mercalli scale—is still an unresolved question. On the other hand, we may decide to define metrics allowing to handle this type of data. For instance, we may compare their ranks, as done in Spearman's rank correlation.

In categorical data, one may use some type of coding, creating a vector with '1', when a feature corresponds to a category (e.g., a blond-haired person), and '0', when the feature does not belong to that category (not brown- or black-haired person). Then a blond receives the code (1,0,0), a brown-haired person has the code (0,1,0), and black-haired people are coded (0,0,1). Having multivariate categorical feature vectors, we may consider the number of coincidences, creating tables of contingency. For instance, besides the hair, we may consider the eye color ('blue', 'green or gray', 'brown') and exploit it in defining specific metrics such as the "Tanimoto distance".

Here we shall focus on numerical features, which are the most common ones in geophysics and the most suitable for the type of analyses discussed. However, recall that the aim of pattern recognition resides in assigning the patterns to a class. Thus at the end of our application, we transform our, often numerical, features to categories! In our applications for clustering, we shall try to identify groups of patterns considering numerical features, but obtain classes 'A', 'B', and 'C', which do not often have numerical meaning. In other applications, we find that patterns belong to a certain, a priori defined, class, such as a type of rock, a meteorological phenomenon, a signal originating from a type of source. Again, these classes are not specified numerically, but from a verbal description, for which we may use some coding as mentioned above.

1.2.2 Feature vectors

The choice of features is essentially based on effectiveness. First of all, features should describe the object in a way that it can be identified and distinguished without ambiguity. As a single feature is often not sufficient for this purpose, a number of features stored in a feature vector are taken into account. Augmenting the number of components of the feature vector, we increase the probability that we can identify our object without confusing it with others. Nonetheless, the dimension of the feature vector and the choice of the components are a tricky issue. First, it is wise to limit the dimensionality of the feature vector for computational issues. It is a general rule of common sense that the robustness of the classifier increases if the number of free parameters is low compared to the number of training patterns. Care has to be taken when choosing the components of the feature vector. All the components should be closely related to the core problem of classification we intend to resolve. For instance, for classifying the composition of a rock sample, the geographical coordinates of the place where it was found are probably irrelevant, but may disturb the discrimination as they are included in the calculation of the discrimination function.

A further issue is the appropriate aggregation of feature vectors in the case of multiparametric and multidisciplinary data. In multidisciplinary analyses, it may be tempting just to combine features of varying origin, for instance, seismic and infrasound signals recorded on a volcano. However, we may run into problems if these signals are not generated by the same source. Our feature choice therefore should include a sound reasoning about the physical processes, which relate our data to a phenomenon.

Culling together multidisciplinary features entails the problem of biases introduced by the differences in the number of components. For example, suppose to have 20 features related to seismic data, but only three related to deformation data; then changes in the first ones will have a stronger effect than those caused by the latter. This happens because the number of feature vectors $||\boldsymbol{x}||^2$ is obtained from the squared sum of the normalized x_i. Thus, the 20 features of seismic data are likely to outweigh the three features of the deformation data.

1.2.3 Feature extraction

1.2.3.1 Delineating segments

In some applications, we may gain features from the data quite directly, such as in the case of weather conditions (temperature during day and night, number of sunshine hours, etc.). In other applications, the direct use of our available data is not possible. Those cases are represented by time series or images, where we face a sequence of samples, which when taken alone do not have a particular meaning. When dealing with waveforms, such as seismograms, infrasound recordings, or ground deformation signals, we have to identify segments of the data set we can relate to an event that forms our object of interest. In the framework of the CTBT (Comprehensive Test Ban Treaty), we look at seismograms, hydro-acoustic, or infrasound signals to recognize earthquakes and explosions. In ground deformation records, we may be interested in signs left behind by processes of rupturing or ground failure.

Looking at waveforms, the first question we pose is "Was there an event?" Questions like this can be tackled by identifying "local" features, which help the identification of segments in the data set containing events of interest. In time series for instance, we shall be interested in the "phase" in a wide sense. In image processing, we may focus on areas with a specific aspect that can be assigned to a given category, such as a forest, a cornfield, lakes, and rivers. Feature extraction in this context consists of browsing through the whole data set, revealing parameters that depend on a few samples in the small region where we focus on.

From intuition, the answer to the question "Was there an earthquake?" can be answered looking at time series where wave trains have amplitudes significantly higher than the background signal observed over long times. In many applications, such as in seismology, we are not able to wait for the whole seismogram. For example, shortage of storage capacity on our recording system leads to the idea to start recording only when we have clear signs that there is an event interesting for us. As signals are often sampled with a high frequency (such as hundreds of samples per second), parameters or features have to be identified in order to recognize the time of the onset of an event automatically. A classical trigger criterion is the STA/LTA parameter. The STA ("short-time average") is calculated from the mean squared sum of seismic amplitudes over a short-time window, say 1 s, whereas the LTA ("long-time average") is a measure of the background noise. It is obtained in the same way as the STA, but considering longer time windows, for instance a minute (Fig. 1.1).

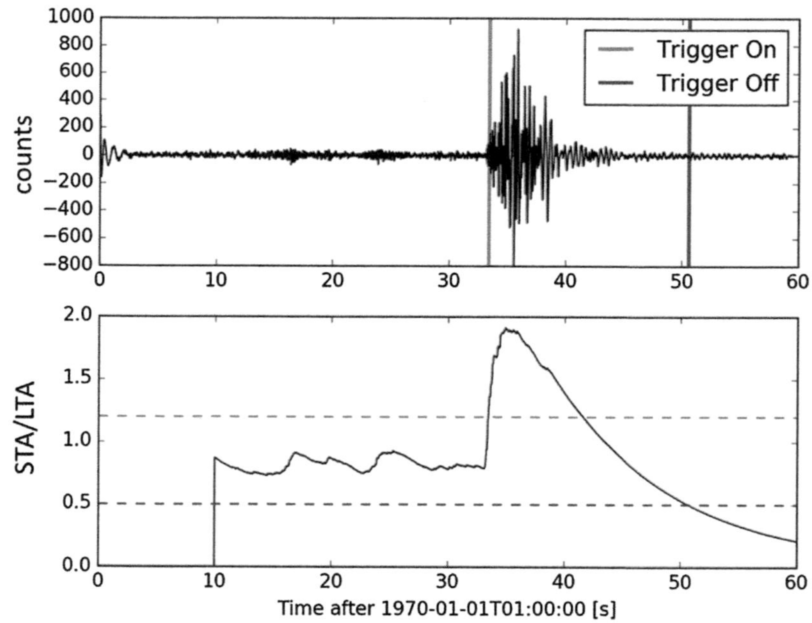

Figure 1.1
Recognition of an event on a seismogram. The "trigger on" line marks the arrival time of the P wave.

The so-called characteristic functions are applied in automatic reading of first arrivals ("P-wave picking"). P-waves are compressional waves. As they travel with higher velocities than other types of waves, they will be the first to be recorded on a seismogram. Typically, their arrival times are easy to identify with the best precision, as the seismogram is not yet blurred by other phases, being shear waves, surface waves, or caused by effects of scattering in a wide sense.

Crampin and Fyfe (1974) defined a set of functions, i.e.,

$$f_l(k) = |x_k - x_{k-l}| \tag{1.1}$$

with x_k being the k-th sample measured on a seismogram and l being a time shift, such as four or eight in their paper. Stewart (1977) proposed a modified characteristic function *f(k)* in the following way

$$d_k = x_k - x_{k-1} \tag{1.2a}$$

$$f(k) = d_{k-1} \text{ if } g(k) \neq g(k-1) \tag{1.2b}$$

$$f(k) = d_k \text{ if } g(k) = g(k-1) \text{ and } h_k = 8 \tag{1.2c}$$

$$f(k) = d_k + d_{k-1} \text{ if } g(k) = g(k-1) \text{ and } h_k \neq 8 \tag{1.2d}$$

with $g(k) = 1$ for $d_k \geq 0$ else $g(k) = -1$ and $h_k = |\Sigma_l g_k|$, l runs from k-7 to k. As shown in Fig. 1.2, this function is sensitive to changes with respect to spectral characteristics, which, besides mere amplitudes, are an important feature for the identification of a seismic phase.

We refer to the seismological literature on this topic for more details, for instance to Diehl and Kissling (2000) who give a detailed description of methods which account for the development of the "signal to noise" ratio; features are obtained from the rise of the signal

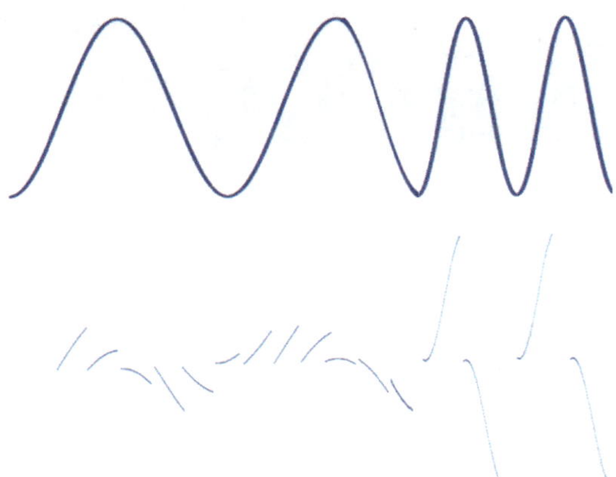

Figure 1.2

A chracteristic funcion obtained after fStewart (1977). The input signal is a sinusoid, where the frequency changes at a certain time, while peak amplitudes remain constant. The lower trace is the response of Stewart's function. We can define an amplitude threshold where we "declare" the arrival of a phase.

and the slope ("ASNR", Amplitude Signal to Noise Ratio approach). In the "FSNR" approach (Frequency Signal to Noise Ratio), changes in the frequency content are taken into account.

1.2.3.2 Delineating regions

In image processing, we rarely use the field of pixels without any preprocessing step. Often we shall analyze an image with respect to a "texture", which can be in the form of "brick", "checker board", "grass", "leaves", etc. In Fig. 1.3, we show simple examples of textures made up of 50% white and black pixels. We recognize the differences between the various images based on the order of pixels.

Various parameters characterize the texture (see, e.g., Shapiro and Stockman, 2000). One of them is the so-called "cooccurrence matrix" for which a simple example is shown in Fig. 1.4.

This matrix considers the relation of neighborhood between pixels. For instance, in the 4×4 image of Fig. 1.4, we construct the matrix for neighborhood **C**(0,1), that is we look along the rows to the right. We find pairs (1,0) two times, thus the value for the position (1,0) in this matrix is 2, whereas it is 0 for position (0,1) as we do not find any '1' to the right of a '0'. Similarly, we may look along columns, where we find four times a '0' below a '0', two times a '0' below a '1', but zero times a '1' under a '0'. Finally, we consider the neighborhood along the diagonals and find the configuration (0,0) two times, (1,1) once, and (2,2) once. Note that we have been considering only direct neighborhoods. We can define similar matrices for larger distances among the pixels.

Figure 1.3
Example for different textures with 50% of white and black pixels.

(A)

1	1	0	0
1	1	0	0
0	0	2	2
0	0	2	2

(B)

i/j	0	1	2
0	4	0	2
1	2	2	0
2	0	0	2

(C)

I/j	0	1	2
0	4	0	2
1	2	2	0
2	0	0	2

(D)

i/j	0	1	2
0	2	0	1
1	2	1	1
2	0	0	1

Figure 1.4
Construction of the cooccurrence matrix (redrawn from Shapiro and Stockman, 2000). (A) 4×4 image, (B) configuration (0,1), (C) configuration (1,0), and (D) configuration (1,1).

Other variants of the cooccurrence matrix are considered. For instance, we may define

$$C_N(i,j) = C(i,j)/\Sigma_i\Sigma_j C(i,j) \tag{1.3}$$

where the $C_N(i,j)$ value fall into a range between 0 and 1. A further variant is the creation of a symmetric cooccurrence matrix $C_S = C_d(i,j) + C_{-d}(i,j)$, where d gives the length of the neighborhood interval (in our example above $d = 1$). With C_S, we consider configurations "right" or "left" together with "below" and "above".

The cooccurrence matrices are still cumbersome for further analysis, such as for comparing two textures. The derivation of numeric features offers a way out of this problem. For instance, we may define

$$\text{"Energy"} = C(i,j)/\Sigma_i\Sigma_j C_N(i,j)^2$$
$$\text{"Entropy"} = \Sigma_i\Sigma_j C_N(i,j)^2 log_2\, C_N(i,j)$$
$$\text{"Contrast"} = \Sigma_i\Sigma_j (i-j)^2 C_N(i,j)$$
$$\text{"Homogeneity"} = \Sigma_i\Sigma_j C_N(i-j)/(1+|i-j|)$$
$$\text{"Correlation"} = \Sigma_i\Sigma_j (i-\mu_i)(j-\mu_j) C_N(i,j)/\sigma_i\sigma_j$$

where μ_i, μ_j are the means over the rows and columns, and σ_i, σ_j are the corresponding standard deviations.

Laws (1980) proposed convolution masks for the texture-based feature generation. We start with the vectors

L5 (level) = (1,4,6,4,1)
E5 (edge) = (−1,−2,0,2,1)
S5 (spot) = (1,0,2,0,1)
R5 (ripple) = (1,−4,6,−4,1)

and create, for instance, the convolution mask E5L5 forming the product $\mathbf{E5}^T\mathbf{L5}$. As a result, we obtain a 5 × 5 convolution matrix, which is applied to the pixels of an image by carrying out a convolution in two dimensions.

$$\begin{vmatrix} -1 & -4 & -6 & -4 & -1 \\ -2 & -8 & -12 & -8 & -1 \\ 0 & 0 & 0 & 0 & 0 \\ 2 & 8 & 12 & 9 & 2 \\ 1 & 4 & 6 & 4 & 1 \end{vmatrix}$$

In the same way, we can form other combinations, such as E5E5, S5S5, R5R5, E5L5, S5L5, R5L5, S5E5, R5E5, and R5S5, which yield nine features suitable for the pattern recognition techniques discussed in the following chapters.

1.2.4 Transformations

The performance of any automatic recognition approach can be only as good as the data used to train it. Thus, the appropriate representation of the raw data provides the basis for a well-working classification system. In order to achieve this goal, we often must "translate" the raw data to a form, which is suitable for learning and allows a generalization for the given problem. The procedure of converting the data in this optimal format is known as feature transformation, at the end of which, the signal space is mapped into a transformed feature space. Viewing the data in this transformed space may allow us to recognize elements—separating functions, data clusters, etc.—that could not be seen before. The choice of the applied transformation is imposed by the specific problem and its practical implementation issues. It might also be necessary to apply several transformations consecutively.

The often-used Principal Component Analysis (PCA) (also called Karhunen–Loève transformation) allows us to represent any multidimensional feature vector to a new system of variables, which are uncorrelated to each other. In that way, redundant information is decreased while the discriminative information of individual components is exposed. The Independent Component Analysis (ICA) follows a strategy similar to PCA. Using ICA, we transform our data into variables independent of each other, which is a stronger condition than uncorrelated variables.

The Fourier transform—often applied in the analysis of (time) sequences—does not directly lead to a data reduction. We first map the raw data into the frequency domain where data reduction can be achieved more easily. By decomposing the raw signal into its individual frequency components, specific constituents not interesting for our problem can be eliminated, which renders the recognition easier. Variants of the Fourier transform, as the Short-Time Fourier Transform (STFT) and spectrograms, allow accounting for local fluctuations within the data set. The wavelet transforms follow a scope similar to the STFT using basis functions of finite length, which limits the effects of window margins.

1.2.4.1 Karhunen–Loève transformation (Principal Component Analysis)

Transforming our original data set to a new feature space, we aim at reduction of dimensionality. Therefore, many of the techniques are based on the creation of an orthogonal vector space, so that we have components that are independent of each other. That way we are able to omit components that are considered of minor relevance, as they contribute little to the total variability of the observations. A frequently applied technique is the PCA, also known as Karhunen–Loève transformation.

Looking at the bivariate Gaussian in Fig. 1.5, we may define a unique variable, which measures a principal axis along the major axes of the ellipse rather than measuring in two

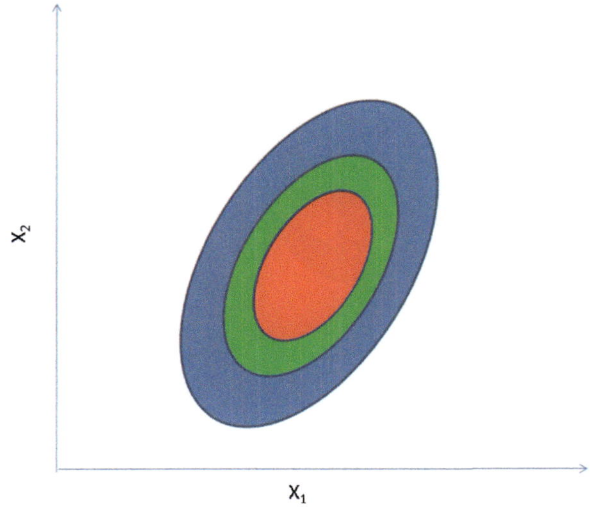

Figure 1.5
A bivariate Gaussian distribution.

directions X_1 and X_2. This new variable represents the majority of the dispersion, and at the same time, it reduces the dimensionality of the problem. We start with

$$y = A^t x \tag{1.4}$$

where both $E(y) = E(x) = \mathbf{0}$

and

$$E(yy^T) = E(A^T xx^T A) \tag{1.5}$$

(dispersion of y in terms of x).

The matrix xx^T is symmetric; hence, the eigenvectors are orthogonal. If A is chosen so that the columns correspond to the eigenvectors, we get $E(yy^T) = E(A^T xx^T)A = \Lambda$, which is a diagonal matrix with elements being the eigenvalues λ_l. We can achieve a reduction of dimensions by considering only a limited number of eigenvectors. These are not necessarily the largest ones as illustrated in Fig. 1.6.

Even though we are tempted at the first glance to use the eigenvector with the largest eigenvalue, as it explains the major part of the dispersion, it may not be the best choice for the separation. In Fig. 1.6, the separation of the two groups is more efficient using the eigenvector corresponding to a smaller eigenvalue. Nonetheless, there is a big advantage of using the Karhunen–Loève transformed data: after the transformation, linear separating elements can be recognized in the marginal distribution, whereas they may be blurred in the original system of axes.

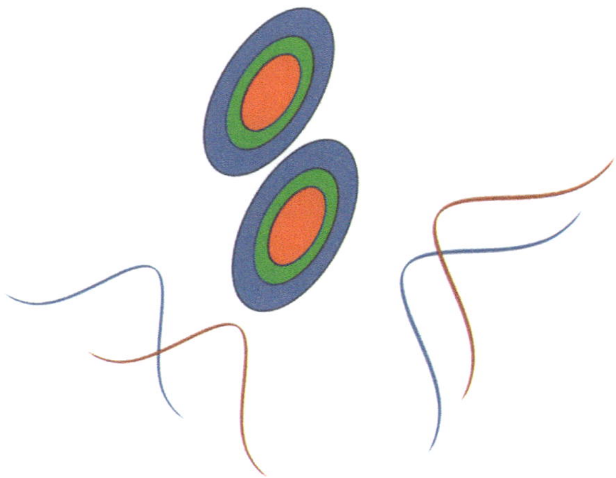

Figure 1.6
Separation of two groups using the eigenvectors.

1.2.4.2 Independent Component Analysis

As seen above, the Karhunen–Loève (PCA) transform creates features, which are mutually uncorrelated. It is an appropriate solution for reducing dimensionality minimizing the approximation (squared) error. In Fig. 1.6, however, we recognize that this minimization may be questionable, for instance, for discrimination purposes. In ICA (see Hyvärien et al., 2001 and references therein), we go beyond the goal of mere decorrelation. Similar to PCA, we define a transform

$$\mathbf{y} = \boldsymbol{W}\boldsymbol{x} \tag{1.6}$$

such that the components of **y** are mutually independent. This is a stronger condition than uncorrelated and applies for data not following a Gaussian distribution. In fact, for our discrimination purposes, the component orthogonal to the eigenvector with the largest eigenvalue is the preferable one. A way to identify these components resides in the analysis of the higher-order cumulants (see Appendix A1.1), such as the skewness and, in particular, the kurtosis $\widetilde{\omega_4}$ (see Eq. A1.10 in the appendix A1.1). ICA based on the kurtosis is carried out in two steps:

- perform PCA on the input data in order to obtain

$$\widetilde{\boldsymbol{y}} = \boldsymbol{A}^t\boldsymbol{x} \tag{1.7}$$

- obtain a second matrix $\widetilde{\boldsymbol{A}}$ and create new components by

$$\boldsymbol{y} = \widetilde{\boldsymbol{A}}\boldsymbol{x} \tag{1.8}$$

such that the sum of the squared fourth-order auto-cumulants, $\sum_i \widetilde{\omega}_4(y_i)^2$, is maximum. This is achieved by a suitable diagonalization of $\widetilde{\boldsymbol{A}}$. At the end, we get the transform

$$y = (A\widetilde{A})^T x = Wx$$

Note that the diagonalization (making that cross-kurtosis vanishes) may succeed only approximately, e.g., if input data do not obey a linear model or are noisy, and cumulants are only approximately known.

The approach consisting in nulling the second- and fourth-order cumulants is the most commonly used ICA. For the sake of completeness, we mention the approach for estimating W based on minimizing the mutual information. First, we write the information entropy

$$H(y) = \int p(y)\ln(p(y))dy \tag{1.9}$$

where $p(y)$ is a joint probability density function. The mutual information between the components of y is given by

$$I(y) = -H(y) + \Sigma_i H(y_i) \tag{1.10}$$

which is the so-called Kullback–Leibler distance. It measures the distance between the joint probability density function $p(y)$ and the product of the respective marginal distributions $\Pi_i p(y_i)$. Thus W is designed with the scope of minimizing $I(y)$, in order that the joint $p(y)$ is close to the product of the marginal distributions $p(y_i)$. In PCA, for Gaussian distributions, this is already achieved by nulling the off-diagonal elements of the covariance matrix. That way, the ICA approach exploiting the Kullback–Leibler distance can be understood as generalization of PCA.[1] We address the interested reader for more details to the textbook on ICA by Hyvärien et al. (2001).

1.2.4.3 Fourier transform

The Fourier transform belongs to the family of integral transformations where a data set is represented by a superposition of the so-called "basis functions". It is defined by the Fourier integral

$$f(x) = \int f(\omega)e^{-i\omega x}d\omega \tag{1.11}$$

or, in the discrete formulation, as Fourier series

$$f(x) = \Sigma_n f(\omega)e^{-in\omega_0 x} \tag{1.12}$$

with i denoting the imaginary part. In the discrete formulation, we use the fundamental frequency $\omega_0 = 2\pi/T$; T gives the length of our series (in time series analysis, we call it a "window"). Higher frequencies are given as multiples, overtones, of ω_0. From the equality

[1] Note, however, that $I(\mathbf{y})$ is not always a metric in a strict sense, as the condition for a metric—$d(\mathbf{a},\mathbf{b}) = 0$ if $\mathbf{a} = \mathbf{b}$—is not always warranted (see Duda et al., 2001).

$$e^{-i\omega x} = \cos(\omega x) + i\,\sin(\omega x),$$

one can immediately recognize that our data series is a sum of sine and cosine functions with an angular frequency ω. We can consider the pair of sine and cosine functions as a basis function of the Fourier transform. The $f(\omega)$ are the weights expressing the degree to which a frequency ω contributes to a signal and are found from

$$f(\omega) = 1/(2\pi) \int f(x) e^{-i\omega x} dx \tag{1.13}$$

and

$$f(n\omega_0) = 1/T\ \Sigma_{0<x<T} e^{-in\omega_0 x} \tag{1.14}$$

The Fourier integral converges for a series of finite length, such as transient phases in a waveform record (seismogram, electrocardiogram, etc.), which has an onset and dies out after some time. Assuming a transient, however, we implicitly add an infinite number of zeros outside the window of interest ("zero-padding"). Thus, our time window T is actually infinite, and the fundamental frequency ω_0 tends to 0. With increasing T, we gain in "frequency resolution" as the frequency steps $n\omega_0$ become smaller.

Following the definition above, $f(\omega)$ and $f(n\omega_0)$ are complex values. Taking the squared sums of the real and imaginary parts, we get the power spectral density

$$\text{Pow}\, f(n\omega_0) = \text{Re}^2 f(n\omega_0) + \text{Im}^2 f(n\omega_0) \tag{1.15}$$

which is a measure of energy of a series of samples (often a time signal) represented by a frequency. We get a spectrum by plotting the power spectral density as a function of ω.

1.2.4.4 Short-time Fourier transform and spectrograms

In case of long-time series, one could opt to consider them all at once. In this case, we lose information about fluctuations in the window. A way out is the Fourier transform over a short series (time windows) of samples. In a gliding window scheme, we shift the short window over a number of samples and carry out the Fourier transform for each step (see Fig. 1.7). As a result, we get an ensemble of spectra, the so-called "spectrograms". Spectrograms are popular as they have a better resolution "in time" (for time series); however, they lack in the representation of low frequencies due to the limited length of the window. Their use requires specific steps of preprocessing, such as tapering, baseline correction (offset and trend removal, high-pass filtering) for which we refer the interested reader to textbooks of signal processing (e.g., Oppenheim and Schafer, 1999; Hamming, 1983; Kanasewich, 1981).

A further advantage of the spectrogram method is the variety of statistical parameters, which can be extracted from an ensemble of STFTs. Besides considering the means of the

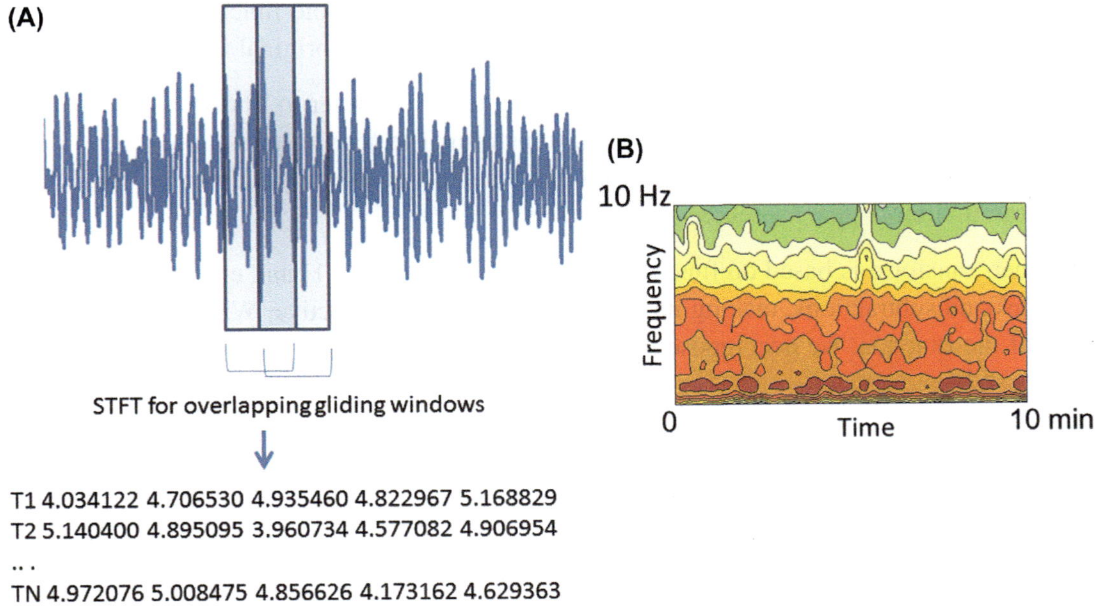

Figure 1.7

(A) Flowchart of STFT applying overlapping gliding windows. For each window T_i, Fourier Power Spectral Densities (PSD) are calculated and stored in a matrix. Each row reports the frequency components (PSD values) obtained in the window T_i in columns. (B) Example of a spectrogram obtained from a seismic signal. It shows the development of the spectral characteristics with time. PSD are represented by the colors (dark brown: high values of PSD, yellow and green: low PSD).

spectral components in the windows T_i, we can calculate standard deviations, peak holds, and consider quantiles, such as the median (50% of all values are lower than the median), 25% or 75% percentiles. For instance, focusing on low percentiles, we may efficiently eliminate short-lived "transient" parts of the original signal, where high amplitudes, and consequently high PSD, are observed (see, e.g., Di Grazia et al., 2006; Langer et al., 2011).

1.2.4.5 Discrete wavelet transforms

The Fourier transform is a widely used concept for data representation for its intuitive understanding. We recognize music not from the crude time series, but we perceive sequences of tones, similar to what we have seen in spectrograms. A drawback, however, is the superposition of infinite sinusoidal functions. This leads to unpleasant margin effects at the beginning and at the end of a data series. Commonly these problems are cured by applying tapering functions with the cost to disrupt the original data.

An alternative to using infinite basis functions, as sines and cosines, resides in the representation of data as a superposition of the so-called "wavelets", i.e., functions with a finite length. Typical wavelets resemble tapered versions of sinusoid functions; unlike the latter, however, their application does not lead to any loss of the original information.

Here we briefly outline the concepts, and for more details we refer to Theodoris and Koutroumbas (2009). We start with the Haar transform. The formalism is quite straightforward and provides an insight into fundamentals of the discrete wavelet transform in a more general sense.

The Haar transform is based on functions in a closed interval [0, 1] and entails only additions and subtractions, making its use in computers very effective. With k being the order of the function, we decompose it into two integers p and q

$$k = 2^p + q - 1 \tag{1.16}$$

$$k = 0, 1, \ldots 2^n - 1; 0 \leq p \leq n - 1; 0 \leq p \leq 2^p \forall\, p \neq 0; q = \begin{cases} 0 \leq q \leq 2^p, & p \neq 0 \\ q \in \{0, 1\}, & p = 0 \end{cases}$$

The Haar functions of order L are given as

$$h_{00}(z) = \frac{1}{\sqrt{L}}, \quad z \in \{0 \ldots 1\} \tag{1.17}$$

and

$$h_{pq}(z) = \begin{cases} \dfrac{1}{\sqrt{L\,2^{p/2}}} & \forall\, (q-1)^{-2p} \leq z \leq \left(q - \dfrac{1}{2}\right)^{-2p} \\ \dfrac{1}{\sqrt{L^{-2p}}} & \forall \left(q - \dfrac{1}{2}\right)^{-2p} \leq z \leq q^{-2p} \\ \mathbf{0} & \end{cases} \tag{1.18}$$

for other z.

Consider the Haar functions of order $L = 8$, i.e., $k = 0,1 \ldots 7$; we get k, p, and q as.

k	0	1	2	3	4	5	6	7
p	0	0	1	1	2	2	2	2
q	0	1	1	2	1	2	3	4

The Haar transform is applied as follows. Let L = 2, then

$$y_1(k) = \frac{1}{\sqrt{2(x(2k) + x(2k+1))}}$$

$$y_0(k) = \frac{1}{\sqrt{2(x(2k) - x(2k+1))}}$$

and k runs from 0 to $N/2-1$ with N being the lenght of the wavelets. Note that in y_1, we form a sum of two adjacent x, and in y_0, we take the difference. The first one is a low-pass filter and the second a high-pass filter.

The transfer function of the two filters is

$$H_1(z) = \frac{1}{\sqrt{2(1+z)}}$$

$$H_0(z) = \frac{1}{\sqrt{2(1-z)}}$$

here z is a complex variable of unit length. Fig. 1.8 shows the impulse response.

The eight order Haar transform $\boldsymbol{H}_8=$

$$1/\sqrt{8}\begin{vmatrix} 1 & 1 & 1 & 1 & 1 & 1 & 1 & 1 \\ 1 & 1 & 1 & 1 & -1 & -1 & -1 & -1 \\ \sqrt{2} & \sqrt{2} & -\sqrt{2} & -\sqrt{2} & 0 & 0 & 0 & 0 \\ 0 & 0 & 0 & 0 & \sqrt{2} & \sqrt{2} & -\sqrt{2} & -\sqrt{2} \\ -2 & -2 & 0 & 0 & 0 & 0 & 0 & 0 \\ 0 & 0 & 2 & -2 & 0 & 0 & 0 & 0 \\ 0 & 0 & 0 & 0 & 2 & -2 & 0 & 0 \\ 0 & 0 & 0 & 0 & 0 & 0 & 2 & -2 \end{vmatrix}$$

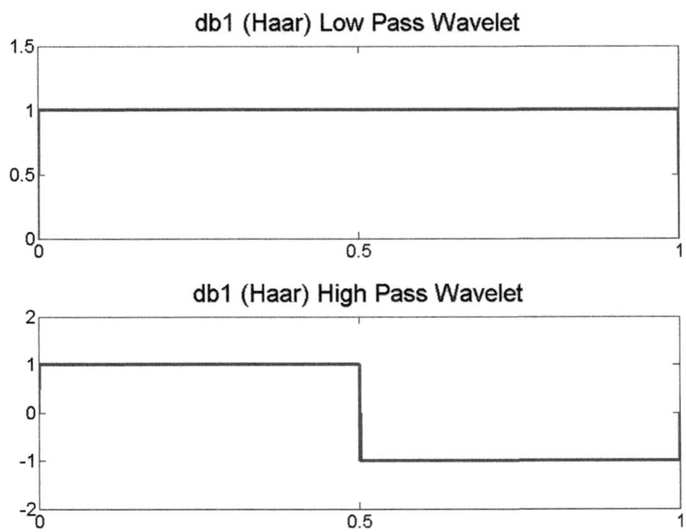

Figure 1.8
The Haar wavelets (low- and high-pass wavelets).

We perform the transform

$$y = H_8 x \tag{1.19}$$

where x is an input series with eight elements ($k = 0 \ldots 7$), and y is the output. The lower four components of y correspond to $y_0(k)$. The upper elements are obtained iteratively. We take the output of our low-pass filter and apply again our two filters

$$F_1(z) = H_0(z^2)H_1(z) = \frac{1}{2}(1+z)(1-z^2) = \frac{1}{2}(1+z+z^2-z^3)$$
$$F_0(z) = H_1(z^2)H_1(z) = \frac{1}{2}(1+z)(1+z^2) = \frac{1}{2}(1+z+z^2+z^3)$$

and

$$y_1(0) = [1/2 \quad 1/2 \quad -1/2 \quad -1/2 \quad 0 \quad 0 \quad 0 \quad 0]x^T$$
$$y_1(1) = [0 \quad 0 \quad 0 \quad 0 \quad 1/2 \quad 1/2 \quad -1/2 \quad -1/2]x^T$$

which corresponds to the output of the third and fourth row in H_8. We further split considering $F_1(z)$ that is the response of the applied low-pass filter in this iteration and obtain

$$y_2(0) = \frac{1}{\sqrt{8}}[1 \quad 1 \quad 1 \quad 1 \quad -1 \quad -1 \quad -1 \quad -1]x^T$$

output with respect to the second row in H_8, and finally the last one

$$y_3(0) = \frac{1}{\sqrt{8}}[1 \quad 1 \quad 1 \quad 1 \quad 1 \quad 1 \quad 1 \quad 1]x^T$$

Given the components of $\mathbf{y}$, we reconstruct the original data sequence by

$$x = H^T y \tag{1.20}$$

As the low- and high-pass filters established in the Haar transform are poor with respect to their frequency characteristics, filters with more satisfying characteristics have been proposed.

With h_0(k) and h_1(k) being the impulse responses of our filters, we may write

$$y_0(k) = \Sigma_l x(l)\, h_0(n-l)_{|n=2k} \tag{1.21a}$$
$$y_1(k) = \Sigma_l x(l) h_1(n-l)_{|n=2k} \tag{1.21b}$$

Similar to the Haar transform, we obtain the impulse responses of the filters

$$y = T_i x \tag{1.22}$$

where T_i is the filter matrix.

$$\left|\begin{array}{ccccccc} .. & .. & .. & .. & .. & .. & .. \\ .. & h_0(2) & h_0(1) & h_0(0) & h_0(-1) & h_0(-2) & .. \\ .. & h_1(2) & h_1(1) & h_1(0) & h_1(-1) & h_0(-2) & .. \\ .. & .. & .. & h_0(2) & h_0(1) & h_0(0) & .. \\ .. & .. & .. & h_1(2) & h_1(1) & h_1(0) & .. \\ .. & .. & .. & .. & .. & h_0(0) & .. \\ .. & .. & .. & .. & .. & h_1(2) & .. \\ .. & .. & .. & .. & .. & .. & .. \end{array}\right|$$

The reconstruction matrix $\boldsymbol{G}$, which restores the vector x, is obtained easily, provided that the $h_i(2k-n)$ are orthogonal. Its elements are obtained from

$$g_i(n) = h_i(-n) \tag{1.23}$$

The "Daubechies" wavelets (Daubechie, 1991) come with the nice property to have filters with maximal flat properties. At the same time, they are orthogonal (Fig. 1.9).

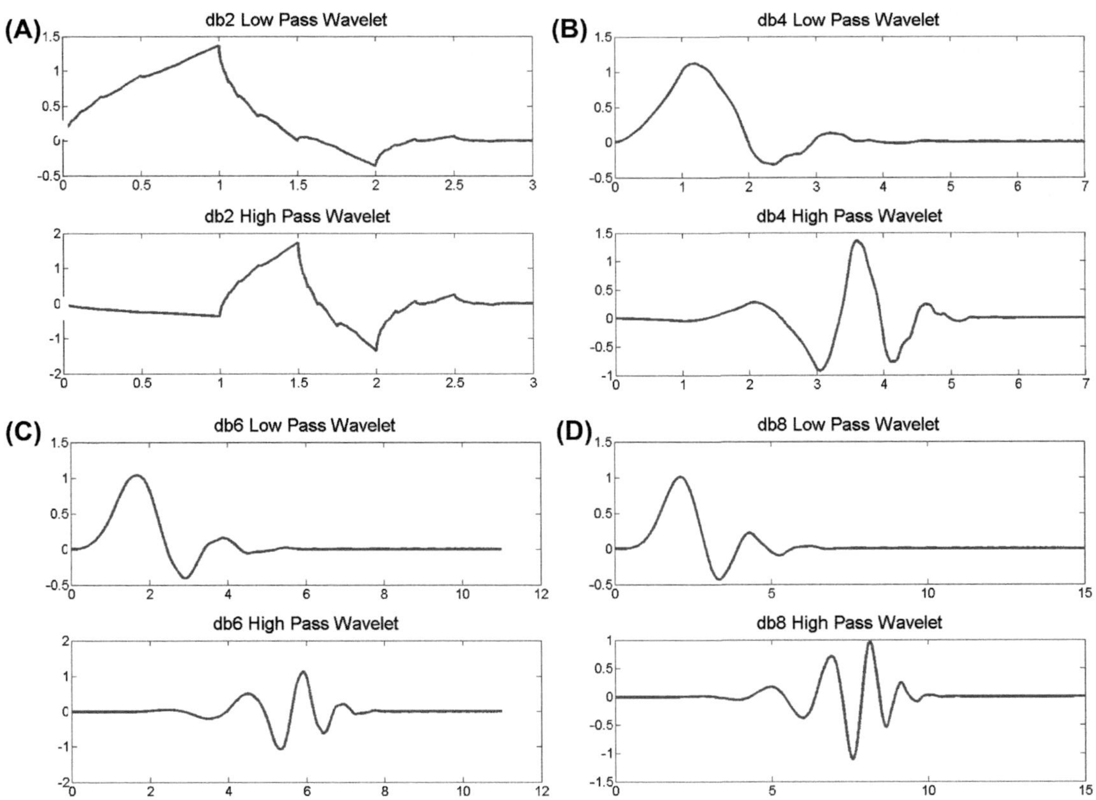

Figure 1.9
Daubechies wavelets (low- and high-pass impulse response) of order 2 (A), 4 (B), 6 (C), and 8 (D).

Their low-pass filter coefficients are

$$\begin{aligned}
&\text{DB2} = h_1(n) = [0.7071\ 0.7071]\\
&\text{DB4} = [0.4830\ 0.8365\ 0.2241{-}0.1294]\\
&\text{DB6} = [0.3367\ 0.8069\ 0.4599{-}0.1350\ {-}0.08544]\\
&\text{DB8} = [0.2303\ 0.7148\ 0.6309{-}0.0280\ {-}0.1870\ {-}0.0308\ 0.0329{-}0.0106]
\end{aligned}$$

The inverse filter coefficients are obtained simply as

$$g_i(n) = h_i(-n)$$

i.e., the mirrored version of $h_i(n)$. Coefficients of the high-pass filters h_0 can be found from

$$h_0(n) = (-1)^n h_1(-n + 2L - 1) \tag{1.24}$$

with L being the length of the filters.

There are various alternative wavelet transforms, such as the "Morlet" wavelet, "Mexican hat", "Gaussian" wavelets, etc. These alternatives are attractive at a first glance as they are symmetric and are described with explicit expressions. However, they come with a number of drawbacks, such as they do not obey orthogonality, reconstruction is not available, and the numerical algorithms are relatively slow. The "Meyer" wavelet (Meyer, 1993, see Fig. 1.10) is orthogonal and symmetric; however, it is not strictly finite.

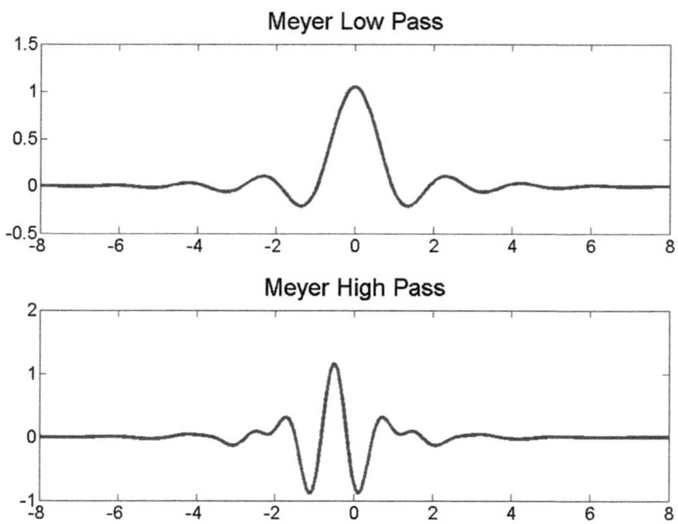

Figure 1.10
The Meyer wavelets.

1.2.5 Standardization, normalization, and other preprocessing steps

Multivariate data are often composed of features, which are measured in different units and ranges. For example, the height of a person can be measured in centimeters, inches, or meters, and the weight can be measured in pounds or kilograms. In order to warrant a balanced weight of these features, we must normalize them appropriately. Typical normalizations are

range: $f(x) = (x - \min(x))/(\max(x) - \min(x))$
standard deviation: $f(x)=(x - \mu)/\sigma$,
σ being the standard deviation of the population and μ its mean
logistic: $f(x) = 1/(1 + e^{-\alpha x})$
logarithm $f(x) = \log(x - \min(x) + 1)$
Hist: sort the x_k, creating a sequence from low to high and use their rank as new values.

1.2.5.1 Comments

Even though the normalizations with respect to range and standard deviation look very similar, there are considerable differences. The first is sensitive to the tails of a distribution, as one single value at the extremes may control the normalization. In the second, the term in the denominator depends on the whole population rather than single extreme values, which brings a larger stability. Still, extreme values have a strong impact on the normalization, as the standard deviation is based on the squared distances of the samples from their average (see Appendix A1.1). Alternatively, one could consider the mean of the absolute distances of the samples from their average.

The logistic function and its variants reserve a specific treatment of values in the tails of a distribution. For small values, the denominator tends to infinity, thus $f(x) \rightarrow 0$. On the other side, for large x, the denominator tends to 1, and $f(x) \rightarrow 1$ as well.

Many processes in geology entail processes of growth and variation that are expressed "by a factor of" rather than "by a value".

Self-similar objects[2] are quite common in geology and related fields. They are at the basis of the "fractal geometry of nature" (Mandelbrot, 1983). Underlying distributions turn out as "exponential" or "log-normal" instead of "uniform" or "normal", i.e., Gaussian (see Appendix A1.2). With a logarithmic normalization, we turn log-normal distributions into a normal ones.

[2] Varying feature components by a factor leads to self-similarity, in the sense that geometrical aspects, besides the size, are maintained. Increasing all sides of a body by a factor, a brick remains a brick as the ratio of length, width, and height is not altered. Geologists know that and always add a reference (e.g., a coin) to a photo taken in the field.

With the "Hist" normalization (see, e.g., Vesanto et al., 2000), we work with ranked data, which implies a considerable loss of information. The rank difference between two neighboring samples does not measure the real difference between them. On the other hand, we get rid of distortions related to the peculiarities of the probability function of the samples (see Appendix A1.1). In fact, many statistical procedures are based on ranked data rather than on their absolute values (e.g., Wilcoxon test, Spearman's rank correlation, etc.).

1.2.5.2 Outlier removal

Outliers are points falling far away from the mean of the corresponding random variable, and they may severely affect the assessment of the error and the quality of our pattern recognition techniques. Often the error is given in the form of squares, which exacerbates the problem with outliers. However, there is no fixed rule for their identification. If we work with few samples and in a low-dimensional feature space, they may be easily identified and discarded. Otherwise, one may rely on percentile-based criteria, for example considering only a high percentile of the data, such as 99% or 99.9%, removing 0.5% or 0.05% of the data at the tails. Such a technique is known as "winsorizing". We might prefer not to touch the data set with any a priori defined criterion by using a cost function, which is not sensitive to the presence of outliers.

1.2.5.3 Missing data

It may happen that some samples were not collected, leaving for instance a gap in a time series. In other cases, the feature vector is incomplete, i.e., there are some missing components. A missing value in a time series can be replaced by the average of the two neighboring ones, i.e., we carry out a linear interpolation. Such a simple estimation of missing values can be justified, as the linear interpolation—compared to other methods to fill a gap—has less degrees of freedom. Following Occam's Razor, we may adopt this solution in the majority of the cases. In the case of incomplete feature vectors, we may follow the strategy of "Expectation Maximization". Suppose our feature vector is

$$\boldsymbol{x}_{com} = [\boldsymbol{x}_{obs}, \boldsymbol{x}_{mis}] \tag{1.25}$$

where subscripts "com", "obs", and "mis" stand for complete, observed, and missing parts of the vector. We ask for the conditional probability that $\boldsymbol{x}_{mis}$ assumes values given by the $\boldsymbol{x}_{obs}$. considering

$$p(\boldsymbol{x}_{mis}|\boldsymbol{x}_{obs}; \boldsymbol{\theta}) = p(\boldsymbol{x}_{obs}, \boldsymbol{x}_{mis}; \boldsymbol{\theta}) / p(\boldsymbol{x}_{obs}; \boldsymbol{\theta}) \tag{1.26}$$

with $\boldsymbol{\theta}$ being an unknown set of parameters of the probability density function (pdf), which are estimated from the $\boldsymbol{x}_{obs}$, and

$$p(\boldsymbol{x}_{obs}; \boldsymbol{\theta}) = \int p(\boldsymbol{x}_{com}; \boldsymbol{\theta}) d\boldsymbol{x}_{mis}$$

In other words, we first infer an estimation of missing components from a set of complete l-dimensional feature vectors, where we can design a pdf of components where missing values occur along with those for which observations exist. For instance, we may have an incomplete vector $\boldsymbol{x}$, where the component j is not recorded. Then we get a most likely guess on the basis of the other complete feature vector considering $p(\boldsymbol{x}_i|\boldsymbol{x}_{1\ \ldots\ j-1,j+1\ \ldots\ L})$.

Preprocessing encompasses all necessary steps for a proper application of pattern recognition tools. For this purpose, we should make sure that all components of the feature vectors have equal weight, regardless of the physical units they may have. This is usually achieved by carrying out a normalization. We also want to represent the data space with our samples as good as we can. Therefore, we request that feature values have a continuous distribution between their extremes. We try to avoid occurrences as large gaps, overcrowded intervals, and outliers. Further steps are specific transformations, which aim at a reduction of the dimensionality of the problem. For the latter, there is a large variety of available techniques; we limit ourselves to the description of a few of them.

1.2.6 Curse of dimensionality

The term "curse of dimensionality" encompasses phenomena that arise in the presence of high-dimensional data spaces. The expression was coined by Bellman (1961) considering problems in dynamic optimization. One of the most striking examples is the phenomenon of data concentrating at the boundaries of a high-dimensional hypercube.

Imagine a circle and a sphere, both having a diameter $2r$ (see Fig. 1.11). They can be inserted in a square or a cube having the same side length $2r$. Assuming samples being uniformly distributed in the area given by the square, then the probability of finding a sample in the circle is given by $\pi\ r^2$, which is ca. 78.5% of the total. In the three dimensional case, we have to compare the volumes of the hypercube and the sphere, the

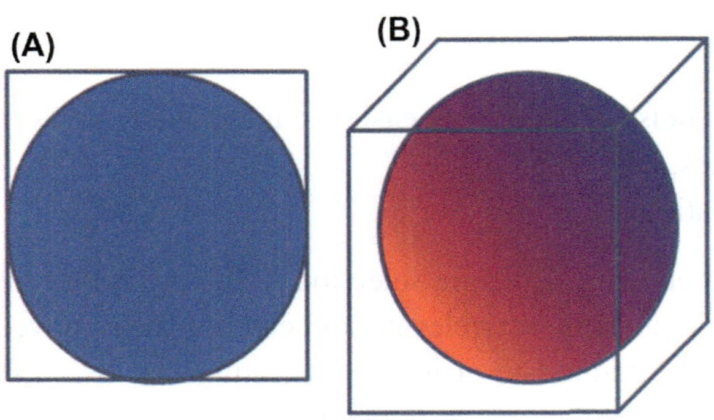

Figure 1.11
With increasing dimensionality, more samples tend to be found at the margins of a feature space.

latter being $4/3\pi\ r^3$, which is 52%. In general, for dimension n, we find a volume of the hypersphere given by

$$V(r) = \pi^{\frac{n}{2}} r^n / \Gamma\left(\frac{n}{2} + 1\right) \tag{1.27}$$

with Γ being the gamma function. The ratio of hypersphere/hypercube volume vanishes rapidly as the dimension n increases.

A related effect is the matter of distances. Suppose randomly distributed samples in an n-dimensional data space. Then the ratio of encountered minimum and maximum distances approaches 1, or

$$\lim_{n \to \infty} \left[(dist_{\text{max}} - dist_{\text{min}}) \,/\, dist_{\text{min}} \right] = 0 \tag{1.28}$$

In other words, distances tend to concentrate around a certain value.

All these effects have a common source, that is, that data become sparse when the dimension increases. In "nearest neighborhood" approaches, the curse of dimensionality becomes a serious drawback. In two dimensions, we need a subcube with a side length of ca. 30% of the total to gather 10% of the samples, and in a ten-dimensional case, we need over 80%. At the same time, the samples have an increasing number of near neighbors, which entails a considerable computational burden in algorithms based on neighborhood considerations. Eventually, we conclude that any effort to limit the dimensionality of the problem is justified. As a minimum condition, we may require that the number of samples is greater than the number of dimensions.

1.2.7 Feature selection

In machine learning and statistics, feature selection (sometimes referred to as variable selection, attribute selection, or variable subset selection) is the process of selecting a subset of relevant features (variables, predictors) for use in model construction. Feature selection techniques are applied for three reasons:

- simplification of models to make them easier to interpret by users,
- shorter training times, and
- enhanced generalization

The central premise when using a feature selection technique is that the data contain many features that are either redundant or irrelevant and can thus be removed without incurring much loss of information. Redundant or irrelevant features are two distinct notions, since one relevant feature may be redundant in the presence of another relevant feature with which it is strongly correlated. Feature extraction and transforms may be a part of the selection process. As we have seen earlier, they may help to reduce the number of components to be considered.

Features should allow the definition of a suitable metric for the difference between two patterns. A metric must have four properties. Given the feature vectors **a**, **b**, and **c**, with **d**(,) being the distance between two vectors, these properties are.

Nonnegativity: $\mathbf{d(a,b)} \geq \mathbf{0}$
Reflexivity: $\mathbf{d(a,b)} = \mathbf{0}$, if and only $\mathbf{a} = \mathbf{b}$
Symmetry: $\mathbf{d(a,b)} = \mathbf{d}(\mathbf{b},\mathrm{a})$
Triangle inequality: $\mathbf{d(a,b)} + \mathbf{d(b,c)} \geq \mathbf{d(a,c)}$.

In the literature, various options are reported, such as the Euclidean distance, the Mahalanobis distance, Manhattan distance (or city block distance), etc. More about these definitions can be found in textbooks like that of Duda et al. (2001). In numerical features, Euclidean distances and modifications are the most common ones.

Typically, a metric is interpreted in a statistical sense. For example, given a certain distance of a pattern A to the centroids of class $\mathbb{B}$ or $\mathbb{C}$, what is the probability that A belongs to $\mathbb{B}$ rather than $\mathbb{C}$? In Appendix A1.1, we show how we can transform a distance d into a probability. A in this sense is a single pattern described by a feature vector **a**, $\mathbb{B}$ and $\mathbb{C}$ are two classes of patterns described by the same type of parameters as A, and the centroids of $\mathbb{B}$ and $\mathbb{C}$ are denoted as **b** and **c**. Knowing **d(a,b)**, d(**a,c**), and the probability density functions of $\mathbb{B}$ and $\mathbb{C}$, we shall be able to answer the question. We may exploit the distance of the two centroids of $\mathbb{B}$ and $\mathbb{C}$, **d(b,c)**, for the selection of the features. Clearly, we wish that classes $\mathbb{B}$ and $\mathbb{C}$ are indeed two different classes. In classical statistics, the answer can be found in statistical tests, which are based on metrics using a normalization with respect to the standard deviation. In univariate statistics, we may apply Student's *t*-test, otherwise we use Hotelling's T^2 test (Hotelling, 1933, see Appendix A1.2), which is based on the Mahalanobis distance (Mahalanobis, 1925). In this metric, the distance vector **d(b,c)** is normalized by the inverse covariance matrix

$$\tilde{d}^2 = (\mathbf{b}-\mathbf{c})\boldsymbol{C}_{B,C}^{-1}(\mathbf{b}-\mathbf{c})^T \tag{A1.29}$$

where $\boldsymbol{C}_{\mathbf{B,C}}$ is the pooled covariance matrix of samples belonging to classes $\mathbb{B}$ and $\mathbb{C}$. On the basis of the Hotelling's T^2 test, we may also derive a criterion for deciding whether some components are relevant for our distinction. Recall Hotelling's T^2

$$T^2 = N_B N_C/(N_B + N_C)\,\tilde{d}^2 \tag{1.30}$$

(see Appendix A1.2). We first obtain T^2 for the full set of components T^2 and $\tilde{T}^2$ for the reduced set with $\tilde{L}$. We define

$$F = (N_B + N_C - L - 1)/\left(L - \widetilde{L}\right)\left(T^2 - \widetilde{T}^2\right)/\left((N_B + N_C - 2) + \widetilde{T}^2\right) \quad (1.31)$$

where N_B and N_C are the number of samples in groups $\mathbb{B}$ and $\mathbb{C}$. F follows an F distribution with degrees of freedom $L - \widetilde{L}$and $N_B + N_C - L - 1$ (see, e.g., Krzanowski, 1988). For small levels of significance, we may decide that working with a reduced feature set is sufficient for our purposes.

The t- and T^2-tests can give useful hints if our problem is linear and our features follow a multivariate Gaussian distribution. Recall that the T^2 test uses a pooled covariance matrix; in other words, we assume that it describes dispersion of both groups $\mathbb{B}$ and $\mathbb{C}$. In many practical applications, this does not hold, and we also have to be aware that features are not normally distributed. Therefore, we may follow more general testing schemes, which rely on the available data themselves.

In the classification techniques with supervision (see Chapter 2), we will first examine whether our classification scheme is able to reproduce the a priori information. For a given feature vector, our classifier will output a '1' (or TRUE) if a pattern belongs to some class $\mathbb{A}$, otherwise the output is '0' (or FALSE). In this phase, we verify whether our classifier is able to "learn" how to distinguish the patterns. If the problem is well conditioned, we get a high rate of success, i.e., '1' assigned to the desired class. Nonetheless, the only success of learning does not suffice to warrant a high quality classifier. In Chapter 2, we shall learn that nonlinear classification can be transformed to a linear one augmenting the dimensionality of the feature vector. This is exploited in the Multilayer Perceptron and the Support Vector Machines. In the extreme case, we may be able to resolve any classification problem, provided that we have an unlimited number of dimensions or features. Even though we may achieve a (close to) 100% percent of success, the punishment is just round the corner. Applying the classifier to other patterns, outside the basket of samples used for learning, we risk considerable error. Therefore, in supervised classification schemes, we perform the so-called "cross-validation". In cross-validation, some data are set aside in the sense that they are not used for learning the classifier criteria. This data set is referred to as the "test set" whereas the data used for learning are called the "training" set. The success of a classifier is assessed by applying it to the test data and checking whether we obtain the desired output. That way we are able to verify whether our whole scheme, comprising the chosen features, is valid.

We shall learn more about testing schemes in the chapters presenting real data applications (in particular, Chapter 4 and 6). Here we just cite some elements, which guide the design of the testing schemes:

- test and training sets are defined by a random selection. As data are always limited, we may decide to use a minor part for testing such as 20%–30%;

- testing has often to be repeated several times so that (almost) all patterns have a chance to be selected as test patterns; and
- testing schemes should go beyond the simple rate of success of matching classifications, as they may achieve a success by chance, particularly when one class has much more samples than the competing ones.

We can exploit our general test schemes for purposes of feature selection in the same way as the conventional statistical test. We may repeat the classification using various configurations of features, for instance by applying a jack-knife approach or nonlinear optimization algorithms such as genetic algorithms where we search for an optimum configuration of features considering the error for the test data set.

Appendix 1 Basic notions on statistics

A1.1 Statistical parameters of an ensemble

Many considerations presented here are based on statistics, where we describe the distribution of ensembles of values and objects. As their number is typically large, we condense the information by using some parameters summarizing the essential properties of the distribution of our data.

The mean of an ensemble of N data $\boldsymbol{x}_i$ is given by

$$\mu = 1/N\ \Sigma_i x_i \tag{A1.1}$$

where i runs from 1 to N. In the same way, we define the mean vector

$$\boldsymbol{\mu} = 1/N\ \Sigma_i \boldsymbol{x}_i \tag{A1.2}$$

which sometimes is referred to as the "centroid" of an ensemble. The second parameter is a measure of the dispersion of the data, which we express by the variance

$$\sigma^2 = \frac{1}{N-1}\Sigma_i (x_i - \mu)^2 \tag{A1.3}$$

The square root σ (standard deviation) is the root mean square of the distances of the samples from the average. We now may ask about probability to find a sample within a certain range. Alternatively, we might consider the rank of our samples and the percentiles of the data found at the limits of the range. This requests ordering the data with respect to their size. The use of ranked data and percentiles makes no assumption on the probability density function of our population of samples. However, it comes with severe drawbacks when we face with multivariate data.

Most commonly, the probability density function is assumed to be Gaussian:

$$f(x) = 1\Big/\sqrt{(2\pi\sigma^2)e^{-\frac{1}{2}(x-\mu)^2/\sigma^2}} \tag{A1.4}$$

Given the mean μ and the variance σ^2, we can calculate the probability of finding a value x_i in the range of $\mu \pm \Delta x$, for instance $\mu \pm \sigma$, i.e., 68%, and $\mu \pm 2\sigma$, i.e., 95%. That is, the distance Δx transforms into a probability using the variance as the parameter of conversion. In pattern recognition, we first search for a distance measure and then use it as a criterion for assigning a pattern to a class.

In the multivariate case with L components, the distance $\boldsymbol{\mu} \pm \Delta \boldsymbol{x}$ can be a criterion for assigning a pattern to a class only if the data are normalized and uncorrelated. In general, the sum of variances $\Sigma_i \sigma_j^2$, j running from 1 to L, does not represent the total dispersion of our data. A generalized measure of dispersion of a multivariate ensemble is given by the covariance matrix $\boldsymbol{C}$.

Considering the random variable X_1 and X_2, then

$$\sigma_1^2 = \frac{1}{N-1}\Sigma_i (x_{1i} - \mu_1)^2 = E(X_1 - \mu_1)^2$$

$$\sigma_2^2 = \frac{1}{N-1}\Sigma_i (x_{2i} - \mu_2)^2 = E(X_2 - \mu_2)^2$$

and

$$\sigma_{12} = \sigma_{21} = E[(X_1 - \mu_1)(X_2 - \mu_2)]$$

where $E(.)$ is the expectation of the argument specified in the parantheses.

For the L-dimensional ensemble, the covariance matrix $\boldsymbol{C}$ reads as

$$\boldsymbol{C} = \begin{bmatrix} \sigma_{11} & \cdots & \sigma_{1L} \\ \vdots & \ddots & \vdots \\ \sigma_{L1} & \cdots & \sigma_{LL} \end{bmatrix} \tag{A1.5}$$

$\boldsymbol{C}$ is symmetric, i.e., $\sigma_{ij} = \sigma_{ji}$. Together with the centroid vector, the covariance matrix allows us to identify position and orientation of the ensemble in the data space. Recall, for instance, the regression coefficient in the two-dimensional case, which is given by

$$\beta_{ij} = \sigma_{ij} / \sigma_{ii}$$

and gives the slope of the regression line. Rather than using the regression coefficient, we may use the eigenvalues and eigenvectors of $\mathbf{C}$, specifying the shape and orientation of the distribution of our ensemble. In PCA, we exploit the eigenvectors for rotating our samples to a new system of axes, facilitating the discrimination of data groups.

Similar to the univariate case, we may use the distance of a sample $\boldsymbol{x}_{ii}$ from the centroid $\boldsymbol{\mu}_A$ as a measure of the probability that the sample belongs to class A. In doing so, we suppose that members of class $\mathscr{A}$ are multivariate normally distributed around $\boldsymbol{\mu}_A$ with $\boldsymbol{C}_A$ being the covariance matrix.

The distance, which can be read as probability, is the so-called Mahalanobis distance, i.e.,

$$d^2 = (\boldsymbol{x}_i - \boldsymbol{\mu}_A)\boldsymbol{C}_A^{-1}(\boldsymbol{x}_i - \boldsymbol{\mu}_A)^T \tag{A1.6}$$

In the case of a univariate Gaussian, we were able to say that 68% of data fell in a range of $\pm\sigma$ around the average, i. e., we consider a normalized distance $(x-\mu)/\sigma$ for the transformation of a, physical, distance into a probability measure. In the multivariate case, this kind of transformation is obtained by calculating the Mahalanobis distance, where $\boldsymbol{C}^{-1}$ appears instead of the σ in the denominator (Fig. A1.1).

For some distributions, the parameters such as mean or variance are not defined. Consider, for instance, the Cauchy probability distribution

$$f(x) = \left(1 + x^2\right)/\pi \tag{A1.7}$$

The theoretical mean is obtained from

$$\mu = \int x f(x) dx \tag{A1.8}$$

Carrying out the integration from $-\infty$ to ∞, we discover that the integral does not converge. A (imaginary) body having a Cauchy mass distribution has no defined center of gravity, as its tails are too heavy.

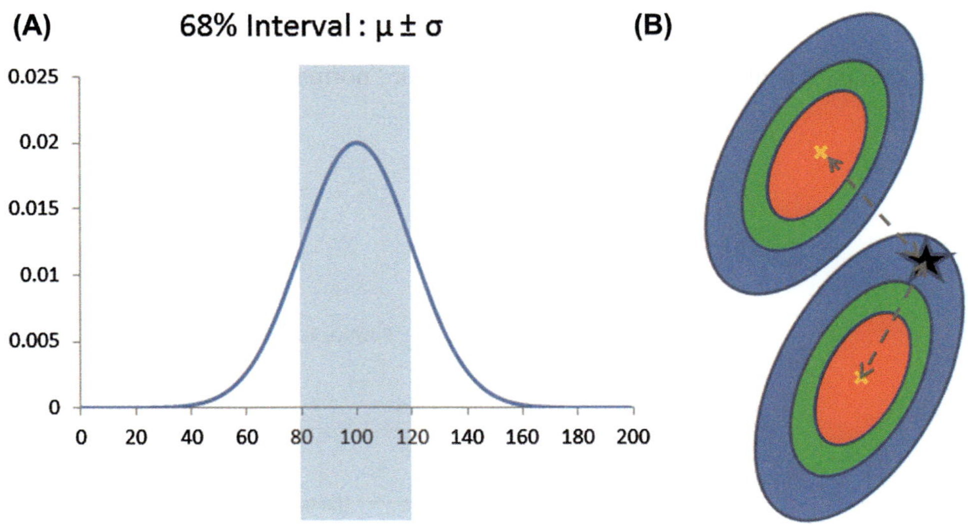

Figure A1.1

Variance and distance. In the univariate case (A), a Euclidean distance measured in multiples of the standard deviation turns into a probability. In the multivariate case (B), we have to account for the direction in which we measure. The colored ellipses depict bivariate Gaussian distributions for two different ensembles. Even though the Euclidean distance of the black asterisk to the centroid of the upper ensemble is smaller than the one to the other centroid, the asterisk belongs more likely to the latter one. The Mahalanobis distance accounts for that.

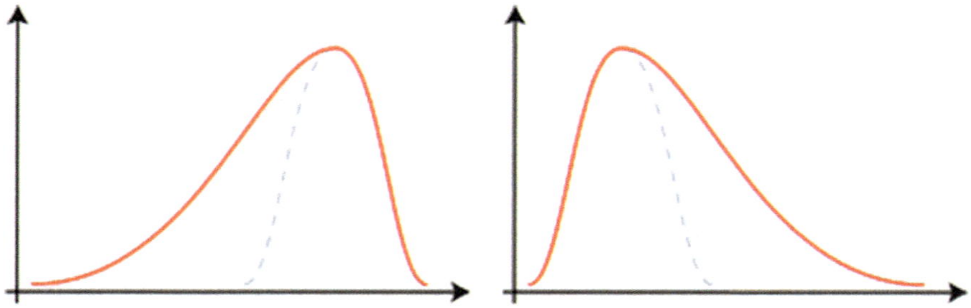

Figure A1.2
Distributions with positive (A) and negative (B) skewness. For reference, a Gaussian distribution is shown as a dotted line.

The standard deviation has been derived from the summed squares of the differences of x with respect to its mean. For distributions not being uniform or Gaussian, the so-called "higher-order cumulants" can be of interest. The "skewness" γ is given by

$$\gamma = E((X - \mu)/\sigma)^3 \tag{A1.9}$$

and is a measure of asymmetry of a distribution. For instance, the log-normal distribution is asymmetric (Fig. A1.2).

The "kurtosis" is obtained from

$$\omega_4 = E(X - \mu)^4 \tag{A1.10}$$

In order to compare it more easily with a Gaussian (or "normal" in Fig. A1.3) distribution, one defines

$$\widetilde{\omega}_4 = \omega_4 \Big/ \left(\sigma^2\right)^2$$

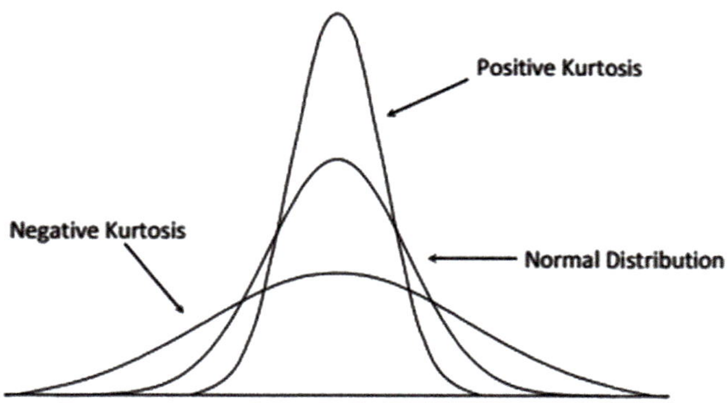

Figure A1.3
Shape of distribution functions with positive and negative kurtosis. A Gaussian (or normal) distribution is depicted as reference.

Having $\widetilde{\omega}_4 = 3$ for a Gaussian distribution, we speak of a positive kurtosis for $\widetilde{\omega}_4 > 3$, and a negative kurtosis for $\widetilde{\omega}_4 < 3$.

A1.2 Distinction of ensembles

In classical statistics, we face with a problem which is closely related to the classification issues discussed in our book. A typical question is whether two ensembles belong to the same parent population. For instance, we check a number of batteries finding that some of them do not match the requested lifetime. Did something deteriorate during the process of fabrication? In our problem of feature selection, we may have the inverse question—are the features appropriate for discrimination purposes?

A first answer is given by the famous "Student's t-test". Suppose we have two groups A and B, for which we estimate the means μ_A and μ_B. We consider the difference $\mu_A - \mu_B = d$ and normalize d for the pooled variance, i.e., $s^2 = (s_A^2\ (N_A - 1) + s_B^2(N_B - 1))/(N_A + N_B)$, with the $s_{A,B}^2$ being the estimated variances of groups A and B. The variable

$$t = \sqrt{[N_A N_B\ (N_A + N_B - 2)/N_A + N_B\ )]d/s} \tag{A1.11}$$

can be used for testing the hypothesis that the two ensembles differ with respect to their mean. For instance, for $t = 1.97$, we can reject the hypothesis "the two means are equal" with a level of significance of 95%.

For a similar testing in the multivariate case with L dimensions, we may again exploit the concept of the Mahalanobis distance

$$D^2 = (\boldsymbol{\mu}_A - \boldsymbol{\mu}_B)\boldsymbol{C}_{A,B}{}^{-1}(\boldsymbol{\mu}_A - \boldsymbol{\mu}_B)^T$$

where $\boldsymbol{C}_{A,B}$ is the pooled covariance matrix of $\mathbb{A}$ and $\mathbb{B}$. Similar to the Student t-test, we define Hotelling's T^2, i.e.,

$$T^2 = N_A N_B/(N_A + N_B)D^2 \tag{A1.12}$$

and test for

$$F = (N_A + N_B - L - 1\ /\ [(N_A + N_B - 2)L]T^2 \tag{A1.13}$$

where F follows the F-distribution. From these, we may get a first idea whether the choice of features is appropriate. Note that we also can formulate a minimum request on the number of samples with respect to the dimension of the feature vector L. If $N_A + N_B \leq L + 1$, the separation of the two groups will be rejected whatever distance we may find.

CHAPTER 2

Supervised learning

2.1 Introduction

In the preceding chapter, we learned about patterns and objects, and the way about how these are characterized. We discussed on how to identify suitable feature vectors and outlined some procedures, which facilitate our task, such as choosing a suitable standardization, carrying out some calculus for the extraction of features, and performing transformations, which further make our task more affordable. In this chapter and the next chapter, we shall learn about techniques on how to handle the complex objects and patterns, allowing one to establish criteria for making decisions or taking actions. The first group of methods, treated in this chapter, belongs to the "supervised learning" and "supervised classification" strategies. These strategies make use of a priori information inferred from examples of patterns, supposing to know to which class they belong. For instance, we look at a rock, and—based on our observations—we assign it to a category—volcanic, plutonic, sedimentary, etc. In the analysis of images, we shall assign certain areas to "woods", "farmland", "water", and other, moving on from texture properties. In the following, we shall present an example problem, which consists of assigning seismic signals to the category "earthquake" or "explosion". For the sake of ease, we keep it simple, considering only two dimensional feature vectors and only two classes, in fact "earthquakes" or "explosions". Nonetheless, being a quite simplified problem, basic concepts become quite clear and generalization to more complex tasks with higher dimensional feature vectors and more classes is straightforward.

Our decision, whether to assign a pattern to a class, exploits features and feature vectors, such as the ones described in Chapter 1. However, this a priori classification—carried out by us as experts—does not always follow well established criteria or rules, in the sense "if condition TRUE then class A". Indeed, being faced with multivariate features—feature vectors—the definition of those rules is rather difficult requiring the construction of multibranched decision trees or being even an unaffordable task at all. So, we often base our decision on some not explicitly stated experience, intuition.

The techniques of supervised classification offer possibilities of identifying reproducible formalisms, i.e., rules or criteria in a wide sense. Those will help us to find an answer, when the expert's opinion is not available, or when we have to make many decisions in little time. An important factor is the reevaluation of a priori established class membership

Advantages and Pitfalls of Pattern Recognition. https://doi.org/10.1016/B978-0-12-811842-9.00002-9

of the patterns. First, we examine whether our adopted scheme is able to provide this formalism, in other words, it is able to learn. Failure during learning may have two reasons:

- the chosen method or features are inappropriate or
- the a priori classification (expert opinion) has flaws.

The first reason can be quite easily discovered, whereas the second one requires further analysis known as the test phase.[1] In Chapter 1, we mentioned, among others, cross-validation tests in the context of feature selection and reduction; here we may exploit them for checking whether our classification scheme is stable. Cross-validation tests are based on data with (presumably) known class membership, which is set aside, i.e., not used for learning. Once the classification formalism has been found, it is applied to the data set not used for learning, in the sense of a blind check. Unsatisfying results mean that we reject the formalism as it "overfits". The error estimated during learning is overly optimistic, because the classifier used has too many degrees of freedom. If the a priori classification is flawed, the classifier may be able anyway to find a solution that is valid only for the specific data used during learning, but fails if applied to new data.

In the following, we shall outline basic concepts of some learning schemes with supervision. We start with linear discrimination, applying it to a famous problem in seismology—the discrimination of earthquakes from nuclear test explosions. Being far from presenting an exhausting solution for the issue, it allows an intermediate understanding of the discrimination problem. Besides using the classical Fisher discrimination technique, we can generalize linear discrimination in the perceptron approach and the Support Vector Machines. The latter have become rather popular as they can be extended to problems, where discrimination with linear functions is not possible. An intriguing property of these techniques is that we do not have to make any a priori assumptions on the distribution of the features of our patterns.

In Hidden Markov Models (HMM) we consider sequences of observations rather than single patterns. In our framework of supervised classification we have to restate the problem, as we have no straight forward metrics allowing to distinguish between the classes. Generative models, such as HMMs and Bayesian Networks (BNs), are based on the solution of the task "find the most likely model explaining the sequence of observations at best". These strategies offer an appealing characteristics, asvariables making up our observations can be of any type: numerical, ordinal or categories. Even class member ships found by other classification methods can be re-used in HMMs or BNs.

[1] The necessity for this kind of testing resides in the fact that techniques like the MLPs and SVMs allow us to classify any data, even consisting of noise. The same holds if the number of features is unlimited.

2.2 Discriminant analysis

2.2.1 Test ban treaty—some history

Back in 1954, 9 years after the "Trinity" test—the world's first nuclear explosion was conducted in Alamogordo (New Mexico, USA), on 16 July, 1945, and the Prime Minister of India, Jawaharlal Nehru, advocated a "standstill agreement" on nuclear testing (http://www.ctbto.org/the-treaty/history). By the mid-1950s, USA and USSR started conducting high-yield thermonuclear weapon testing in the atmosphere. As the radioactive fallout from these tests increased international concern, the Partial Nuclear Test-Ban Treaty (PTBT) was signed in 1963 banning nuclear testing in outer space, the atmosphere, and underwater.

However, underground testing was not banned. Therfore, the number of tests greatly increased. The reason for not banning underground nuclear tests was partly due to verification difficulties. Nuclear test ban discussions held in 1958 provided a strong boost to fill that need. A panel of US experts considered research needs for improving the national capability in the detection and discrimination of underground nuclear explosions (see Peterson and Hutt 2014). One of the panel recommendations was the installation of standardized seismographs with accurate clocks at 100 to 200 existing seismograph stations. Even though the needs for the detection of underground nuclear tests provided a decisive kick, the boost of the monitoring systems achieved a broader target, producing valuable data needed for fundamental research in seismology.

The deployment of the "World Wide Standard Seismic Network" (WWSSN) by the USA in the 1960s formed the backbone of the surveillance system for nuclear testing carried out by the USSR and other countries, such as China and India. Many countries around the world participated in the WWSSN, making it the first true global seismic network. The stations were equipped with two types of three component seismographs: short-period sensors with a natural period of $T_0 = 0.8$ s and long-period sensors with $T_0 = 15$ s. Records on the short-period seismometers were commonly used for the estimation of the body wave magnitude m_b, whereas the surface magnitude M_S was defined on the amplitude of surface waves, which—for teleseismic earthquakes—have a dominant period of ca. 20 s and can be well read on the long-period records of the WWSSN network.

2.2.2 The M_S–m_b criterion for nuclear test identification

The distinction between nuclear tests and earthquakes exploits differences in their sources, which mirror in the characteristics of the radiated seismic signals. In a first approximation, nuclear shots are explosive events causing compression in all directions and a radiation pattern of seismic energy showing spherical symmetry. Typically, the geometrical dimensions of the sources are small and the duration of the radiated signal, which mainly travels in the form of compressional (P-) waves, is short.

Earthquakes are the result of a sudden release of energy accumulated in rock during long-lasting tectonic deformation. At some point, the material fails in a shear fracture and seismic energy radiates with a prevailing content of shear (S-) waves and, to a minor degree, in the form of P-waves. Besides, earthquake sources can reach considerable geometrical dimensions, such as tens or hundreds of km. The duration of the source signal is proportional to the dimensions of the seismic source, about 3 s for each 10 km of rupture length.

Let us suppose that an earthquake and explosion emit equal-amplitude P-waves. The earthquake also emits large shear waves, which together with the compressional waves are converted to Rayleigh waves at the earth's surface (see Blanford, 1977). Thus, the surface magnitude M_S based on Rayleigh waves from an earthquake is larger than that from an explosion even in the case of equal m_b. From these considerations, we may infer that the evaluation of M_S and m_b should provide a valid criterion for this discrimination problem. Nevertheless, there are occasional earthquakes (and explosions) for which the M_S/m_b discriminant fails.

2.2.3 Linear Discriminant Analysis

The discrimination between earthquakes and nuclear explosions is a two-class distinction problem with two variables: M_S and m_b (see Fig. 2.1). In classical univariate statistics, we can base our distinction criterion between the two groups on average and variances, asking for instance whether the average value of the group $\mathbb{A}$ is higher than that of the group $\mathbb{B}$, and the significance of the distinction resulting. The Student *t*-test is based on this consideration. Such reasoning, however, is quite meaningless for our purposes, as

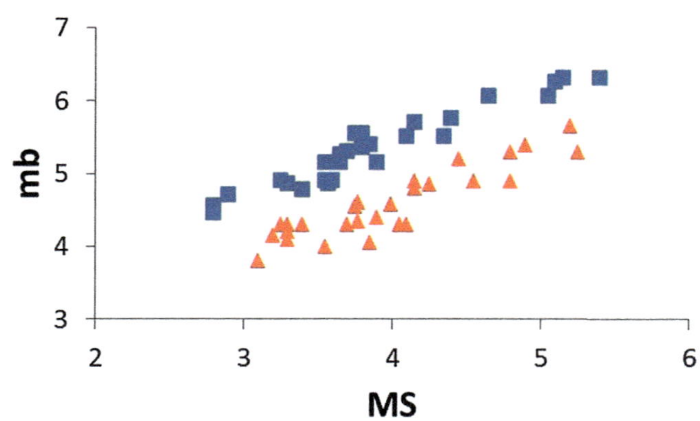

Figure 2.1
M_S—m_b relation for earthquakes (*red triangles*) and nuclear tests (*blue squares*). Values correspond to Marshall and Basham (1972).

earthquakes are not simply "larger" or "smaller" than nuclear explosions. In other words, we cannot solve the problem considering M_S or m_b only.

An example of distribution of the magnitudes Ms and m_b is shown in Fig. 2.1. In this example, group $\mathbb{A}$ corresponds to earthquakes with N_A samples and group $\mathbb{B}$ covers N_B samples of nuclear tests. Components X_i are the magnitudes. In univariate problems, the dispersion of the data is defined by its variance. As discussed in Box 2.1, the variance with respect to the single components may not suffice to describe properly the dispersion of these multivariate data sets. We use instead the covariance matrix $\boldsymbol{C}$, in which the variance (with respect to the single components) is found in the diagonal elements, whereas the covariance is given in the off-diagonal elements (see Appendix 2.1). We notice that a distinction between the two groups in Box 2.1 is enabled by a rotation of the axis. The angle of rotation is chosen so that the direction of one of the new axes coincides with the direction where the elongation, or spread, of the data clouds is maximum. The other axis is perpendicular. For the identification of the orientation of the new axes, we can exploit the covariance matrix $\boldsymbol{C}$, which is symmetric and positive semidefinite. The identification of the axis where the maximum elongation/spread is measured is an eigenvalue problem, where the largest eigenvalue describes the maximum elongation of the data cloud, and the corresponding eigenvector gives its orientation with respect to the original axes. On the contrary, the minimum dispersion is encountered along the axis corresponding to the smallest eigenvector, and this is the one where the distinction can be seen more clearly.

In our specific problem of distinction between earthquakes and nuclear tests, we describe the position of the two data sets—earthquakes and nuclear tests—in the two dimensional M_S—m_b data space by the coordinates of the centroids, which are obtained by

$$\overline{\boldsymbol{X}}_{A,B} = \sum_{i=1}^{N_{A,B}} \boldsymbol{x}_i$$

where $N_{A,B}$ indicates the number of earthquakes or nuclear test. In our example the centroids or average vectors are

$$\overline{\boldsymbol{X}}_{\mathrm{A}} = \left(\overline{M}_S, \overline{m}_b\right) = (3.99,\ 4.58)$$
$$\overline{\boldsymbol{X}}_{\mathrm{B}} = \left(\overline{M}_S, \overline{m}_b\right) = (3.89, 5.33)$$

The extension and orientation of the data sets in the M_S—m_b system of axes are given by the covariance matrices $\boldsymbol{C}_A$ and $\boldsymbol{C}_B$ of the two groups, here

$\boldsymbol{C}_A$	0.3866	0.2706
	0.2706	0.2311
$\boldsymbol{C}_B$	0.4922	0.3603
	0.3603	0.2865

Box 2.1 Marginal Distributions.

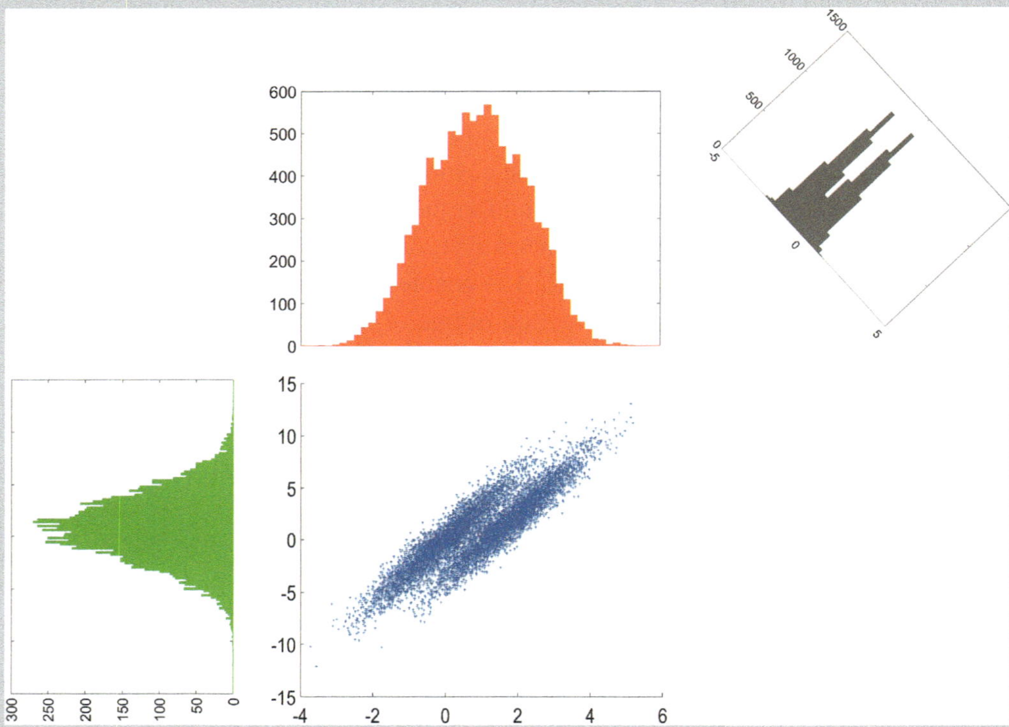

The above-mentioned reasoning is strongly related to a general problem encountered with the so-called marginal distributions. They come as a projection of a multivariate distribution to a low-dimensional representation. In the figure shown here, we demonstrate the issue with a hypothetical two-dimensional (2D) test set, created randomly from two Gaussian distributions. In the central panel of the figure, we clearly recognize a separation of the two data groups. In the projection of the distribution to one of the axes (either X1 or X2, see panels at the left and bottom), the distinction is essentially eclipsed. The drawback of marginal distributions pops up every time there is a plot for a number of variables less than those originally present in the data set. The problem can be fixed by a suitable data transformation—here a simple rotation, which allows us to distinguish the two groups—which is illustrated in the panel in the upper left corner. The user may reproduce the figure using the MATLAB™ script "S2_2" coming along with the book.

On a sheet of paper, we can easily create 2D plots, but the representation of higher dimensional distributions is a complicated issue—and this difficulty is one of the motives for applying pattern recognition techniques!

we form the pooled covariance matrix

$$C_P = \frac{(N_A \cdot C_A + N_B \cdot C_B)}{(N_A + N_B)}$$

where N_A and N_B are the number of earthquakes and nuclear tests, respectively.

We now calculate the eigenvalues λ and eigenvectors $\boldsymbol{w}$ of C_P, getting

$$\begin{aligned}\lambda_1 &= 0.6773, \lambda_2 = 0.021 \\ \boldsymbol{w}_1 &= (0.602, \; -0.7985) \\ \boldsymbol{w}_2 &= (-0.7985, \; -0.602)\end{aligned}$$

Finally, we rotate our original vectors multiplying them with the eigenvector $\mathbf{w_i}$, getting a new set of now univariate values, i.e.,

$$\begin{aligned}Z_1 &= \boldsymbol{X}^T_{\mathrm{A,B}} \cdot \boldsymbol{w}_1 \\ Z_2 &= \boldsymbol{X}^T_{A,B} \cdot \boldsymbol{w}_2\end{aligned}$$

For the sake of distinction, we should establish a threshold between the two data sets. As we use a pooled covariance matrix, we assume implicitly that $C_A \sim C_B \sim C_P$, and the separating line should pass at equal distance from the two average vectors $\overline{\mathbf{X}}_{\mathrm{A}}$ and $\overline{\mathbf{X}}_{\mathrm{B}}$. We take the global mean

$$\overline{\boldsymbol{X}}_{\mathrm{glob}} = 0.5 \cdot \left(\overline{\boldsymbol{X}}_{\mathrm{A}} + \overline{\boldsymbol{X}}_{\mathrm{B}}\right)$$

and rotate to get

$$\begin{aligned}z_1 &= \overline{\boldsymbol{X}}^{\mathrm{T}}_{\mathrm{glob}} \cdot \boldsymbol{w}_1 \\ z_2 &= \overline{\boldsymbol{X}}^{\mathrm{T}}_{\mathrm{glob}} \cdot \boldsymbol{w}_2\end{aligned}$$

and find $z_1 = -1.5874$ and $z_2 = -6.1355$. In our example, all elements of Z_1 belonging to group $\mathbb{A}$ (earthquakes) are above $z_1 = -1.5874$ and those belonging to group $\mathbb{B}$ (nuclear tests) are below (Fig. 2.2). In the original system of axes, the discriminating line passes through $\overline{\boldsymbol{X}}_{\mathrm{glob}} = (3.9435, 4.9611)$ and has a slope of 0.75. A MATLAB™ script "S2_1"reproducing Fig. 2.2 is included.

The approach outlined here exploits the concept of PCA and can be easily understood from the underlying geometrical concepts. A more formal description of FLDA is given in Appendix 2.1, where we also discuss the case where $C_{\mathrm{A}} \neq C_{\mathrm{B}}$ and the discrimination function becomes quadratic.

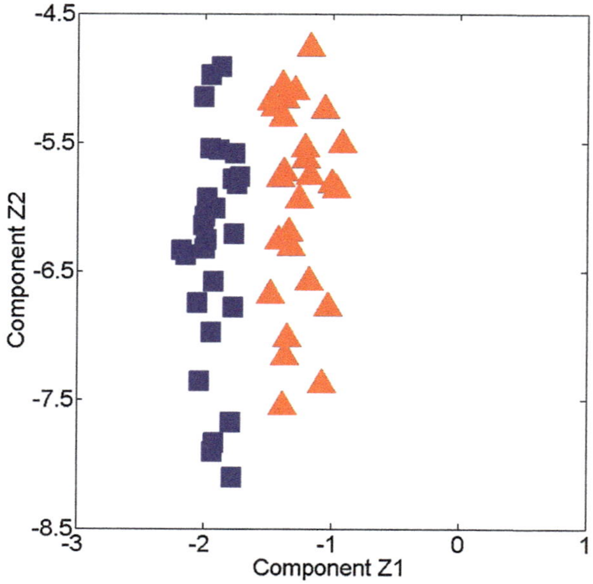

Figure 2.2
Earthquakes and nuclear tests after rotation.

2.3 The linear perceptron

The perceptron concept was invented by Rosenblatt (1958) as a probabilistic model for information storage in the brain. Fig. 2.3 outlines the basic characteristics of a simple linear perceptron. In the jargon of artificial neural networks, $\boldsymbol{x}$ is referred to the input layer. Interconnections between the neurons are activated or inhibited according to the stimulus arriving at a sensorial layer. The degree of activation (or weights) of the interconnection can be adjusted by comparing the response of the network to some known information ('target'). In the 'untrained' network, the weights of the interconnections have some random values and are successively adjusted in a manner that the network finally responds in the desired way.

Rosenblatt was able to proof the convergence of the learning, i.e., that—in linear problems—it is possible to adjust the network weights such that their response is close to the desired one. For the intriguing similarity of the computer-based processing scheme to learning process encountered in brains, those systems of interconnected neurons are often referred to as "Artificial Neural Networks". However, these somewhat mystical terms lead to overly optimistic expectations, finally reflecting discredit on the whole field of research. Minsky and Papert (1969) reformulated the perceptron approach in rigorous terms of prediction calculus. Their harsh criticism, in particular with respect to limits of the approach in nonlinear classification problems (impossibility to resolve the simple XOR

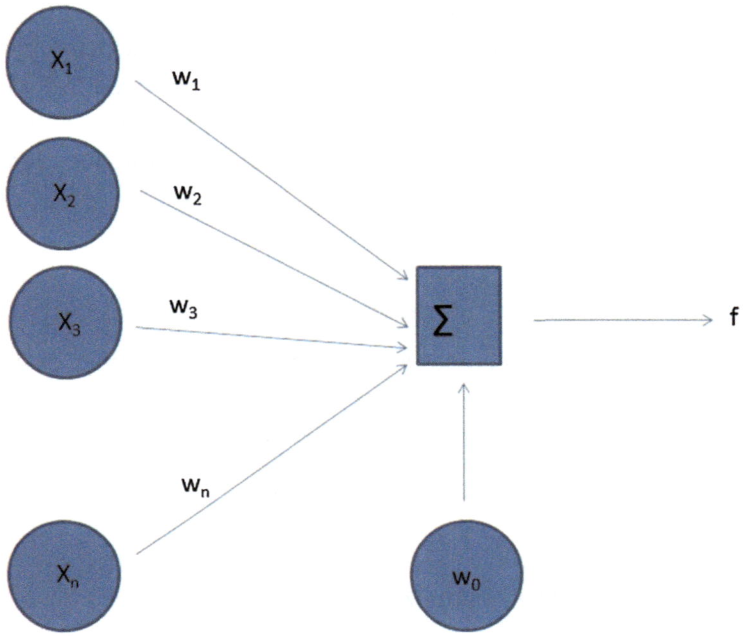

Figure 2.3
Flow chart of the basic perceptron model. We distinguish an input layer where the data vectors $\boldsymbol{x}$ with components x_i are stored. A weight w_i is applied to each of the x_i. Lines representing the weights w_i are often referred to as "synapses", whereas the nodes (both the input nodes x_i as well as the processing units, such as the "Σ") are "neurons". Finally, the sum of the weighted x_i is further transformed applying an activation function f (in this context f is linear). During the learning phase, the output of the perceptron is compared to a desired output; weights w_i are adjusted in order to minimize the difference of the perceptron's output and the target.

problem), brought the perceptron concept almost to an eclipse. In the following decades, it turned out that the limits can be overcome by considering more complex configurations and nonlinear activation functions. Consequently, the method remains on the screen of researchers working on classification problems.

Here we outline the perceptron concept in terms of prediction calculus and underscore the link to the more conventional, statistics-based approaches outlined in Section 2.1. In the PCA and FLDA (see Appendix 2.1) example, we solved the discrimination problem by reducing the multivariate variable $\boldsymbol{X}$ to a one-dimensional variable Z, carrying out a suitable linear combination of the components. We have discussed generalizations allowing for differences in the covariance matrices between the groups. In both cases, we assume that the distance of a sample $\boldsymbol{x}_i$ to the average $\boldsymbol{\mu}_{\mathrm{k}}$ of the k-th group or class is a measure for the probability that the sample belongs to this group. Finally, we assign a class membership to the group for which the highest probability is achieved.

In the perceptron approach, we follow a slightly different idea (see, e.g., Bishop, 1995; Theodoridis and Koutroumbas, 2009). We ask for a function $g(\boldsymbol{x})$ which separates the members of two groups $\mathbb{A}$, $\mathbb{B}$ to the highest degree.[2] Once established, we wish to use it as a decision criterion for the class membership of the samples. Ideally, it will be able to separate all samples, that is, the class memberships using our distinction element are the ones originally defined.

We start with the linear case:

$$g(\boldsymbol{x}) = \boldsymbol{w}^T\boldsymbol{x} + \boldsymbol{w}_0 \tag{2.1}$$

which is a line for a 2D case and becomes a (hyper)plane if the number of dimensions is greater. Having the so-called biases $\boldsymbol{w}_0 \neq 0$, $g(\boldsymbol{x})$ will not pass through the origin. We extend the dimension of the vector $\boldsymbol{x}$ considering $\widetilde{\boldsymbol{x}} = (\boldsymbol{x}, \mathbf{1})$ and search for $g(\widetilde{\boldsymbol{x}})$ such that

$$\boldsymbol{w}^T\widetilde{\boldsymbol{x}} > 0 \text{ for all } \widetilde{\boldsymbol{x}} \; \varepsilon \; \mathbb{A} \tag{2.2a}$$

$$\boldsymbol{w}^T\widetilde{\boldsymbol{x}} < 0 \text{ for all } \widetilde{\boldsymbol{x}} \; \varepsilon \; \mathbb{B} \tag{2.2b}$$

The solution is searched in terms of optimization; to this purpose, we choose a cost function

$$J(\boldsymbol{w}) = \sum_{\widetilde{\boldsymbol{x}} \in \mathbb{Y}} \left(\delta_x \boldsymbol{w}^T\widetilde{\boldsymbol{x}}\right) \tag{2.3}$$

with $\mathbb{Y}$ being the set of misclassified samples. $\delta_x = -1$ for$\widetilde{\boldsymbol{x}} \; \varepsilon \; \mathbb{A}$ (but classified as $\mathbb{B}$), $\delta_x = 1$, and $\widetilde{\boldsymbol{x}} \; \varepsilon \; \mathbb{B}$ (but classified as$\mathbb{A}$), making sure that J is positive at any time and zero if no samples were misclassified. Compared to PCA/FDLA or quadratic discrimination, the solution for $\boldsymbol{x}$ is not unique. Typically, iterative procedures have to be applied for its identification (see Appendix 2.2).

The iteration will formally not converge if the two groups cannot be separated by a linear element. We could accept a number of misclassifications but this tolerance had to be fixed a priori, which is a severe drawback of such an approach. As we have to live with errors and noise, we may attempt to design a linear classifier where the desired output is again ± 1, but the output of our classifier may not be exactly equal to the desired one.

We redefine our cost function

$$J(\boldsymbol{w}) = E\left(\left(y - \boldsymbol{x}^T\boldsymbol{w}\right)\right)^2 \tag{2.4}$$

[2] It is an intriguing property of the perceptron approach that the degree of separation is a user-defined measure. Rather than depending on the statistical properties of the data sets—as the PCA or the FLDA–it is based on the a priori defined target classes. It also allows the application of specific cost functions that weigh individually the contribution of each sample to the error.

$E(.)$ is the expectation. $J(\boldsymbol{w})$ corresponds to the mean square error (MSE). Taking the derivative and setting to 0 (looking for the minimum of J), we find

$$2E\left(\boldsymbol{w}\left(\boldsymbol{x}\left(y-\boldsymbol{x}^{T}\boldsymbol{w}\right)\right)\right)=0 \tag{2.5}$$

and solve for $\boldsymbol{w}$

$$\boldsymbol{w}=E\left(\boldsymbol{x}^{T}\boldsymbol{x}\right)^{-1}E(\boldsymbol{x}y) \tag{2.6}$$

The term $E(\boldsymbol{x}^{T}\boldsymbol{x})$ is the so-called "autocorrelation matrix" of $\boldsymbol{x}$ (corresponds to the covariance matrix if averages of $\boldsymbol{x}$ are zero), and $E(\boldsymbol{x}y)$ is the cross-covariance of $\boldsymbol{x}$ and y, i.e.,

$$E(xy)=E\left(\begin{pmatrix} x_1\, y \\ \vdots \\ x_l\, y \end{pmatrix}\right)$$

The mean square optimal vectors are obtained by solving the linear equation above. We can apply this scheme to our m_b/M_S example of earthquakes and nuclear explosions. For this purpose, we have created a small MATLAB script named "S2_3", which starts with the steps:

1. Augment the dimension of the data vectors by adding a column of '1's.
2. Form the expectation matrix $E(\boldsymbol{x}^{T}\boldsymbol{x})$ and take its inverse.
3. We need the vector $\boldsymbol{y}$ for the target output, and calculate $\boldsymbol{w}$ as above.
4. Calculate the vector of outputs from the product $\boldsymbol{x}^{T}\boldsymbol{w}$, considering only positives.
5. Sum over outputs multiplied with the targets. Here we consider only negative products, i.e., misclassifications. As we noticed earlier, earthquakes and nuclear explosions can be perfectly separated with a linear function. Therefore, we shall get a zero error.

We can also create a test data set. This can be achieved by generating values randomly, assuming a multivariate Gaussian with the averages and covariance of our original data sample (for instance, using the MATLAB "mvnrnd" command). As the earthquakes and nuclear explosions were well separated, we get zero error also for the random test set. We may simulate a situation in which we have a stronger noise in the test data (Fig. 2.4). This helps us to understand whether our distinction criterion is robust, i.e., whether we can trust the result when it is applied to new data (that is, the target output—earthquake or nuclear explosion—is not known a priori). We shall come back to the test issue repeatedly in our book as it forms a critical issue, especially in Chapter 4.

In our example, we just multiplied the covariance matrix of our random set by a factor 4, simulating a stronger presence of noise. In fact, we get some small error, as for some samples of the test data set, linear separation is not possible (Fig. 2.5).

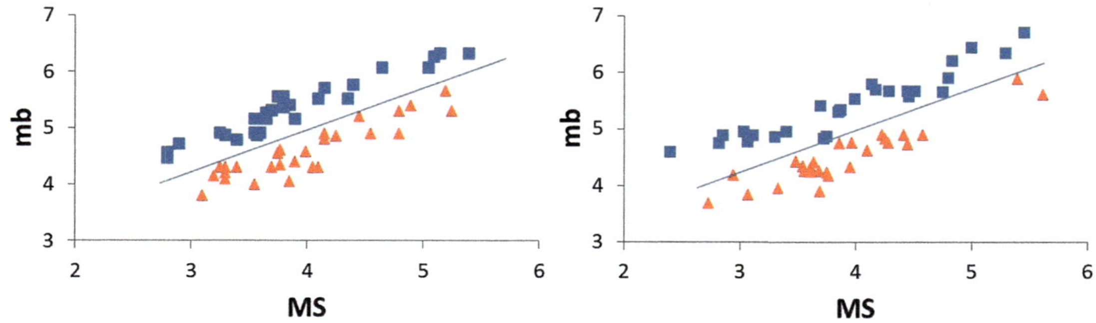

Figure 2.4
Left panel: original data set—blue squares are the nuclear explosions and red triangles are the earthquakes. Right panel: randomly generated test set using the averages and covariance estimated from the original data set. Note that the test set was created using the pooled covariance matrix of the two data groups, which causes slight differences in the aspect of the data clouds. Nonetheless, our distinction rule yields zero error for the test set.

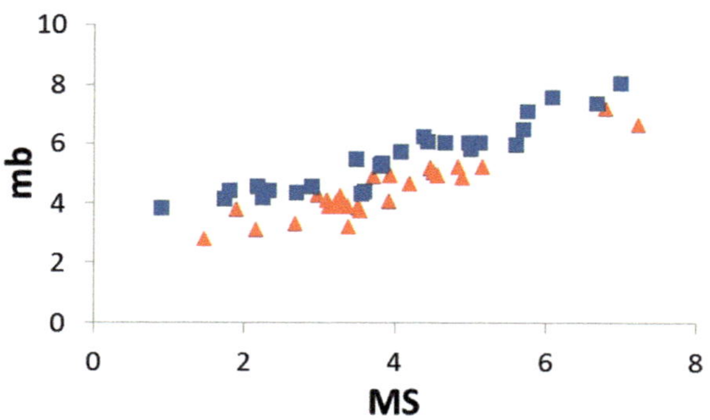

Figure 2.5
Randomly generated test set, using the averages from the original data set and covariance four times greater than those of the original data.

Fig. 2.6 shows the calculated class membership values, $\boldsymbol{x}^T\boldsymbol{w}$. The blue circles are those obtained for our random simulation of M_S–m_b pairs (see Fig. 2.4, left panel). From the figure, we can see that 27 correspond to randomly generated M_S–m_b pairs of earthquakes (index 1–27) and 27 correspond to those of nuclear tests (index 28–54). All class memberships for the simulated earthquake data are found with class membership below 0 and all nuclear tests have class memberships above 0. In other words, all samples fall into the red fields in Fig. 2.6, which indicate the range of matching classification. We obtain a 100% separation of the two data groups. We can apply our discrimination criterion if the averages and the pooled covariance matrix are of the same order as in the original data set. The red triangles show the scores when we simulated M_S–m_b pairs with a covariance matrix four times greater than the original one (see Fig. 2.5). Here we get a

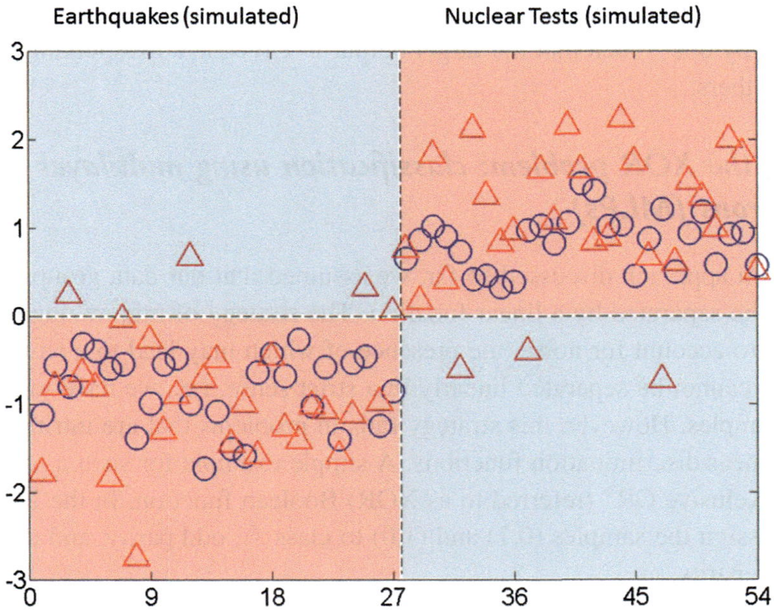

Figure 2.6

Classification scores applying the M_S—m_b discrimination criterion to randomly generated test data. The blue circles stand for data where averages and covariance matrix correspond to the original M_S—m_b pairs of earthquakes and nuclear tests. Red triangles stand for data generated with covariance four times greater than in the original data. All blue circles fall into the red areas, i.e., no misclassification is observed. Some of the red triangles fall into the blueish areas, i.e., are misclassified.

few positive class membership values for the simulated earthquakes (the desired output is negative) and some negative for the nuclear tests (where the target is positive). In other words, having more noise in the test data set than in the original one, we get some misclassifications, which fall into the blueish fields shown in Fig. 2.6.

The treatment of the multiclass problem with *M* classes is straightforward. We modify our cost function

$$J(\boldsymbol{w}) = E\Big(\big(y - \boldsymbol{x}^T\boldsymbol{w}\big)\Big)^2 \tag{2.7}$$

such that

$$E\big(\|\boldsymbol{y} - \boldsymbol{W}^T\boldsymbol{x}\|\big) = E\left(\sum_i \big(y_i - \boldsymbol{w}_i^T\boldsymbol{x}\big)^2\right) = min$$

Here $\boldsymbol{y}$ is the vector of class memberships of dimension *M*, and has a '1' for the class *i* to which **x** is supposed to belong and '0's for all other classes. $\boldsymbol{W}$ is the matrix of weight vectors $\boldsymbol{w}_i$, i running from 1 to M. It is sufficient to design each of the discriminant functions—one by one—such that the target output is 1 for the corresponding class and -1 for all the others.

2.4 Solving the XOR problem: classification using multilayer perceptrons (MLPs)

In the perceptron approach discussed so far, we assumed that our data groups—at least in principle—can be separated by a linear function. The strategy of minimizing the squared error allows us to account for noise, the presence of which may lead to a situation where the data groups cannot be separated linearly in a strict sense, i.e., we have to accept some misclassified samples. However, this strategy fails in problems that are intrinsically not solvable with linear discrimination functions. A simple example for such a problem is the well-known "eXclusive OR" (referred to as XOR) Boolean function. In the XOR problem (Fig. 2.7), we assign the samples (0,1) and (1,0) to class $\mathbb{A}$, odd parity, and (0,0) and (1,1) to class $\mathbb{B}$, even parity.

It is evident that the two classes cannot be separated by a single linear element, but we have to insert a second element as shown in Fig. 2.8. In the architecture of the perceptron, this is achieved by adding a new layer of nodes ("neurons") and appropriate "synaptic" weights, obtaining a "Multilayer Perceptron" (MLP). Here we have one input layer, a second layer, called the "hidden" layer, and an output layer (see Fig. 2.9). The hidden nodes can be referred to as "processing units", which become active under specific conditions. For the moment, we assume a threshold or step activation function and we assign -1 for negative values and 1 for positive values. The formalism of such an MLP consists of two steps. In the first step, the input vector $\boldsymbol{x}$ is transformed ("mapped") to a new one, $\boldsymbol{y} = (y_1, y_2)^T$.

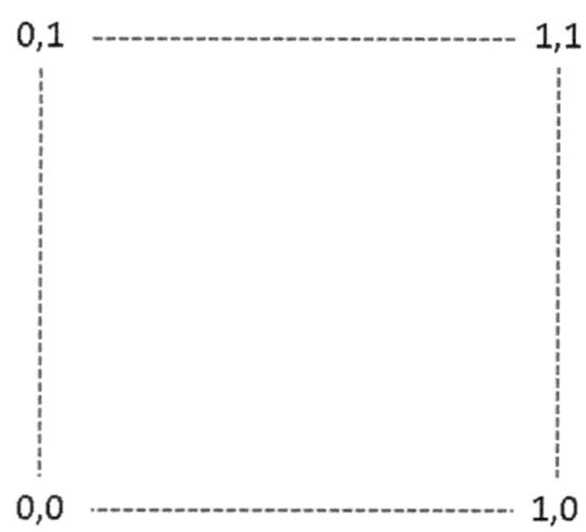

Figure 2.7
The XOR problem.

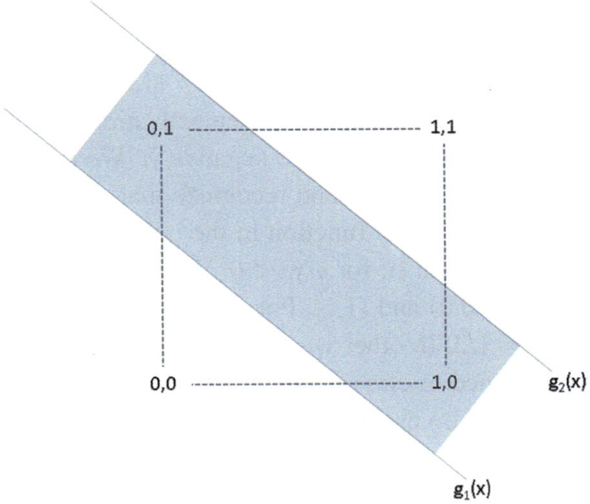

Figure 2.8
Decision area for the XOR problem.

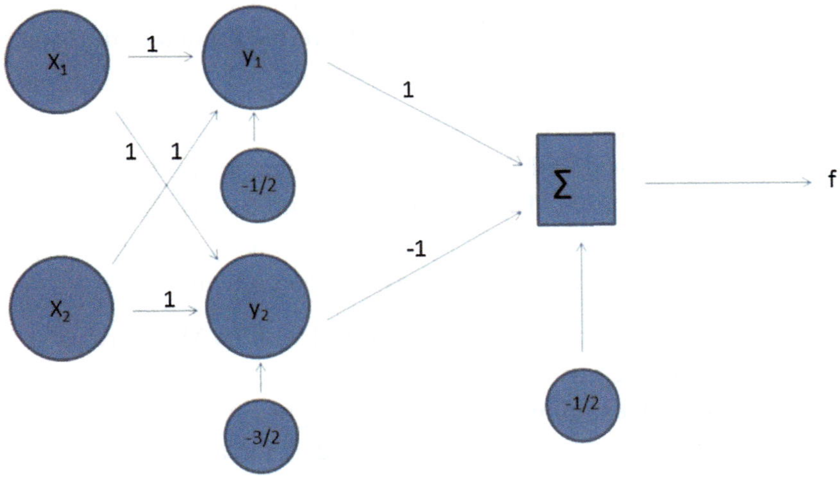

Figure 2.9
Perceptron with one hidden layer. Biases (constants) are represented in the smaller circles.

In the next step, we define the decision element separating the new sample vectors $\mathbf{y}_i$. Suppose the configuration of the MLP shown in Fig. 2.9, which transforms to the equations for the decision elements:

$$y_1 = g_1(\boldsymbol{x}) = x_1 + x_2 - 1/2 = 0$$
$$y_2 = g_2(\boldsymbol{x}) = x_1 + x_2 - 3/2 = 0$$
$$g(\boldsymbol{y}) = y_1 - y_2 - 1/2 = 0$$

(see, e.g., Bishop, 1995). The y_i depends on the input values x_i and some constants that are referred to as "biases" in the neural network jargon. The sample $\boldsymbol{x} = (0,0)$ transforms to $\boldsymbol{y} = (-1/2, -3/2)$, $\boldsymbol{x} = (1,1)$ becomes $\boldsymbol{y} = (3/2,1/2)$, $\boldsymbol{x} = (1,0)$ is mapped to $\boldsymbol{y} = (1/2,-1/2)$, and $\boldsymbol{x} = (0,1)$ to $\boldsymbol{y} = (1/2, -1/2)$. Now apply the step activation function (or sign function) to the y_i, i.e., $z_i = 1$ for positive and $z_i = -1$ for negative y_i. We may plot the transformed values in a new system with the y_i as axes and recognize that we get a linearly separable problem. We use again a step activation function in the output, i.e., we set $z = 1$ for $g(\boldsymbol{y}) > 0$, class $\mathbb{A}$, and $z = -1$, class $\mathbb{B}$, for $g(\boldsymbol{y}) < 0$. We find $g(\boldsymbol{y}) = -1/2$ for the two samples with even parity, $\boldsymbol{x} = (0,0)$ and $(1,1)$. For our samples with odd parity, $\boldsymbol{x} = (1,0)$ and $\boldsymbol{x} = (0,1)$, we find $g(\boldsymbol{y}) = 1/2$. In other words, our even-parity samples are mapped to a negative range and are assigned to class $\mathbb{B}$ ($z = -1$), and the odd parities are mapped to a positive range, i.e., $z = 1$, and are assigned to class $\mathbb{A}$.

In our XOR case, the solution of the classification problem resides in the addition of a second decision element, i.e., the function $g_2(\boldsymbol{x})$. In the MLP scheme, this corresponds to the addition of the second layer (the hidden layer). We can add more units defining more complex decision boundaries. Note that each hidden unit divides the input space with a line or hyperspace. That way, the decision boundaries surround a convex region. The classification capabilities of the two-layer MLP are therefore limited to decision bodies with a convex hull (see Bishop, 1995).

In Fig. 2.10, we reproduce a figure by Bishop (1995), illustrating a few decision boundaries that can be generated with MLPs having no, one, or two hidden layers. It can be demonstrated (see Bishop, 1995; Theodoridis and Koutroumbas, 2009) that, by applying the step activation function in the hidden units, at least two hidden layers are necessary for solving an arbitrarily complex classification problem.

Even though any classification problem could be solved with an MLP having at least two hidden layers (applying the step activation function in the nodes), those configurations are not very popular yet, as they entail many nodes and weights. A further problem is the step activation function, which is not differentiable and hinders the application of gradient methods for the search of optimum weights. For a more detailed discussion, we refer the interested reader to the already mentioned textbook by Bishop (1995) and to Theodoridis and Koutroumbas (2009) (Box2.2)

2.4.1 Nonlinear perceptrons

The identification of the weights using iterative schemes, such as gradient search methods, is facilitated when activation functions in the units are continuous and differentiable. A well-known example is the sigmoid function

$$f(x) = 1/\left(1 + e^{-\alpha x}\right) \tag{2.10}$$

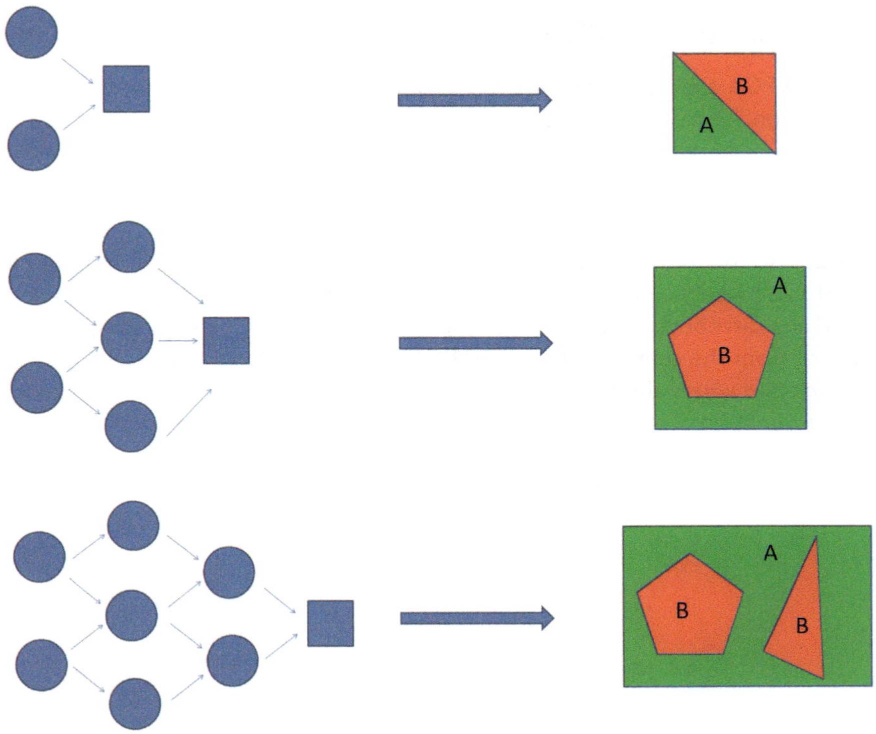

Figure 2.10
Decision boundaries created with perceptrons of varying configuration. Note that this holds for nodes with a step activation function (see Bishop, 1995). Biases are omitted for the sake of simplicity.

Such a sigmoidal function can be considered as a generalization of the step activation function, allowing for smooth flanks. A variation of this is given by a function like

$$f(x) = 2\left(1 + e^{-\alpha x}\right) - 1 \tag{2.11}$$

which varies between −1 and 1 (see Fig. 2.11).[3]

[3] A frequently used alternative to the sigmoid is the hyperbolic tangent *(tanh)* function, which has a shape similar to the curve shown in Fig. 2.11. Besides, these so-called "radial base functions" (RBF) have gained some popularity. They have the general form $f(||\boldsymbol{x} - \boldsymbol{c}||)$, where $\boldsymbol{c}$ is some center. The most widely used form of f is a Gaussian function. Perceptrons with RBF activation functions are reported to learn more rapidly than their multilayer counterparts but have a minor performance with respect to their generalization properties. Centers $\boldsymbol{c}$ may not be known and have to be identified with a clustering method, typically K-means (see Chapter 3). In that case, RBF networks represent a scheme where a combination of supervised and unsupervised learning is used. For more details, see e.g., Skorohod (2017).

Box 2.2 XOR and statistical discrimination.

In the FLDA and PCA approach, we have been using statistical properties of the groups for their discrimination. These parameters are namely the averages and the covariance. In both the FLDA and PCA approach, we have been assuming that both groups have the same covariance matrix, which is estimated by calculating the so-called pooled covariance matrix. This leads to a linear discrimination function by lines and (hyper)planes.

In Appendix 2.1, we discuss on how to use a likelihood approach for discrimination. In that case, we used the covariance matrix for the definition of a metric, which—once we assume a certain underlying distribution—can be interpreted as a probability that a sample placed at a certain distance from the centroid of a group belongs to that group. The likelihood strategy also allows for cases where the covariance matrices of the data group differ. This case leads to nonlinear discrimination functions. In the example shown here, we have two ensembles where diag $\mathbf{C}_1$ = diag $\mathbf{C}_2$, but off-diagonal elements are of opposite sign. Averages of both groups coincide and the distribution of both classes is bivariate Gaussian.

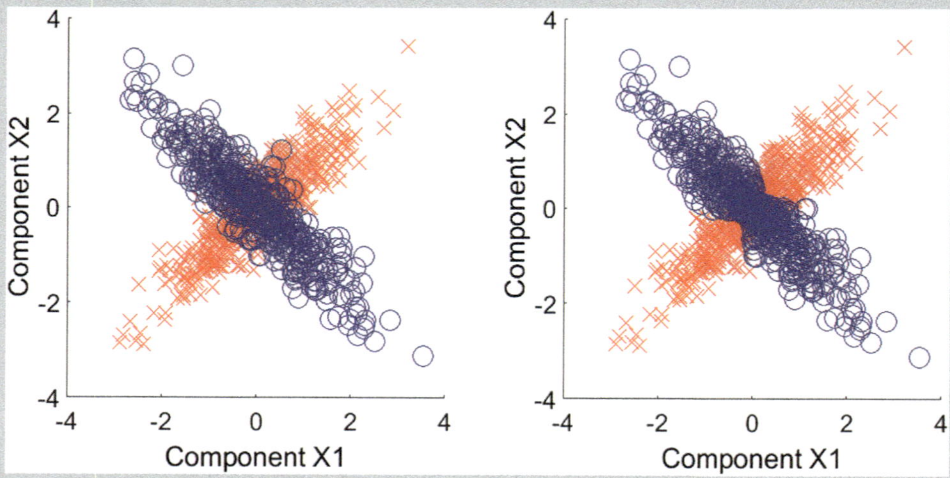

In the figure, to the left, we show the two data groups with red and blue symbols indicating their target classification. On the right, we see the same data, but class membership has been assigned a posteriori based on a likelihood approach. There are a few differences around the centroids. This is no surprise as the two Gaussians are very similar around the centroids. In general, we notice that the a posteriori classification matches the target pretty well. We may identify the data lying along the diagonal as those with more or less even parity, such as the elements (0,0) and (1,1) in the XOR problem. The other data in perpendicular direction represent the odd parity data (i.e., they correspond to the elements (0,1) and (1,0) in the XOR problem). The reasoning outlined here follows a Bayesian strategy, i.e., we assume that the probability that a sample belongs to a class can be inferred from the Mahalanobis distance to the centroid of the class. In the perceptron approach, we avoid this a priori assumption. In unsupervised classification, however, the application of a likelihood strategy can be very successful.

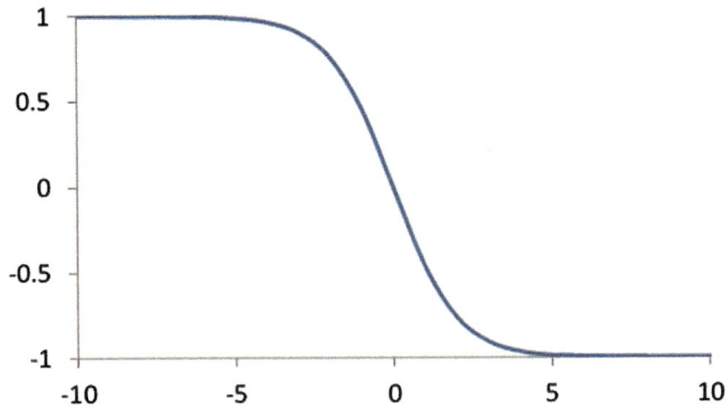

Figure 2.11
The sigmoid function.

The clue of using those sigmoidal activation functions resides in the theorem of Cybenko (1989), which says:

Denoting the n-dimensional unit cube as I_n and the space of the continuous functions as $C(I_n)$, then the final sums of

$$G(\boldsymbol{x}) = \sum_{i=1}^{n} \beta_i \sigma\left(\boldsymbol{w}_j^T \boldsymbol{x} + \vartheta_j\right) \tag{2.12}$$

where $\boldsymbol{w} \in R^n$, β_j, $\vartheta_j \in R$, and $\sigma(.)$ a sigmoidal function, are dense in $C(I_n)$.

In other words, it is possible to approach any function of arbitrary degree of complexity by a two-layer neural network, whose hidden neurons apply a nonlinear weighting function. Thus classes, whose envelopes are of complicated shape, can be anyway distinguished from each other, and it is possible to establish any kind of regression function between input and output vectors.

We can denote the output of an MLP with one hidden layer (see, e.g., Hornik et al., 1989) as

$$y_k(\boldsymbol{x}) = \sum_{j=1}^{N_H} c_j \left[\sigma\left(\boldsymbol{w}_j^T \boldsymbol{x}\right) + t_j\right] + c_k \tag{2.13}$$

(see Fig. 2.12), where N_H is the number of nodes in the hidden layer. The output consists of the weighted sum of sigmoidal functions applied to the units of the hidden layer. We can understand this operation as a kind of transform, similar to the well-known Fourier transform in Fourier analysis, where we represent our original data as the sum of sine and

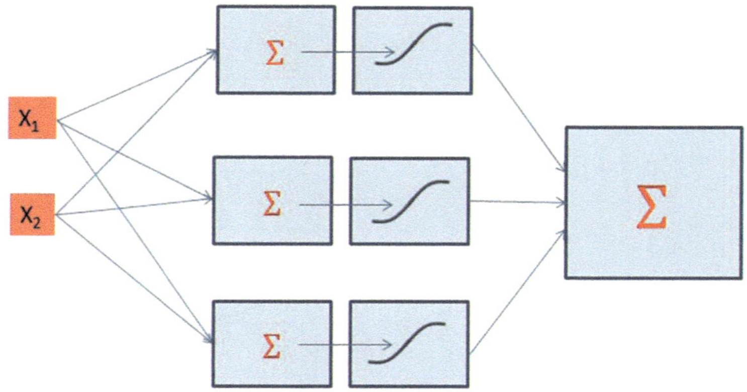

Figure 2.12

A perceptron with one hidden layer and nonlinear activation functions in hidden nodes. This configuration represents the minimum scheme for a perceptron able to resolve general nonlinear classification problems.

cosines. The accuracy that can be achieved depends on the number of elements—hidden nodes and number of sigmoidal functions. The higher the number, the better the accuracy. As before, the determination of the weights is an optimization problem. Among the most popular ones, there are the backpropagation algorithm (originally proposed by Werbos, 1974), simulated annealing (Kirkpatrick et al., 1983), and genetic algorithms (e.g., Holland, 1975, 1992; Goldberg, 1989). Even though being very powerful, the latter are highly time consuming; this prevents their application to larger networks. Backpropagation is considerably faster, but one may be trapped in local minima. However, this problem can often be fixed by starting over with a new set of initialization weights or a slight change to the number of hidden neurons. Despite that backpropagation does not always lead to a global optimum of the weights, it is the most frequently applied method.

The essence of backpropagation relies on the fact that the error encountered in the output can be traced back (backpropagated) to the weights in a simple way following the Generalized Delta Rule. During the training, the data vectors are passed one by one to the network updating the weights by a fraction η of the backpropagated error. The choice of the learning parameter depends on the problem, a small value leads to a better but slower convergence whereas too high values will make the iteration bounce between solutions with an unsatisfying error. The interested reader may find more details and practical hints on training problems in textbooks like Freeman and Skapura (1992). Here we outline the backpropagation algorithm in Appendix 2.2.

2.5 Support vector machines (SVMs)

In SVMs, we try to separate the examples belonging to different a priori known classes by a gap that is as wide as possible. The elements that delineate the margins of the gap are the so-called "support vectors" from which the name of the techniques is derived. In the linear case, the identification of the widest gap leads to a unique solution, which is not guaranteed by a linear perceptron approach (see Section 2.3). Though being essentially based on the identification of a linear discriminating function, nonlinear classification problems can be tackled by applying the so-called "Kernel Trick". The trick consists in applying a suitable mapping function of the original feature space to a higher dimensional one, at the same time maintaining numerical efficiency. In the following, we give a short description of the SVM concept, starting from the linear problem and outlining the "Kernel Trick".

SVMs were invented by Vapnik in 1995 (see Vapnik, 1999; Vapnik and Kotz, 2006) and are by now one of the most popular methods in supervised machine learning. Among the rich literature on the subject, we mention here Campbell and Ying (2011), Han et al. (2011), Schöllkopf and Smola (2002), Theodoridis and Koutroumbas (2009).

2.5.1 Linear SVM

Similar to the perceptron, we start with the linear separable problem, i.e., we search a linear function given in Eq. (2.1)

$$g(\boldsymbol{x}) = \boldsymbol{w}^T\boldsymbol{x} + w_0 = 0$$

which separates our two groups at best. In the linear perceptron, we considered the term "at best" simply defined by the number of misclassified samples, i.e., samples ending up on the wrong side of the discrimination function. That way the solution of the discrimination problem is not unique, however. In Fig. 2.13, we recognize various lines separating the two groups without any misclassification, for instance the two dotted lines, the thin solid and the fat line right in the middle of the two groups. In the SVM approach, we raise the request on the quality of separation. Indeed, we look for the elements that separate the convex hulls of the two groups with the largest margin.

In Fig. 2.13, we delineate this margin with two thin solid lines. These lines touch the border elements of the two groups, highlighted as circles with orange and blue rings, which are addressed to as "support vectors". The fat line represents the optimum discrimination element. From conventional math, we calculate the width of the margin by $2/||\boldsymbol{w}||$.

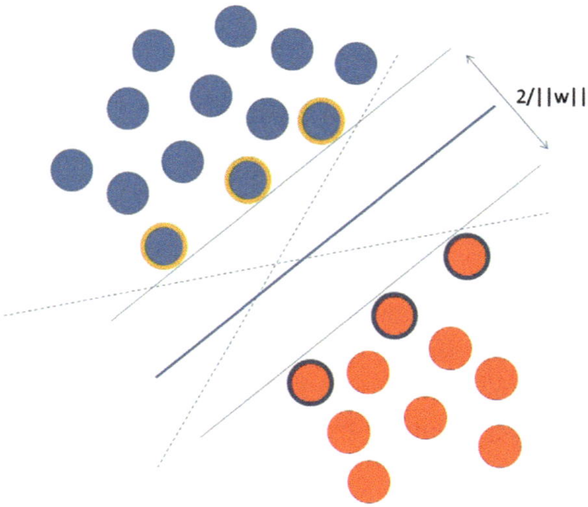

Figure 2.13
The (linear) SVM problem. The support vectors are marked with blue and orange rings around the corresponding circles.

As in the linear perceptron, we require

$$\mathbf{w}^T\mathbf{x} + w_0 \geq 0 \ \forall \mathbf{x} \in \mathbb{A}$$
$$\mathbf{w}^T\mathbf{x} + w_0 \leq 0 \ \forall \mathbf{x} \in \mathbb{B}$$

or

$$y_i\left(\mathbf{w}^T\mathbf{x}_i + w_0\right) \geq 1 \tag{2.14}$$

for all $i = 1\ldots$N.

In this case, we define the cost function as

$$J(\mathbf{w}, w_0) = 1 / 2\|\mathbf{w}\|^2 \tag{2.15}$$

which has to be minimized. This corresponds to maximizing the width of the margin between the two groups. The two elements delineating the margins are defined by the vectors for which

$$\mathbf{w}^T\mathbf{x} + w_0 = \pm 1 \tag{2.16}$$

Those vectors are referred to as "support vectors".

The identification of the minimum of the cost J thus becomes a nonlinear (quadratic) optimization problem with linear constraints. Optimization problems with constraints are typically tackled using the Lagrange multipliers, see Appendix 2.3.

Finally we have to identify the optimum Lagrange multipliers such that

$$max_\lambda = \left(\sum_i \lambda_i - 1/2 \sum_{ij} \lambda_i \lambda_j y_i y_j \boldsymbol{x}_i \boldsymbol{x}_j \right) \tag{2.17}$$

subject to

$$\Sigma_i \lambda_i y_i \geq 0$$
$$\lambda \geq 0$$

In the real world, we are faced with noisy data. Suppose the gross of our data remains separable with a linear element. We want to maintain the concept of linear classification, despite some few samples behaving as outliers and falling on the wrong side of the hyperplanes. For SVMs, where separation relies critically on the marginal samples—support vectors are at the margins—those outliers may become a nasty phenomenon. For this reason, we introduce the so-called slack variables ξ.

We distinguish between points where

$$0 \leq y_i \left(\boldsymbol{w}^T \boldsymbol{x}_i + w_0 \right) \leq 1 \tag{2.18a}$$

i.e., points which fall between the hyperplanes but are anyway correctly classified,and

$$y_i \left(\boldsymbol{w}^T \boldsymbol{x}_i + w_0 \right) < 0 \tag{2.18b}$$

i.e., definitely misclassified samples.

We generalize our criterion to

$$y_i \left(\boldsymbol{w}^T \boldsymbol{x}_i + w_0 \right) \geq 1 - \xi_i \tag{2.19}$$

where ξ_i is the slack variable. For $0 < \xi_i \leq 1$, samples are inside the space between the two hyperplanes, but are still correctly classified. Samples, for which the slack variable $\xi_i \geq 1$, are misclassified.

The cost function reads as

$$J\left(\boldsymbol{w}, w_0, \xi = 1/2 \|\boldsymbol{w}\|^2 + C \sum_i I(\xi_i) \right) \tag{2.20}$$

with I being 1 for positive ξ_{ii}. That means that we wish to maximize the distance between the two hyperplanes (as before), minimizing the number of samples for which $\xi_i > 0$. C is a parameter that weights to which degree the number of samples $\xi_i > 0$ is considered to be important.

Finally, we end up with

$$max_\lambda = \left(\sum_i \lambda_i - 1/2 \sum_{ij} \lambda_i \lambda_j y_i y_j \boldsymbol{x}_i \boldsymbol{x}_j \right) \tag{2.21}$$

subject to

$$0 \le \lambda_i \le C$$

and

$$\sum_i \lambda_i y_i \ge 0$$

More details on this can be found, e.g., in Theodoridis and Koutroumbas (2009). Corresponding standard solutions can be found in libraries like MATLAB™. The MATLAB script "S2_4" coming along with this book exploits routines downloaded from the companion website of Theodoridis et al. (2010) (http://booksite.elsevier.com/9780123744869). These routines are based on the so-called Sequential Minimal Optimization (Platt, 1999; Mattera et al., 1999) and we address the interested reader to this material.

2.5.2 Nonlinear SVM, kernels

In the perceptron approach, we were able to overcome the original limitations to linear separable classes by applying specific activation functions in the nodes, in particular the sigmoid activation function. In SVMs, we can follow a strategy of trying to transform the nonlinear distinction problem to a linear one. This can be achieved by applying a mapping

$$\boldsymbol{y} = [f_1(\boldsymbol{x}), f_2(\boldsymbol{x}) \ldots f_k(\boldsymbol{x})] \tag{2.22}$$

and then work with the $\boldsymbol{y}$, asking for functions $f(.)$ which assure that

$$\begin{aligned} w_0 + \boldsymbol{w}^T \boldsymbol{y} &> 0 \; \forall \boldsymbol{x} \in \mathbb{A} \\ w_0 + \boldsymbol{w}^T \boldsymbol{y} &< 0 \; \forall \; \boldsymbol{x} \in \mathbb{B} \end{aligned}$$

In terms of the original feature vector **x**, the discrimination function has the form

$$g(\boldsymbol{x}) = w_0 + \sum_i w_i f_i(\boldsymbol{x}) \tag{2.23}$$

i.e., an approximation in terms of preselected interpolation functions $f_i(.)$. For instance, in regression we may generate a nonlinear functional from a sequence of polynomials. Here let us consider again the XOR problem discussed earlier. Choosing a mapping

$$\boldsymbol{y} = [x_1, x_2, \; x_2 x_2]$$

we map [0,0] to [0,0,0], [1,1] to [1,1,1], [1,0] to [1,0,0], and [0,1] to [0,1,0]. The vertices of the XOR-square in the original system of axes $[x_1, x_2]$ are mapped to vertices of a cube in the new system of axes $[y_1, y_2, y_3]$. which can be separated by a plane

$$y_1 + y_2 - 2y_3 - 0.25 = 0$$

thus the distinction function is

$$g(\boldsymbol{x}) = -0.25 + x_1 + x_2 + 2x_1x_2$$

and we assign class $\mathbb{A}$ if $g(\boldsymbol{x}) > 0$ and $\mathbb{B}$ vice versa (Box 2.3).

Recall that the linear separation problem is solved by

$$g(\boldsymbol{x}) = \boldsymbol{w}^T\boldsymbol{x} + w_0$$

In the SVM approach, we find the weights from the (support) vectors $\boldsymbol{x}_i$

$$\boldsymbol{w} = \sum_i \lambda_i y_i \boldsymbol{x}_i$$

(see Appendix 2.3) so that

$$g(\boldsymbol{x}) = \sum_i \lambda_i y_i \boldsymbol{x}_i^T \boldsymbol{x} + w_0 \tag{2.24}$$

i.e., depending on the inner product of the two vectors, $\boldsymbol{x}_i^T\boldsymbol{x}$.

Now suppose we are using the mapping function $f(\boldsymbol{x})$,

$$f(\boldsymbol{x}) = \left[x_1^2, \sqrt{2}\,x_1x_{2,}x_2^2\right] = z$$

and apply the 3D z for our linear separation problem. Then, the term z^Tz. transforms into

$$\left[x_{1i}^2x_1^2, 2x_{1I}x_1x_{2i}x_2, x_{2i}^2x_2^2\right]$$

but this corresponds to the product $\left(\boldsymbol{x}_i^T\boldsymbol{x}\right)^2$

In other words, choosing a suitable mapping function, $\Phi(\boldsymbol{x})$, we are able to express the product $\Phi(\boldsymbol{x}_i)\Phi(\boldsymbol{x}_j)$ by a kernel $K(\boldsymbol{x}_i, \boldsymbol{x}_j)$ which itself is a simple function of the original feature vectors $\boldsymbol{x}_i, \boldsymbol{x}_j$. In our decision function, we can directly use the kernels $K(\boldsymbol{x}_i, \boldsymbol{x}_j)$, so our task is limited to specify the kernels instead of the mapping function $\Phi(.)$. Most commonly used kernels are:

Linear : $K(\boldsymbol{x}_i, \boldsymbol{x}_j) = \boldsymbol{x}_i^T\boldsymbol{x}_j$

Polynomial : $K(\boldsymbol{x}_i, \boldsymbol{x}_j) = \left(\boldsymbol{x}_i^T\boldsymbol{x}_j + 1\right)^q,\ q > 0$

Radial Basis Functios (RBF) : $K(\boldsymbol{x}_i, \boldsymbol{x}_j) = exp\left(-\|\boldsymbol{x}_i - \boldsymbol{x}_j\|^2 / \sigma^2\right)^q$

Hyperbolic Tangent : $K(\boldsymbol{x}_i, \boldsymbol{x}_j) = \tanh\left(\beta\boldsymbol{x}_i^T\boldsymbol{x}_j + \gamma\right)$

Box 2.3 Cover's theorem—linear dichotomies.

In our discussion of the XOR problem, we have applied a transformation of a 2D feature vector $\boldsymbol{x}$ to a 3D $\boldsymbol{y}$, which enabled us to resolve the classification task applying a linear separation element. There is a deeper reason behind that as found by Cover (1965). Given the number of points in an l-dimensional space, we have

$$O(N, l) = 2\sum_i P(i)$$

possibilities to arrange separations of the N points in two classes using a linear distinction element. $P(i)$ is obtained as $P(i) = (N-1)!/(N-1-i)!i!$, with i running from 0 to l. At the same time, we have 2^N possible configurations of our samples in the two groups (including configurations for which linear separation is not possible. An example is shown in the figure below, where we indicate eight configurations. In reality, this number has to be doubled, as a group of samples may be assigned to either of the two classes. The configuration in red is not separable with a line.

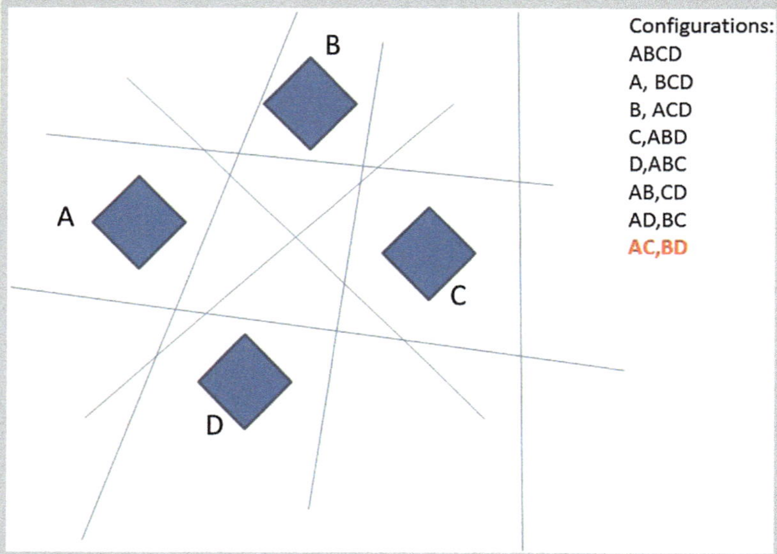

The probability to meet linear separable configurations among all possible ones is

$$p(l, N) = O(N, l) \cdot 2^{-N} \sum_i \frac{(N-1)!}{(N-1-i)!i!}$$

with i running from 1 to l. For large dimension l and small N ($N < 2l + 2$), the probabilities of finding linear separable configurations approach 1. Given N samples (feature vectors), a mapping to a higher dimensional space increases the probability of locating them in spaces which can be separated linearly from each other. This phenomenon is exploited in nonlinear SVMs, where the original feature vectors undergo a transformation applying the so-called kernel functions, which lead to variables with a higher dimension than the original feature vectors. The treatment of the XOR case in the text above highlights this fact.

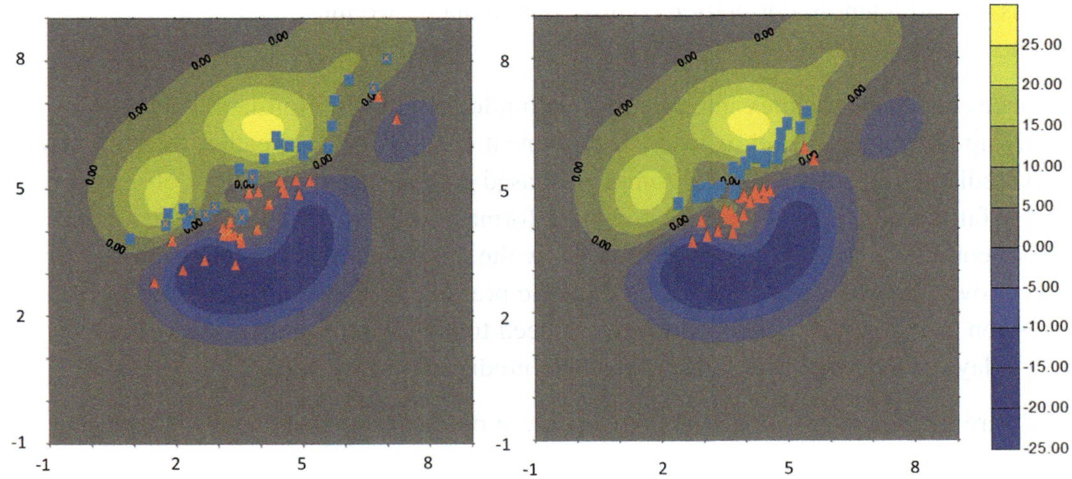

Figure 2.14

Application of an SVM using the radial base functions (RBF) kernel. Target classes are represented by the red diamonds and black triangles. In the training set (left panel), the classifier creates a nonlinear decision function shown by the isolines. Isolines can be obtained with the MATLAB script "S2_4". The support vectors—marked by yellow crosses—are placed around the "0" isoline, which forms the final discrimination element, whereas the nonlinear margins can be inferred from the distribution of the support vectors. On the right panel, we see possible problems when applying such a decision function to a test set not used during training. Some elements in the center of the plot turn out as clearly misclassified.

The choice of the kernel functions is a matter of convenience. No doubt that, as they bring along a nonlinear mapping, their behavior (and performance) may vary, in particular when the number of feature vectors is limited. This is shown in Fig. 2.14, where we use the data set used in Fig. 2.5 for the SVM training and apply it to the data in Fig. 2.4 (right panel).

2.6 Hidden Markov Models (HMMs)/sequential data

2.6.1 Background—from patterns and classes to sequences and processes

In the previous sections, we have considered objects and classes being independent from each other. In Geophysics as in other scientific fields, however, we often deal with processes characterized by a sequence of observations. The objects we are interested in cannot be characterized by single patterns but require to consider their interrelations. For example, in speech analysis, we first characterize sounds or vowels, and then we form words and expressions as a sequence of such vowels. Sentences are again a sequence, a sequence of words in this case. Numbers can be understood as a sequence of digits, such as 293 being composed of '2 ' '9', and '3'. No doubt that identifying the single digits alone is not sufficient to correctly recognize the number—we must read them in the right sequence.

Sequences can have complex interrelations with a lot of hidden information about the underlying phenomenon. Let us take the prominent example of the weather forecast. If we try to predict if it is going to snow tomorrow based on a snapshot of today's observations, we may fail because we miss all the context information. Besides observing the actual weather conditions at our site, we need to know the past: what was the temperature gradient over the last days? Does the barometric pressure history tell us that there is a depression crossing? In weather forecast, we need to take into account the patterns over the last days or even weeks to make a realistic prediction for the future.[4]

HMMs are the techniques that allow us to analyze observations in a context. The name "Markov model" comes from the Russian mathematician Andrey Andreyevich Markov. The underlying principle of Markov models is that they express chains of observations, being them patterns or classes of patterns. We do not know the underlying process; rather we attempt to derive the hidden information considering the chain of interrelations between the observations.

The first application of HMMs was in speech recognition beginning in the 1970s (e.g., Baker, 1975; Jelinek et al., 1975; Huang et al., 1990). In the subsequent decades, HMMs were applied to the analysis of biological sequences (Bishop and Thompson, 1986), bioinformatics (Durbin et al., 1999; Pachter and Sturmfels, 2005), face identification (Nefian and Hayes, 1998), and electrocardiogram classification (Koski, 1996). Motivated by the similarities of acoustic and seismic waveforms, HMMs were introduced in the field of seismology about 20 years ago (Ohrnberger, 2001). Besides their application in seismic signal classification (e.g., Dammeier et al., 2016), HMMs have also been employed for volcano state description (Bebbington, 2007), snow fall prediction (Joshi et al., 2017), and avalanche forecasting (Joshi and Srivastava, 2014).

In a general sense, HMMs are the so-called "double stochastic processes", which can be described by a set of parameters θ. In the following, the evolution of weather conditions and avalanche release is used to illustrate the concepts of single and double stochastic processes. In meteorology, the weather of consecutive days can be described as a process. Fig. 2.15 shows the weather example with the three states, $\mathbb{S} = \{\text{sunny } s_1, \text{ snowy } s_2, \text{ cloudy } s_3\}$. The weather observations on each single day can be classified into one of these three states. However, we neglect the information whether a low-pressure front is to arrive or whether we are in a regime of a stable high-pressure phase. In other words, overall meteorological

[4] In principle, we can apply the philosophy of sequential data also to the spatial distribution of observations. In weather forecast, we consider indeed the distribution of relevant parameters over an extended area besides their temporal development.

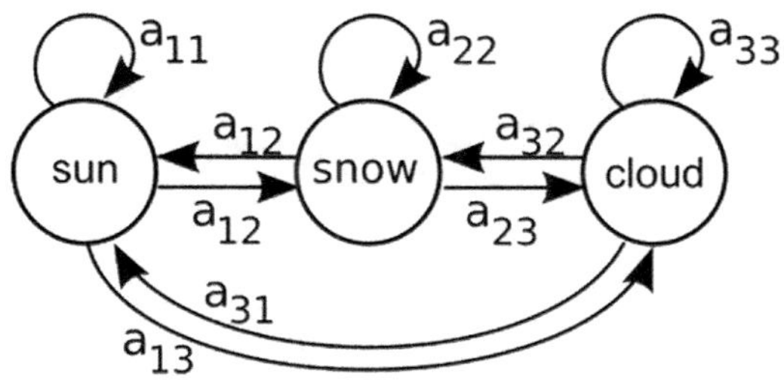

Figure 2.15

Sketch of a Markov model describing the weather. The model consists of three states (sunny, cloudy, and snowy). Transition probabilities a_{ij} are indicated by arrows.

conditions can change over time, meaning that the process will change between the states generating a sequence of visited states $\boldsymbol{q} = q_1, q_2,.. q_T$, with $q_T \in \mathbb{Q}$. The change between state i and state j is a random process which is described by a so-called transition probability a_{ij}. The first-order Markov assumption presumes that today's weather can always be predicted solely given the knowledge of the weather of the past day. In other words, the choice of state is made purely on the basis of the previous state, which is assumed to summarize all the past information. Hence, transition probabilities are given by

$$a_{ij} = P(q_t = s_j | q_{t-1} = s_i)$$

Classically, these transition probabilities are collected in a state transition matrix **A**

		sun	*snow*	*cloud*
A =	*sun*	0.5	0.25	0.25
	snow	0.375	0.125	0.375
	cloud	0.125	0.625	0.375

In terms of the weather example that translates to the following: we observe a sunny day today, which has the initial or a priori probability π_1, i.e., the probability to see a sunny day at all. With a transition probability of

$$a_{11} = P(q_2 = s_1 | q_1 = s_1) = 0.5$$

it will be sunny again tomorrow with a probability of 50%. With a probability of

$$a_{13} = P(q_2 = s_1 | q_1 = s_1) = 0.25$$

it will be cloudy the day after tomorrow. The probability for the total observation sequence, here 'sun' and 'cloud', will then be $\pi_1 a_{11} a_{13}$.

Now we extend the Markov chain above to an HMM. The observation sequence is now the number of avalanche releases in the Swiss Alps. On three consecutive days $T = 3$, we observe $\boldsymbol{o} = (o_1 = 10, o_2 = 5, o_3 = 8)$. Now we use an HMM whose states correspond to

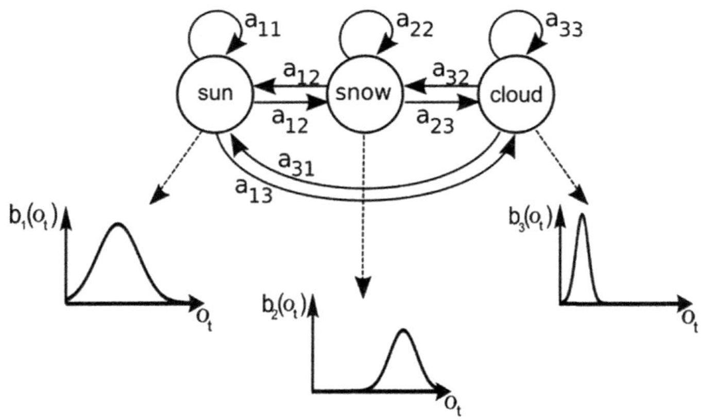

Figure 2.16
Sketch of a hidden Markov model. Transition probabilities a_{ij} are indicated by the arrows. For each hidden state, a probability density function b_i exists which determines the output of the corresponding state.

the weather conditions relating the observed number of avalanche releases. For each weather state, there exists a probability density function $b_i(o)$ for the number of avalanche releases. The connections between hidden states and observations are called "emission probabilities". Fig. 2.16 shows such an example of HMM. If it is cloudy, a reduced number of avalanche releases is observed, while when it snows, the number of avalanche releases increases. Therefore the hidden state sequence $\boldsymbol{q} = (q_2, q_3\ q_1)$ might have generated/might be the reason for the observed number of avalanche releases.

An HMM is a double stochastic process. The first process describes the transitions between hidden states which gives the hidden state sequence $\boldsymbol{q} = (q_1, q_2,..\ q_T)$. The second process determines the output of the HMM. Based on the emission probability of the current state, an observation symbol is issued. Thus, we can completely describe an HMM with N states by the triple $(\boldsymbol{\Pi}, \boldsymbol{A}, \boldsymbol{B})$:

- initial state probability vector $\boldsymbol{\Pi} = (\pi_1, \pi_2, \pi_3, \ldots, \pi_N)$,
- transition probability matrix $\boldsymbol{A}_{NxN}$, where $a_{ij} = P(q_t = s_j | q_{t-1} = s_i)$, and
- emission probability matrix $\boldsymbol{B}_{NxM}$, where $b_j(k) = P(o_k \text{ at time } t | q_t = s_j)$ (Box 2.4)

2.6.2 The three problems of HMMs

When using HMMs for any pattern recognition problem, we have to solve the following tasks (see, e.g., Duda et al., 2001):

a. Evaluation—given an HMM (i.e., transition and emission matrix are known): what is the probability that it has created a particular sequence of observations?

Box 2.4 Tossing coins behind a curtain.

Consider an experiment where coins are tossed behind a curtain, producing a sequence of 'heads' (H) and 'tails' (T), with the probabilities of getting heads or tails given by $P(\mathrm{H})$ and $P(\mathrm{T})$. With a single coin we have $P(\mathrm{T}) = (1 - P(\mathrm{H}))$. This is a process with observable states i and j, as they coincide with the outcomes H and T. Consider now two coins and we again see the outcomes, whereas the experimenter tossing the coins is hidden behind the curtain. That entails, when seeing a head H, we do not know which of the coins produced it. Consequently, a sequence being 'HT', 'HH', 'TH', or 'TT' can be obtained either by tossing coin 1 and then 2 or by tossing the same coin two times. The full number of necessary parameters is 6, i.e., we need to know $P_1(\mathrm{H})$, $P_2(\mathrm{H})$, $P(1,1)$, $P(2,2)$, $P(1,2)$, and $P(2,1)$, with the $P(i,i)$ being the transition probabilities from i to j and vice versa. As probabilities of an event add to 1, the number of parameters reduces in effect to 4. In general, we report all transition probabilities in the matrix $\boldsymbol{A}_{N\mathrm{x}N}$, as above in the text. On the other hand, we can identify the $P_i(\mathrm{H})$, $P_i(\mathrm{T})$ with the emission probability matrix $\boldsymbol{B}$.

b. Decoding—given an HMM: which sequence of hidden states leads most likely to the given sequence of observations?
c. Training—we know the principal structure of the HMM, i.e., size of the transition and emission matrices, but ignore their elements. Based on a given set of observations, we determine the triple $(\boldsymbol{\Pi}, \boldsymbol{A}, \boldsymbol{B})$ that most probably created the set.

In Box 2.5, we present a simple example of an HMM and outline the solution for all three tasks. A detailed description can be found in Appendix 2.4.

In general, classification considers the probability of seeing class $\mathbb{C}_k$ given an observation $\boldsymbol{o}, P(\mathbb{C}_k|\boldsymbol{o})$. The classification schemes discussed in the previous sections consider some measures allowing us to establish whether some patterns belong to a certain class. In FLDA (see Chapter 2.1), such a measure is obtained from the distance of an object $\boldsymbol{o}_i$ to the centroid of a class of objects $\mathbb{C}_k$. In its generalization (Appendix 2.1, Eq. A2.1.10 ff.), we transform such a distance into a probability of an object belonging to a class, assuming a certain probability density for the classes. In the linear SVM problem (see Fig. 2.13 in Chapter 2.5.1), the probability of seeing class $\mathbb{C}_k$ given an observation $\boldsymbol{o}, P(\mathbb{C}_k|\boldsymbol{o})$, can be related to the distance between observations and the separating element.

In HMMs, we have neither a "centroid" of a class nor a straightforward metrics to measure the similarity of an object to a prototype of some category; also, we do not have linear or nonlinear discriminating elements. This is certainly a consequence of the characteristics of the objects we deal with in HMMs, which are formed by a sequence of observations whose order is decisive. We therefore have to restate our strategy in the sense

Box 2.5 Going through an HMM.

Suppose we have a sequence of observations, say the result of rolling two dice. The first, red die is fair, i.e., the results {1, 2, 3, 4, 5, 6} are equally probable. The second, green die is loaded. The behavior of the two dice is given by the "emission matrix" [1/6 1/6 1/6 1/6 1/6 1/6; 1/10 1/10 1/10 1/10 1/10 ½]. The experimenter (behind the curtain) has two coins of red and green color. The outcome of the coins is weighted, with probabilities of getting 'head' of 0.9 and 0.95, respectively. Correspondingly, the probabilities of getting 'tail' are 0.1 and 0.05. We write down this as a matrix [0.9 0.1; 0.95 0.05]. The sequence of observations is composed of the outcomes of dicing. At any time of the experiment, we toss the coin in order to decide which die to roll.

We start with the experiment, rolling the red die, i.e., we begin the state 1, and notice the outcome ("emission").

We toss the red coin. If the result is 'head'—which happens in 90% of the cases—we keep on with the red die. If the result is 'tail'—this is to happen in 10% of the cases—we switch to the green die. In other words, we switch to state 2 in this case.

We continue tossing now the green coin and roll the green die, adding its outcome to the sequence of our observations. We continue with the green die, unless the coin gives 'tail', which makes switch to state 1 (red die).

As observers from outside, we ignore how the experimenter acted behind the curtain. Nonetheless, we are curious to understand, for instance, whether dicing was performed with fair dice and what could be the parameters controlling the outcome of the experiment. For this purpose, we try to get an estimate on the two matrices, i.e., the transition matrix describing the switch from one state to another and the emission matrix, which tells us the probability of outcomes at each state. An algorithm doing this is known as the 'Baum–Welch' algorithm. Having gained a guess for the two matrices, we can also reconstruct the most probable sequence of states. For this purpose, we may use the 'Viterbi' approach. The reader can play with the MATLAB code 'S2_5' coming along with this book (see also Theodoridis et al., 2010).

In our avalanche problem, we may put us in the position of the observer, i.e., just monitoring the occurrence of avalanches. How are they related to the states, here given by the weather conditions? There are now two possibilities: we have a rough idea about the structure of the HMM, but do not know either the sequence of states or the values in the transition end emission matrices.

On the other hand, we may have observed also the weather conditions, i.e., we know the sequence of states, but still ignore the matrices. We can formulate the problem like this: find the best matrices relating the occurrence of avalanches to the weather conditions.

that we ground our decision on the probability that the observation sequence $\boldsymbol{o}$ was created by a model with specific characteristics. In other words, its model belongs to class $\mathbb{C}_k$. We state the problem as

seek the class k *which has the maximum probability given the observation sequence* **o**, *i. e.,*

$$argmax_k(P(\mathbb{C}_k|\boldsymbol{o}) \tag{2.25}$$

To solve Eq. (2.25), we use the Bayes' rule for conditional probabilities

$$P(A|B) = \frac{P(B|A)P(A)}{P(B)}$$

where $P(A|B)$ is the likelihood of event A given that B has occurred and $P(B|A)$ is the likelihood of event B given that A has occurred. $P(A)$ and $P(B)$ are the probabilities of observing A and B independently of each other (see statistical textbooks, e.g., Kreyszig, 1982). For our problem, we have

$$P(\mathbb{C}_k|\boldsymbol{o}) = P(\boldsymbol{o}|\mathbb{C}_k)P(\mathbb{C}_k) \,/\, P(\boldsymbol{o}) \tag{2.26}$$

Maximizing $(\mathbb{C}_k|\boldsymbol{o})$ can now be replaced by maximizing $P(\boldsymbol{o}|\mathbb{C}_k)$, if the a priori probabilities $P(\mathbb{C}_k)$ and $P(\boldsymbol{o})$ are known. Being the relative frequency of class $\mathbb{C}_k$, $P(\mathbb{C}_k)$ is assumed uniform if not known. Therefore, it does not affect the maximization of Eq. (2.26). In addition, $P(\boldsymbol{o})$ is independent of class $\mathbb{C}_k$ and also does not affect the maximum argument of Eq. (2.26). Hence $P(\boldsymbol{o}|\mathbb{C}_k)$ is the driving factor for maximizing the left side of Eq. (2.26). The estimation of $P(\boldsymbol{o}|\mathbb{C}_k)$ can be obtained by a multiplication of the sequence of probabilities (i.e., transition and emission probabilities) that generated the observed data sequence, summed over all possible state sequences (see Fig. 2.17).

However, calculating the probability in this manner is computationally expensive, particularly with large models or long sequences. The introduction of the so-called forward or backward probabilities (also called partial probabilities) reduces the complexity of the problem. Details on the procedure can be found in Appendix 2.4. Hence, while a discriminative approach does not make any assumption about the data-generating process, a generative approach turns the problem around. Instead of finding the class given the observations, we aim to maximize the likelihood of our observations given the class. In other words, we picture individual observation samples to be generated from the class distribution $P(\boldsymbol{o}|\mathbb{C}_k)$ and match a new observation against all available models to see which class is most likely to generate the observed signal.

The optimal state sequence is the combination of hidden states that maximizes $P(\mathbb{O},\mathbb{Q}\,|\theta)$, where $\mathbb{Q}$ is the set of all possible states. Unlike the former problem, we are not summing up over all path, but we are looking for the maximum. We can find the most probable sequence of hidden states by listing all possible sequences and finding the probability for

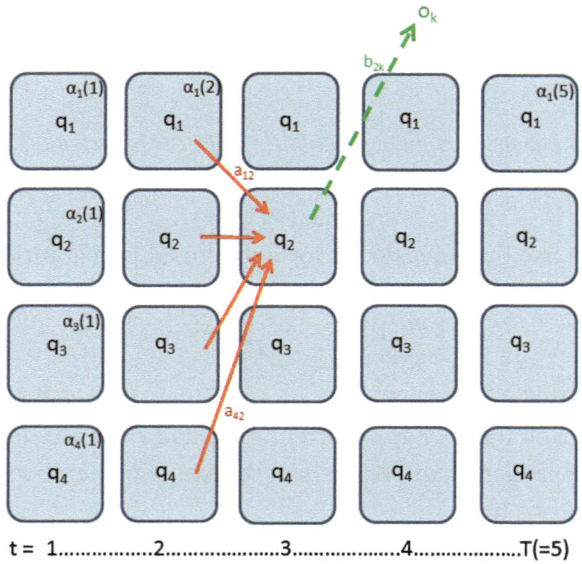

Figure 2.17

Trellis representation of HMM. Suppose we are the q_2 at $t=2$. Then, at $t=3$, the probability of $q_j(t=3)$ is found by adding over a_{j2}. From that state, an observation will be emitted with a probability b_{jk}.

each combination. As with the forward algorithm, we can reduce the complexity of the problem using partial probabilities. Having calculated those partial probabilities, it is possible to record in what state the system must have been at time $t-1$, if it is to arrive optimally at state i at time t. This recording is done by holding for each state a back pointer f, which points to the previous state that optimally provoked the current state. Besides a reduction of complexity, this so-called Viterbi algorithm provides the best interpretation given the entire context of observations—it looks at the whole sequence before deciding on the most likely final state and then backtracking through the f pointers to indicate how it might have arisen. A mathematical description and more details are given in Appendix 2.4.2.

As in HMMs, the hidden states are not observable and a direct training approach cannot be adopted. The parameters describing an HMM can only be inferred from the available observations. Hence, we would like to find the HMM parameters λ which maximize the "Maximum Likelihood" function

$$L_{HMM}(\lambda) = \log P\left(\mathbb{O}|\theta) = log \sum_{q \in \mathbb{S}} P(\mathbb{O}, q|\theta\right)$$

Here, $\mathbb{O}$ is the set of all observations. There is no analytical way to maximize the probability of an observation sequence. However, one way to achieve this goal is to estimate the unknown parameters, so that the output for each model becomes locally maximized using an iterative procedure such as the Baum–Welch algorithm (Baum and Sell, 1968).

2.6.3 Including prior knowledge/model dimensions and topology

There are a number of HMM characteristics and appropriate parameters to be chosen by the user to achieve a well-working system. While the number of observation samples is normally a fixed quantity, the number of states comprising an HMM is defined by the user. The states might correspond to certain physical quantities. For example, in speech recognition, the number of states depends on the number of distinct sound phenomena within one word. Another example comes from seismology. When applied to seismic waveforms, our visible observation is the raw time series or sequence of features obtained by some transformation. The hidden states could be the distinct seismic phases. For local earthquakes, an HMM with three states—P-phase, S-phase, and the so-called "coda"—might be appropriate. For regional or teleseismic events, which occur at a larger distance from our receiver, there may be more phases, and our HMM with only three states is inadequate.

In addition to the number of states, the model topology can be determined based on the underlying process. The most common model topology is the ergodic one. In this configuration, the model can change from each state to every other state without any restriction. However, causality can be introduced, for example modeling purposes a left-to-right topology (Fig. 2.18), which prevents transitions in previous states. Taking again the example of an HMM modeling the seismogram of a local tectonic event: the P-phase will always be followed by the S-phase that is followed by the coda. The left-to-right model is

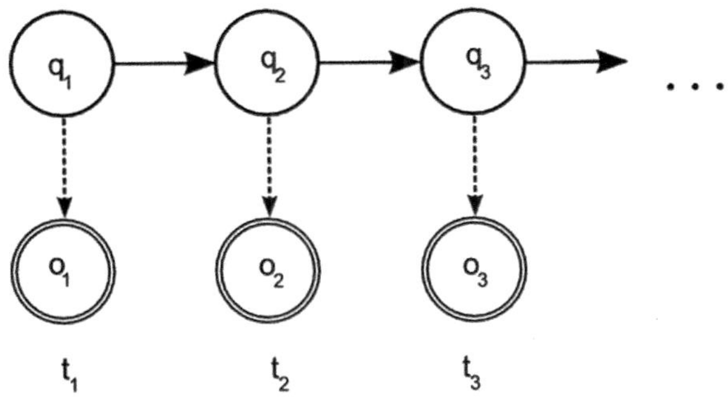

Figure 2.18
HMM with left-to-right topology. For clarity, the sketch shows the simplest version of a left-to-right model. Transitions from state 1 to 3 might also be possible but are not shown.

the most appropriate description of the process and this causality is implemented by restrictions in the transition matrix $\mathbf{A}_{NxN}$. Whenever appropriate, it is advisable to introduce such restrictions as they also reduce the number of parameters to estimate.

The last point concerns the time dependence of transition and emission probabilities. The probabilities in the transition and emission matrix are independent of time, which means that the probabilities do not change in time as the system evolves. This so-called homogenous Markov model is one of the most unrealistic assumptions of Markov models about real processes as data tend to change in time.

2.6.4 Extension to conditional random fields

As we have seen, there are many advantages but also some limitations when using HMMs for labeling sequential data. If the observations are truly sampled from the learned model, it provides a correct description of the corresponding class. However, as in most real-world applications, the trained models are only approximations of the true underlying processes. Consequently, a number of assumptions and/or simplifications might not be correct. For instance, it is not granted that the segmentation of the signal in a number almost stationary parts is appropriate to describe a given class. Furthermore, HMMs evaluate the joint probability of hidden state sequence and observations. To model the joint probability tractably, only dependencies between state variables are modeled while observations are represented as isolated units independent of all other observations in the sequence. Observations depend only directly on the current state. In some cases, it might be advantageous to consider additional information. For instance, when classifying seismic events, information such as seismic phases preceding the current observation or the signal onset quality (e.g., impulsive or emergent) could be considered in the labeling process. Thus, although the approach performs well in many applications (e.g., Koski, 1996; Nefian and Hayes, 1998; Benitez et al., 2007), the question remains: "Is the signal properly described by this model topology or might another less restrictive approach even perform better?"

Discriminative learning strategies for HMMs have also been introduced. A review is given by Jiang (2010). For labeling sequential data, another promising discriminative approach called "Conditional Random Field" (CRF) has been introduced by Lafferty et al. (2001). CRFs are widely used in various research areas, such as text processing (Settles, 2005) or bioinformatics (Sato and Sakakibara, 2005). Instead of using the joint probability of observations and hidden state sequence, CRFs maximize the conditional probability of hidden state sequence given the observations. Arbitrary dependencies between variables are allowed providing a much larger flexibility than classical HMMs. As shown by Sutton and McCallum (2007), HMMs can be easily generalized to CRFs. Similar to HMMs, there are two problems for CRFs: learning the classifier parameters and classifying an unknown observation sequence to one of multiple classes. Both tasks can be performed efficiently by modified versions of the standard dynamic-programming algorithms for HMMs (Rabiner, 1989). The parameters describing a CRF are learned from available training data using the principle of maximum likelihood. The forward–backward algorithm is applied to iteratively solve this task. To label an unknown sequence of observations, we can compute the most likely labeling using the Viterbi algorithm. For that reason, CRFs might provide a valuable alternative for labeling sequential data.

2.7 Bayesian networks

BNs, also known as "Bayesian Belief Networks" or simply "Belief Networks" deal with problems where a distribution cannot be simply described by a parameter vector, but also requires probabilistic or causal dependencies. For instance, we may define the weather by observations like clouds, rainfall, snow, temperature, etc. Certainly, precipitation depends in general on the presence of clouds, and the temperature will condition the type of precipitation, i.e., rain or snow.

The probabilistic or causal dependencies are typically represented in graphs, where we have 'nodes', i.e., the variables, and arrows, which show the relation among the nodes. In Fig. 2.19, we show a graph for our avalanche problem. The figure has a node "snow fall", which is causal for "avalanche" and accumulation of snow with a thickness of ">2m".

In Fig. 2.19, we imagine a problem with snow fall (perhaps somewhere in the high parts of a mountain region), the release of avalanches along the flanks of the mountain, and the accumulation of snow at some site (perhaps a chalet or mountain hut). We analyze the conditions of having a "snow height above 2 m" (for the sake of brevity, we rename this as "Snow > 2 m") at our hut. In this example, the snow height depends on two

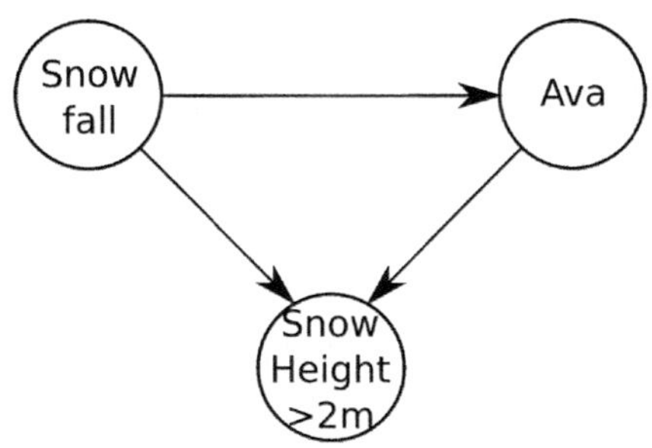

Figure 2.19
A Bayesian network consisting of the three nodes "Snowfall", "Avalanche", and "Snow Height> 2 m", which influence each other as indicated by the arrows. Whether or not the snow height is larger than 2 m depends on the status of its parents< "Snow fall" and "Avalanche"; the occurrence of an avalanche depends on the snow fall.

Table 2.1: Conditional ("local") probabilities for the Snow–Avalanche example.

Snow fall: True/False			
P(Snow = True): 0.4		P(Snow = False): 0.6	
Parent (child) Avalanche|Snow:			
Avalanche|Snow = True		Avalanche|Snow = False	
P(Ava = True): 0.6	P(Ava = False): 0.4	P(Ava = True): 0.2	P (Ava = False): 0.8
Child Snow > 2 m|(Snow, Avalanche):			
Snow	Avalanche	P(Snow > 2 m|Snow, Avalanche)	
True	True	P(True): 0.9	P(False): 0.1
True	False	P(True): 0.5	P(False): 0.5
False	True	P(True): 0.7	P(False): 0.3
False	False	P(True): 0.1	P(False): 0.9

parameters. First of all on meteorological conditions: if it snows, the measured snow thickness at the site will increase. The second factor influencing the snow height is a possible avalanche release: if an avalanche occurs, its deposits will increase the snow height. Besides, avalanche occurrence depends on snow fall. In our example, the phenomenon "snow fall" is a parent process for both "snow height" and "avalanche". "Avalanche" is a parent process for a "snow height > 2 m", but a child of "snow fall". The relation between these processes can be captured by a BN, which can be represented as a ***D****irected* ***A****cyclic* ***G****raph* (DAG). DAGs are defined by arcs and nodes. In addition to the DAG, we need a table of the conditional probabilities (see Charniak, 1991), as shown in Table 2.1. Each node has a potential value, which can be 'true'/'false', probabilities 'low'/'medium'/'high' or an integer. Fig. 2.19 shows an example where all nodes have a 'true' or 'false' value. Each node has a local probability that describes the likelihood of its value without any additional information. If a node has one or more parent, then its prior probabilities are conditioned on the values of the parents. Thus, an arc between two nodes represents some kind of influential relationship between corresponding variables. These relationships correspond to probabilistic dependencies and are specified for each variable X_i by its probability distribution conditioned on its parents $P(X_i|X_{\text{parent(i)}})$. For example, given a BN as shown in Fig. 2.19, we may have the following dependencies.

Parent snow:

The joint probability of all variables is factorized into the local conditional probabilities, i.e.,:

$$\mathrm{P(Snow > 2m, Ava, Snow)} = \mathrm{P(Snow > 2m|Ava, Snow)P(Ava|Snow)P(Snow)}$$

Conditional local probabilities are given applying the Bayes law:

$$\mathrm{P(Snow|Snow > 2m)} = \mathrm{P(Snow, Snow > 2m)/P(Snow > 2m)}$$

For instance, we are interested in the total probability of having Snow > 2 m. Going step by step, we consider all four scenarios leading to a snow height of more than 2 m: P(Snow>2) = P(Snow>2m, Ava = True, Snow = True) + P(Snow>2m, Ava = True, Snow = False) + P(Snow>2m, Ava = False, Snow = True) + P(Snow>2m, Ava = False, Snow = False). Every single joint probability is now decomposed into the local conditional probabilities:

1. P(Snow>2m, Ava = True, Snow = True) = P(Snow>2m|Ava = True, Snow = True) • P(Ava = True|Snow = True) *P(Snow = True)
2. P(Snow>2m, Ava = True, Snow = False) = P(Snow>2m|Ava = True, Snow = False) • P(Ava = True|Snow = False) • P(Snow = False)
3. P(Snow>2m, Ava = False, Snow = True) = P(Snow>2m|Ava = False, Snow = True) • P(Ava = False|Snow = True) • P(Snow = True)
4. P(Snow>2m, Ava = False, Snow = False) = P(Snow>2m|Ava = False, Snow = False) • P(Ava = False|Snow = False).

Given the conditional probabilities in Table 2.1, we find the total probability from 0.9 • 0.6 • 0.4 + 0.1 • 0.8 • 0.6 + 0.7 • 0.2 • 0.6 + 0.5 • 0.4 • 0.4 = 0.216 + 0.048 + 0.084 + 0.08 = 0.428.

In this way, not only the joint probability distribution of all variables but also any other conditional distribution of interest can be derived from the provided local distributions. Thus, one can infer and reason into any direction. For instance, given our example shown in Fig. 2.19, we can answer any question like "What is the probability to have less than 2 m of snow, given the fact that it is snowing?" Using Bayes' law, the answer is:

P(Snow>2m = False|Snow = True) = P(Snow>2m = False, Snow = True)/ P(Snow = True) = [P(Snow>2m = False, Ava = True, Snow = True) + P(Snow>2m = False, Ava = False, Snow = True)]/ P(Snow = True).

After again decomposing into the local conditional probabilities, we end up with

$$\mathrm{P(Snow > 2m = False|Snow = True) =}$$

$$\mathrm{P(Snow > 2m = False|Ava = True, Snow = True)\cdot P(Ava = True|Snow = True) +}$$

$$\mathrm{P(Snow > 2m = False|Ava = False,\ Snow = True)\cdot P(Ava = False|Snow = True)}$$

which yields, inserting the values in Table 2.1:

$$0.1\cdot 0.6+0.5\cdot 0.4=0.26$$

Suppose now, we have noticed a snow height of more than 2 m. What is the probability that there was an avalanche? That is, we ask for probability of a node being both child and parent of other nodes, here P(Avalanche|Snow>2m). We calculate

$$\begin{aligned} \mathrm{P(Ava = True|Snow > 2m = True)} &= \mathrm{P(Ava = True, Snow > 2m = True)} \\ &\mathrm{/\ P(Snow > 2m = True)} = \\ &\mathrm{[P(Ava = True, Snow = True, Snow > 2m = True)} \\ &\mathrm{+\ P(Ava = True, Snow = False,\ Snow > 2m = True)]\ /\ P(Snow > 2m = True).} \end{aligned}$$

Again decomposing into local conditional probabilities, we have

$$\mathrm{P(Ava = true|Snow > 2m) =}$$

$$\begin{aligned} &\mathrm{[P(snow > 2m|Ava = True, Snow = True) \bullet P(Ava = True|Snow = True) \bullet P(Snow = True)} \\ &\quad \mathrm{+ P(Snow} \\ &\quad \mathrm{> 2m = True|Ava = True, Snow = Flase) \bullet P(Ava = True|Snow = False) \bullet P} \\ &\mathrm{(Snow = False)]/P(Snow > 2m)} \end{aligned}$$

Inserting the local values from Table 2.1 and the total probability of having a snow height of more than 2 meters:

$$\mathrm{P(Ava = True|Snow > 2m = True)} = (0.9 \cdot 0.6 \cdot 0.4 + 0.7 \cdot 0.2 \cdot 0.6)/0.428 = 0.70$$

In other words, from the observation of a snow height >2 m, we deduce the probability of an avalanche of ca. 70%.

If we further know that it is snowing, we have just

P(Ava = True, Snow>2m = True, Snow = True)/P(Snow>2m = True, Snow = True)

$$\begin{aligned} &\mathrm{with\ P(Snow > 2m, Snow = True) =} \\ &\mathrm{P(Snow > 2m, Ava = True, Snow = True) + P(Snow > 2m = True,} \\ &\mathrm{Ava = False, Snow = True)\ =} \\ &\mathrm{P(Snow > 2m = True|Ava = True, Snow = True) \cdot P(Ava = True|Snow = True) \cdot} \\ &\quad \mathrm{P(Snow = True)\ +} \\ &\mathrm{P(Snow > 2m = True|Ava = False, Snow = True) \cdot P(Ava = False|Snow = True) \cdot} \\ &\quad \mathrm{P(Snow = True) =} \\ &\mathrm{we\ have} \\ &0.4 \cdot 0.6 \cdot 0.9/(0.9 \cdot 0.6 \cdot 0.4 + 0.5 \cdot 0.4 \cdot 0.4) = 0.7297 \end{aligned}$$

i.e., ca. 73%. In both cases, we should be concerned about the possibility that someone was struck by an avalanche. When it is snowing, our level of alert is even increased.

BNs have a number of interesting properties. Every time new information enters the BN (which can be at any node), it is propagated through the network and the probabilities of all unobserved variables are updated. This new information can be of any type: objective data as well as subjective belief. If no new information arrives, the BN assumes the prior distributions. This allows predictions for missing elements in an incomplete data set. We have seen that using probabilities instead of single point estimates allows us to handle missing data in an elegant way. Uncertainties are modeled explicitly. By learning the BN from data, it is possible to identify the variables that are relevant to the system. Finally, any prior knowledge available (e.g., expert knowledge) about the features can be included.

In the simplest case, the structure of a BN—the elements making up the DAG—is specified by an expert and only the parameters of the conditional (or "local") probability distributions $P(X_i|X_{\mathrm{Parent(i)}})$ (such as the ones given in Table 2.1) are learned from available data. Those networks mostly aim at supporting a decision analysis or at predicting a specific target variable. Traditionally, the learning is based on a score optimization algorithm (e.g., steepest gradient descent, see Russell et al., 1995; Han et al., 2011, or maximum likelihood approach and expectation–maximization algorithm, see Dempster et al., 1977). Alternatively, we may seek to discover internal relationships or causalities in the system. Thus, we are interested in finding the dependency structure of the considered variables that is most probably responsible for generating the observed data, i.e., we chase the network structure and parameters that best explain the data. Vogel et al. (2014) proposed a procedure to learn the network structure (DAG) and the parameters θ (the local probabilities). The algorithm searches for the most probable BN (DAG,θ), given the observed data by aiming to maximize the a posteriori probability P(DAG,θ|data), as suggested by Riggelsen (2008).

Evolving systems in time (i.e., sequential data) can be modeled by the so-called dynamic BNs (DBNs) (Murphy, 2002; Riggelsen et al., 2007; Runge et al., 2014). DBNs relate parameters to each other over one or more adjacent time frames. DBNs form BNs which are unrolled in time and their probability distributions change with time.

Appendix 2

Appendix 2.1 Fisher's linear discriminant analysis

In FLDA (Fisher, 1936), we tackle the problem of distinction by reducing the multivariate problem with m components to a one-dimensional component. For this purpose, we define

$$Z = w_1X_1+w_2X_2\cdot w_mX_m \qquad \text{(A2.1.1)}$$

For the sake of simplicity, we work with two components, but the formalism outlined here can be easily generalized. The w_i coefficients are chosen with the scope of maximizing the degree of separation of the two groups. As a proxy, we use the distance between the averages of the groups A and B. In the 2D case, this is

$$\mu_A = w_1\overline{X}_{1A} + w_2\overline{X}_{2B} \quad \text{(A2.1.2a)}$$
$$\mu_B = w_1\overline{X}_{1B} + w_2\overline{X}_{2A} \quad \text{(A2.1.2b)}$$

where the averages are expressed in terms of Z.

First, we assume that the variance/covariance of both groups are the same and can be estimated as their weighted means. The pooled variances of the groups, in terms of the new variable Z, are given by

$$s_Z^2 = \sum_i (Z_{Ai} - \mu_A)^2 + \sum_j (Z_{Bj} - \mu_B)^2 = w_1^2 s_1^2 + w_2^2 s_2^2 + 2w_1 w_2 s_{12} \quad \text{(A2.1.3)}$$

(s_1^2 and s_2^2 are the pooled variances in the original system of axes, and s_{12} is the pooled covariance).

The coefficients $\boldsymbol{w}_{ii}$ are now identified with the conditions that

$$d^2 / s^2 = |\mu_A - \mu_B|^2 / s^2 = max \quad \text{(A2.1.4)}$$

which can be solved by partial differentiation with respect to a_i and searching for the extremes.

Finally, we have

$$d_1 = \overline{X}_{1A} - \overline{X}_{1B} = w_1 s_{11} + w_2 s_{12}$$
$$d_2 = \overline{X}_{2A} - \overline{X}_{2B} = w_1 s_{21} + w_2 s_{22}$$

where s_{ij} are the elements of the pooled covariance matrix **C** of the two data sets A and B. In matrix form, we write briefly

$$\boldsymbol{d} = \boldsymbol{C}\boldsymbol{w} \quad \text{(A2.1.5)}$$

and resolve for the coefficients of $\boldsymbol{w}$

$$\boldsymbol{w} = \boldsymbol{C}^{-1}\boldsymbol{d} \quad \text{(A2.1.6)}$$

where $\boldsymbol{w}$ is the vector of coefficients and $\boldsymbol{d}$ the distance vector of the two groups of averages. The discriminant function is then given by

$$Z = \boldsymbol{w}\boldsymbol{X}^T \quad \text{(A2.1.7)}$$

We can express a critical value for Z just by considering the averages of groups A and B in terms of Z

$$\underline{Z}_A = \Sigma_i \boldsymbol{w}\, \boldsymbol{X}^T / N_A$$
$$\underline{Z}_B = \Sigma_j \boldsymbol{w}\, \boldsymbol{X}^T / N_B$$

(i, j running from 1 to N_A and N_B) and by taking the averages of Z_A andZ_B.

$$Z_{Crit} = (\underline{Z}_A + \underline{Z}_B)/2$$

The formalism we presented here corresponds to Fisher's approach of discriminant analyses (Fisher's Discriminant Relation). Recall that it is based on the assumption that both groups follow the same distribution around their averages and the covariance matrices are the same, i.e., the pooled covariance matrix is a reasonable estimation of the true one.

We can obtain a more general formulation using a likelihood approach. We consider the ratio

$$L = f_1(\boldsymbol{x})/f_2(\boldsymbol{x}) \tag{A2.1.8}$$

where the f_i are the probability densities of $\boldsymbol{x}$ to belong to one of the groups. Assuming normal distributions for the two groups

$$f = \frac{1}{(2\pi|\boldsymbol{C}^{-1}|)}\exp\left(\left(\frac{1}{2}(\boldsymbol{x}-\boldsymbol{\mu})^T\boldsymbol{C}^{-1}(\boldsymbol{x}-\boldsymbol{\mu})\right)\right) \tag{A2.1.9}$$

we get the log-likelihood ratio

$$\ln L = \frac{1}{2}\ln\left(\frac{|\boldsymbol{C}_1|}{|\boldsymbol{C}_2|}\right) - 1/2(\boldsymbol{x}-\boldsymbol{\mu}_1)^T\boldsymbol{C}_1^{-1}(\boldsymbol{x}-\boldsymbol{\mu}_1) + 1/2(\boldsymbol{x}-\boldsymbol{\mu}_2)^T\boldsymbol{C}_2^{-1}(\boldsymbol{x}-\boldsymbol{\mu}_2) \tag{A2.1.10}$$

The terms of the form $(\boldsymbol{x}-\boldsymbol{\mu})^T\boldsymbol{C}^{-1}(\boldsymbol{x}-\boldsymbol{\mu})$ are known as "Mahalanobis Distance", which expresses the squared distance d^2 of the vectors $\boldsymbol{x}$ and $\boldsymbol{\mu}$.

For $\boldsymbol{C}_1 = \boldsymbol{C}_2 = \boldsymbol{C}$

$$\ln L = \frac{1}{2}\boldsymbol{x}^T\boldsymbol{C}^{-1}(\boldsymbol{\mu}_1-\boldsymbol{\mu}_2) = \boldsymbol{x}^T\boldsymbol{w} \tag{A2.1.11}$$

where

$$\boldsymbol{w} = \boldsymbol{C}^{-1}\boldsymbol{d}$$

For L = 1 (same probability of $\boldsymbol{x}$ to belong to group 1 or 2)

$$\ln L = 0$$

and we get the same solution for $\boldsymbol{w}$ as before. Compared to the formalism above, the likelihood approach allows for $\boldsymbol{C}_1 \neq \boldsymbol{C}_2$. The discrimination then is a quadratic problem of the form

$$\ln L = 0 = \ln\left(\frac{|\boldsymbol{C}_1|}{|\boldsymbol{C}_2|}\right) - d_1^2 + d_2^2 \tag{A2.1.12}$$

(d_i^2 being the Mahalanobis Distance mentioned above). The class membership of a vector $\boldsymbol{x}$ can be assigned for $\ln L > 0$ or $\ln L < 0$.

Appendix 2.2 The perceptron

A function $g(\boldsymbol{x})$ which minimizes the cost function J can be identified using a scheme known as the "gradient descent" method. At each iteration step $t+1$, the vector $\boldsymbol{w}$ is modified with respect to the preceding step t,

$$\boldsymbol{w}(t+1)=\boldsymbol{w}(\mathrm{t})\ -\rho(t)\partial J(\boldsymbol{w})/\partial(\boldsymbol{w}) \tag{A2.2.1}$$

here $\rho(\mathrm{t})$ is the degree to which $\boldsymbol{w}$ at the $(t+1)^{\text{th}}$ iteration is updated with respect to the preceding step. Typically, ρ is a small number and decreases during the iterative process.

Having

$$J(\boldsymbol{w})=\sum_{\boldsymbol{x}\in\mathbb{Y}}\left(\delta_x\boldsymbol{w}^T\boldsymbol{x}\right)$$

we obtain

$$\partial J(\boldsymbol{w})\,/\,\partial(\boldsymbol{w})=\sum_{\boldsymbol{x}\in\mathbb{Y}}(\delta_x\boldsymbol{x})$$

and

$$\boldsymbol{w}(t+1)=\boldsymbol{w}(t)-\rho(t)\sum_{\boldsymbol{x}\in\mathbb{Y}}(\delta_x\boldsymbol{x}) \tag{A2.2.2}$$

Backpropagation

The backpropagation algorithm was proposed by Werbos (1974). Here, we briefly outline the derivation given by Freeman and Skapura (1992). For the sake of simplicity, we limit ourselves to a perceptron with one hidden layer and a sigmoidal activation function in the hidden and output nodes.

Denote the values being passed to the hidden units as

$${}^{h}y={}^{h}wx+{}^{h}w_o \tag{A2.2.3}$$

Then, we get in the output

$${}^{O}\mathbf{y}={}^{O}\mathbf{w}^{T\,h}\mathbf{y}+{}^{O}w_0 \tag{A2.2.4}$$

(the superscripts h and O indicate units in the hidden and output layer, respectively). When we apply an activation function different from linear, we write instead

$${}^{h}\mathbf{y}=f({}^{h}\mathbf{wx}+{}^{h}w_0) \tag{A2.2.5}$$

and

$$^{O}\mathbf{y} = f(^{O}\mathbf{wx} + {}^{O}w_0) \tag{A2.2.6}$$

with f being, for instance, our sigmoidal function.

Let us express the error encounter from each pattern as

$$E = \frac{1}{2}\sum_{ik}(y_k - o_k)^2 \tag{A2.2.7}$$

Here we denote the target output with y_k and the calculated output in the k-th output node with o_k, i.e.,

$$o_k = {}^{O}f\left(\sum_j {}^{o}w_{kj}h_j + {}^{O}w_{k0}\right) \tag{A2.2.8}$$

where $^{O}w_{kj}$ are the weights for the transfer from the j-th hidden node to the k-th output node.

We notice that o_k depends on the output of the hidden nodes h_j

$$h_j = {}^{h}f\left(\sum_i {}^{h}w_{ij}x_i + {}^{h}w_{j0}\right) \tag{A2.2.9}$$

and the h_j depends on the sum of the weighted input x_i, which represents the components of the feature vector $\boldsymbol{x}$. $^{h}w_{ij}$ are the weights for the transfer from the i-th input node to the j-th hidden node.

The error E then reads as

$$E = \frac{1}{2}\sum_{jk}\left(y_k - {}^{O}f\left(\sum_j {}^{O}w_{kj}h_j + {}^{O}w_{k0}\right)\right)^2 \tag{A2.2.10}$$

The activation functions in the output nodes ^{O}f and in the hidden nodes ^{h}f are supposed to be of linear and sigmoidal form. For the sake of simplicity, we use the abbreviations for the arguments of the activation functions in the hidden and output layers, i.e.,

$$u_j = \sum_i {}^{h}w_{ij}x_i + {}^{h}w_{j0}$$
$$v_k = \sum_j {}^{O}w_{kj}h_j + {}^{O}w_{k0}$$

The gradient of error E with respect to the weights is obtained by applying the chain rule

$$\partial E \,/\, \partial^{O}w_{kj} = (y_k - o_k)(\partial f(v_k) \,/\, \partial v_k)(\partial v_k \,/\, \partial^{O}w_{kj})$$

and for the weights between input and hidden

$$\partial \mathrm{E} / \partial^{\mathrm{o}} \mathrm{w}_{\mathrm{ij}} = (\mathrm{y}_{\mathrm{k}} - \mathrm{o}_{\mathrm{k}})\ (\partial^{\mathrm{o}} \mathrm{f} / \partial^{\mathrm{o}} \mathrm{net\_o}_{\mathrm{k}}) (\partial^{\mathrm{o}} \mathrm{net\_o}_{\mathrm{k}} / \partial^{\mathrm{h}} \mathrm{f})\ (\partial^{\mathrm{h}} \mathrm{f} / \mathrm{net\_h}_{\mathrm{j}}) (\partial\ \mathrm{net\_h}_{\mathrm{j}} / \partial^{\mathrm{h}} \mathrm{w}_{\mathrm{ij}})$$

(A2.2.12)

$$\partial E / \partial^h w_{ij} = \frac{1}{2} \sum_k \partial (y_k - o_k)^2 \Big/ \partial^h w_{ij} = -\sum_k (y_k - o_k) \frac{\partial o_k}{\partial v_k} \frac{\partial v_k}{\partial f(u_j)} \frac{\partial f(u_j)}{\partial u_j} \frac{\partial u_j}{\partial^h w_{ij}}$$

We upgrade the weights taking the negative portions of the error gradients:

$${}^{O}w_{kj}(t+1) = {}^{O}w_{kj}(t) + \eta (y_k - o_k) \frac{\partial f(v_k)}{\partial v_k} \tag{A2.2.13a}$$

$${}^{h}w_{kj}(t+1) = {}^{h}w_{kj}(t) + \eta \left(\frac{\partial f(u_j)}{\partial u_j} \sum_{ik} (y_k - o_k) x_i \right) \tag{A2.2.13b}$$

(see also Freeman and Skapura, 1992). The parameter η is addressed to as a "learning parameter" and is typically small with respect to w(t).

Appendix 2.3 SVM optimization of the margins

Recall the definition of the cost function as

$$J(\boldsymbol{w}, w_0) = \frac{1}{2} \|w\|^2$$

which has to be minimized. This corresponds to maximizing the width of the margin between the two groups. The two elements delineating the margins are defined by the support vectors for which

$$\boldsymbol{w}^T \boldsymbol{x} + w_0 = \pm 1$$

and impose the constraints of the optimization of J. The research of minimum of the cost J thus becomes a nonlinear (quadratic) optimization problem with linear constraints.

There the Lagrange function is given by

$$L(\boldsymbol{w}, w_0, \boldsymbol{\lambda}) = \frac{1}{2} \boldsymbol{w}^T \boldsymbol{w} - \sum_i \lambda_i \left(y_i \left(\boldsymbol{w}^T \boldsymbol{x}_i + w_0 \right) - 1 \right) \tag{A2.3.1}$$

The first-term of the Lagrange function corresponds to the cost function J, and the second is the constraint

$$y_i\left(\boldsymbol{w}^T\boldsymbol{x}_i + w_0\right) - 1 \geq 0 \tag{A2.3.2}$$

multiplied by the Lagrange multipliers λ_i. In addition, we require that all $\lambda_i \geq 0$. Taking the derivative of $L(\boldsymbol{w}, w_0, \boldsymbol{\lambda})$ with respect to $\boldsymbol{w}$ and w_0 and equating to 0 yields

$$\boldsymbol{w} = \sum_i \lambda_i y_i \boldsymbol{x}_i \tag{A2.3.3}$$

and

$$\sum_i \lambda_i y_i = 0 \tag{A2.3.4}$$

Consider now the two linear elements that define the margins. Their position is given by

$$\boldsymbol{w}^T\boldsymbol{x} + w_0 = \pm 1$$

Consequently, the support vectors mentioned earlier obey that relation.

We now search for the λ. This can be solved using the Lagrange function above, i.e.,

$$L(\boldsymbol{w}, w_0, \boldsymbol{\lambda}) \tag{A2.3.5}$$

under the constraints

$$\boldsymbol{w} = \sum_i \lambda_i y_i \boldsymbol{x}_i$$
$$\sum_i \lambda_i y_i = 0$$

and

$$\lambda \geq 0$$

ending up with

$$\max \lambda = \left(\sum_i \lambda_i - \frac{1}{2}\sum_{ij} \lambda_i \lambda_j y_i y_j \boldsymbol{x}_i \boldsymbol{x}_j\right) \tag{A2.3.6}$$

subject to

$$\sum_i \lambda_i y_i \geq 0$$
$$\lambda \geq 0$$

Appendix 2.4. Hidden Markov models

Appendix 2.4.1. Evaluation

In the evaluation task, we aim to find the probability $P(\boldsymbol{o}|\theta)$ that the observation sequence $\boldsymbol{o}$ has been generated by the model θ. The estimation of $P(\boldsymbol{o}|\theta)$ can be obtained by a multiplication of the sequence of probabilities (i.e., transition and emission probabilities) that generated the observed data sequence, summed over all possible state sequences.

$$P(\boldsymbol{o}|\theta) = \sum_{q} P(\boldsymbol{o}, \boldsymbol{q}|\theta) \tag{A2.4.1}$$

However, calculating the probability in this manner is computationally expensive, particularly with large model sequences with a long d length T. In a model with N states, it would require $2\ T\ N^T$ multiplications.

By sorting individual sums by time and factorizing, we can reduce the complexity of the problem. The forward-probability α is defined as the probability of the first part of the observation sequence.

$$\boldsymbol{o} = o_1, \ldots\ldots. o_t$$

ending at time t in state $q_t = j$

$$\alpha_t(j) = P(o_1, \ldots., o_t, q_t = j|\theta) \tag{A2.4.2}$$

Thus, the computation of $\alpha_t(j)$ can be carried out by mathematical induction

$$\text{Initialization}: \alpha_1(j) = \pi_j b_j(o_1) \ \text{ for } j = 1, \ldots, N \tag{A2.4.3a}$$

$$\text{Induction}: \alpha_t(j) = \left[\sum_{i=1}^{N} \alpha_{t-1}(i)\alpha_{ij}\right] b_j(o_t), \ \text{ for } j = 1\ldots N, \ \ \mathrm{t} = 2, \ldots T \tag{A2.4.3b}$$

$$\text{Termination}: \ P(\boldsymbol{o}|\theta) = \sum_{j=1}^{N} \alpha_T(j) \tag{A2.4.3c}$$

In contrast, we can also start the sorting with the latest terms. The backward probability $\beta_t(i)$ is defined as the probability of the partial observation sequence starting at time $t+1$, given that at time t, the model is in state $q_t = i$

$$\beta_t(i) = P(o_{t+1}, \ldots.. o_T|q_t = i, \theta). \tag{A2.4.4}$$

Then the following induction scheme can be used to calculate $\beta_t(i)$

$$\text{Initialization}: \beta_T(j) = 1, \text{for } j \ = 1, \ldots, N \tag{A2.4.5a}$$

$$\text{Induction}: \ \ \beta_t(i) = \sum_{j=1}^{N} a_{ij} b_j(o_{t+1})\beta_{t+1}(j), \text{for } i \ = 1, \ldots, N \ \text{ and } \ t = T-1, \ldots, 1 \tag{A2.4.5b}$$

$$\text{Termination}: \quad P(\boldsymbol{o}|\theta) = \sum_{j=1}^{N} \pi_j b_j(o_1)\beta_1(j) \tag{A2.4.5c}$$

Thus, for the evaluation problem either the forward or the backward variable can be used. In contrast, both are needed for the training problem Box 2.6.

Box 2.6 HMM representation as a trellis.

Consider an HMM given by the two matrices.

Transition matrix between states

$$\boldsymbol{A}_{ij} = \begin{vmatrix} 1 & 0 & 0 & 0 \\ 0.2 & 0.3 & 0.1 & 0.4 \\ 0.2 & 0.5 & 0.2 & 0.1 \\ 0.8 & 0.1 & 0.0 & 0.1 \end{vmatrix}$$

Emission matrix for observables given the state j

$$\boldsymbol{B}_{jk} = \begin{vmatrix} 1 & 0 & 0 & 0 & 0 \\ 0 & 0.3 & 0.4 & 0.1 & 0.2 \\ 0 & 0.1 & 0.1 & 0.7 & 0.3 \\ 0 & 0.5 & 0.2 & 0.1 & 0.2 \end{vmatrix}$$

(see Duda et al., 2001). We now wish to know the probability of observing a particular sequence

$$O = \{o_1, o_3, o_2, o_0\}$$

Suppose that our initial state is q_1 (our indices start with 0). We shall obtain the probability of getting o_3 at time $t = 1$, given that we have o_1 at time $t = 0$. In our example, the α_j, i.e., the probability that the model generated the sequence up to t, can be obtained as follows. For instance, we know that at $t = 0$, our system was in a hidden state q_1.

Thus $\alpha_1(0) = 1$ and

$$\alpha_j(0) = 0 \;\; \forall \;\; j \neq 1$$

Next, we calculate $\alpha_j(1)$. Because the visible state o_1 was emitted at $t = 1$, we get $\alpha_0(1) = \alpha_1(0)a_{10}b_{01} = 1 \cdot 0.2 \cdot 0 = 0$. In an analogous way we get $\alpha_1(1) = \alpha_1(0)$ $a_{11}1 = 1 \cdot 0.3 \cdot 0.3 = 0.09$. In a similar way, we can calculate all the other α's along the red arrows shown in the HMM trellis below. The calculus here is simple as we have to account only for the transitions from the known hidden state q_1 (all the other transitions have contribution 0 to $\alpha_j(1)$). For the subsequent time steps, things get a bit more nasty as we have to sum over all hidden states at the previous time steps, as described in Appendix 2.4.1.

Continued

Box 2.6 HMM representation as a trellis.—cont'd

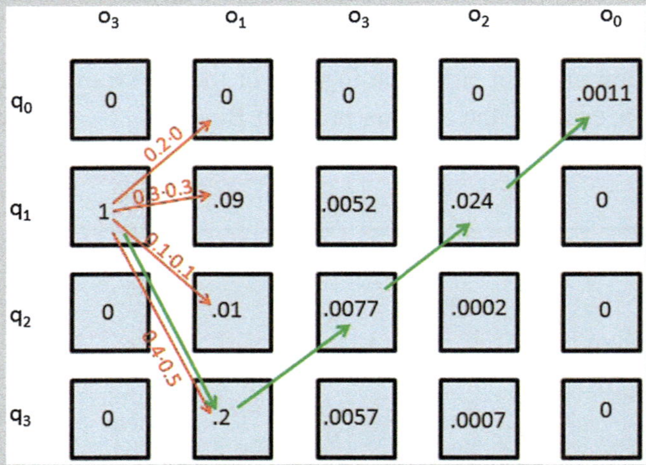

In HMM decoding, we may consider only the path where the α's are maximum. For instance, being started from q_1 and o_3, the highest probability for getting o_1 goes via q_3. Step by step, we go through the trellis moving along the green arrows. This is a simplification as we avoid considering all the α's in the trellis. This simplification has a price: the path delineated by the green arrows includes a transition from q_3 to q_2, which is not possible from the transition matrix **A**. Indeed, the corresponding element for this transition is 0. For more details with respect to this example, see Duda et al. (2001).

Appendix 2.4.2. Decoding—the Viterbi algorithm

The probability of the sequence of hidden states $\boldsymbol{q}$ is defined as

$$P(\boldsymbol{q}|\boldsymbol{o},\theta) = P(\boldsymbol{o},\boldsymbol{q}|\theta)/P(\boldsymbol{o}|\theta) \quad \text{(A2.4.6)}$$

The best hidden state sequence $\boldsymbol{q}^*$ is defined by

$P(\boldsymbol{o}, \boldsymbol{q}^*|\theta) = max_{q\varepsilon\mathbb{Q}}P(\boldsymbol{o}, \boldsymbol{q}|\theta)$, where $\mathbb{Q}$ is the set of all possible state sequences.

Instead of the forward probability $\alpha_t(j)$, we now compute only the maximum probability for generating the partial observation sequence

$$\boldsymbol{o} = o_1, \ldots\ldots.o_t$$

ending at time t in state $q_t = j$

$$\psi_t(j) = \max[(P(o_1, \ldots o_t, q_1, \ldots q_t|\theta)|\boldsymbol{q} \in \mathbb{Q}, q_t = j] \quad \text{(A2.4.7)}$$

The $\psi_t(j)$ are computed recursively as follows. In doing so, the best path is kept by the matrix $\Psi_t(j)$ and the best single path is obtained via backtracking after termination

$$\text{Initialization}: \psi_1(j) = \pi_j b_j(o_1), \Psi_1(j) = 0, \text{for } j = 1, \ldots, \text{N} \tag{A2.4.8a}$$

$$\text{Recursion}: \psi_t(j) = max_i(\psi_{t-1}(i)a_{ij})b_j(o_t), \Psi_t(j) = argmax_i(i)\psi_{t-1}(i)a_{ij}, \text{for } t > 1 \text{ and } j = 1, \ldots, N \tag{A2.4.8b}$$

$$\text{Termination}: P^*(\boldsymbol{o}|\theta) = max_j \psi_T(j),\ q^* = argmax_j \psi_T(j) \tag{A2.4.8c}$$

The optimal single state sequence can then be derived via backtracking

$$q_T^* = \Psi_{t+1}\left(q_{t+1}^*\right), \text{for } t = T-1, \ldots, 1 \tag{A2.4.9}$$

Appendix 2.4.3. Training—the expectation—maximization /Baum—Welch algorithm

The "Expectation—Maximization" (EM) algorithm is an iterative process, which can be used to estimate the parameters of normal density function of an unlabeled training data set. In other words, given the joint distribution $P(c,\boldsymbol{o}|\boldsymbol{\theta})$ of observations $\boldsymbol{o}$ and unknown class labels c, and the model parameters $\boldsymbol{\theta}$, the EM algorithm aims to maximize the likelihood function $P(c|\boldsymbol{\theta})$ with respect to $\boldsymbol{\theta}$.

The iterative process can be summarized in the following four steps, where each cycle increases the log likelihood of the data until it reaches a local maximum

1. Choose initial model parameters $\boldsymbol{\theta}$.
2. Evaluate $P(c|\boldsymbol{o},\boldsymbol{\theta})$—the "E-step" in the EM process
3. Evaluate $\boldsymbol{\theta}^{new} = argmax_{\boldsymbol{\theta}} \sum_c \left(P\left(c|\boldsymbol{o}, \boldsymbol{\theta}^{old}\right) lnP\left(c \middle| \boldsymbol{o}, \boldsymbol{\theta}\right)\right)$—the "M-step" in the EM process.
4. Check if the convergence criterion is satisfied. If not, replace $\boldsymbol{\theta}$ by $\boldsymbol{\theta}^{new}$ and return to the E-step.

The Baum—Welch algorithm is a special case of the EM algorithm and it is used to estimate the parameters of an HMM. The algorithm is based on the forward and backward variables. The parameters to learn are $\boldsymbol{\theta}$, the emission matrix $\boldsymbol{B}$, and the transition matrix $\boldsymbol{A}$. Each emission probability of an HMM can be described as a multivariate density function defined by

$$b_j(\boldsymbol{o}_t) = \sum_{m=1}^{M} c_{jm}\mathcal{N}\left(\boldsymbol{o}_t, \boldsymbol{\mu}_{jm}, \boldsymbol{C}_{jm}\right) \tag{A2.4.10}$$

where c_{jm} is the mixture coefficient of the mth mixture in state j. $\boldsymbol{\mathcal{N}}$ is a Gaussian function with mean $\boldsymbol{\mu}_{jm}$ and covariance matrix $\boldsymbol{C}_{jm}$

Hence, to be more precise, we aim to learn the mean, covariance matrix, and the mixture coefficient of each mixture component at each state and the transition probabilities between states. In the following, N denotes the number of states.

Using the forward and backward variables, the probability of changing from state s_i to s_j, given the observation sequence $\boldsymbol{o}$ and the model $\boldsymbol{\theta}$, can be calculated from

$$\gamma_t(i,j) = P(q_t = s_i, q_{t-1} = s_j|\boldsymbol{o},\boldsymbol{\theta}) = \frac{P(q_t = s_i, q_{t+1} = s_j|\boldsymbol{o},\boldsymbol{\theta})}{P(\boldsymbol{o}|\theta)} \tag{A2.4.11a}$$
$$= \alpha_t(i)a_{i,j}b_j(o_{t+1})\beta_{t+1}(j)/\Sigma_{n=1}^{N}\alpha_t(n)\beta_t(n), 1 \le t \le T$$

From Eq. A.2.4.11a we can get the probability of being in state i, given the observations and the model parameters

$$\delta_t(i) = P(q_t = i|\boldsymbol{o},\boldsymbol{\theta}) = \sum_{j=1}^{N}\gamma_t(i,j) = \alpha_t(i)\beta_t(i) \bigg/ \sum_{n=1}^{N}\alpha_t(n)\beta_t(n) \tag{A2.4.11b}$$

In addition, we need the probability of being in the mth Gaussian mixture component $\boldsymbol{\mathcal{N}}_m(\boldsymbol{o}_t|\boldsymbol{\mu}_m, \boldsymbol{C}_m)$ given by its mean $\boldsymbol{\mu}_m$ and its covariance $\boldsymbol{C}_m$ at time t

$$\varepsilon_t(i,m) = P(q_t = i, m_t = m|\boldsymbol{o},\boldsymbol{\theta}) = P(q_t = i, m_t = m, \boldsymbol{o}|\boldsymbol{\theta})\Big(P(\boldsymbol{o}|\boldsymbol{\theta})$$
$$= \sum_{j=1}^{N}\pi_i c_{im}\boldsymbol{\mathcal{N}}_m(\boldsymbol{o}_1|\boldsymbol{\mu}_m, \boldsymbol{C}_m)\beta_t(i) \bigg/ \sum_{j=1}^{N}\alpha_t(j)\beta_t(j),\ t = 1 \tag{A2.4.12}$$
$$= \sum_{j=1}^{N}\alpha_{t-1}(j)\alpha_{ij}c_{im}\boldsymbol{\mathcal{N}}_m(\boldsymbol{o}_t|\boldsymbol{\mu}_{im}, \boldsymbol{C}_{im})\beta_t(i) \bigg/ \sum_{j=1}^{N}\alpha_t(j)\beta_t(j),\quad t > 1$$

Now, we can reestimate the model parameters using $\alpha_t(i)$ and $\beta_t(i)$

$$\pi_i^{new} = \delta_1(i) = \alpha_1(i)\beta_i(i) \bigg/ \sum_{n=1}^{N}\alpha_1(n)\beta_i(n) \tag{A2.4.13}$$

$$\alpha_i^{new} = \sum_{t=1}^{T-1}\gamma_t(i,j) \bigg/ \sum_{t=1}^{T-1}\delta_t(i) \tag{A2.4.14}$$

$$c_{jm}^{new} = \sum_{t=1}^{T}\varepsilon_t(j,m) \bigg/ \sum_{t=1}^{T}\delta_t(j) \tag{A2.4.15}$$

$$\boldsymbol{\mu}_{jm}^{new} = \sum_{t=1}^{T} \varepsilon_t(j,m)\boldsymbol{o}_t \Big/ \sum_{t=1}^{T} \delta_t(j) \tag{A2.4.16}$$

$$\boldsymbol{C}_{jm}^{new} = \sum_{t=1}^{T} \varepsilon_t(j,m)\left(\boldsymbol{o}_t - \boldsymbol{\mu}_{jm}^{new}\right)\left(\boldsymbol{o}_t - \boldsymbol{\mu}_{jm}^{new}\right)^T \Big/ \sum_{t=1}^{T} \delta_t(j) \tag{A2.4.17}$$

We can explain the Baum–Welch algorithm in terms of EM. For example, the E-step first calculates the number of expected transitions from state i to state j, i.e., $\sum_{t=1}^{T-1} \gamma_t(i,j)$, and the overall number of transitions from state i, i.e., $\sum_{t=1}^{T-1} \delta(i)$. The M-step then calculates the new transition probability a_{ij}^{new} from the number of expected transitions from state i to j normalized by the expectation of being in state i (Eq. A.2.4.14). Analogously, the coefficient c_{jm}^{new} is calculated from the expectation of being in state j with the mixture component m, normalized by the expectation of being in state j (Eq. A.2.4.15). Finally, the mean $\boldsymbol{\mu}_{jm}^{new}$ and the covariance $\boldsymbol{C}_{jm}^{new}$ are calculated from the expectation of the observation $\boldsymbol{o}_t$ and the covariance $\left(\boldsymbol{o}_t - \boldsymbol{\mu}_{jm}^{new}\right)\left(\boldsymbol{o}_t - \boldsymbol{\mu}_{jm}^{new}\right)^T$ in state j with the mixture component m, normalized of the expectation of being in state j (Eqs. A.2.4.16 and A.2.4.17).

CHAPTER 3

Unsupervised learning

3.1 Introduction

In supervised learning, we deal with labeled patterns for the identification of formalisms that relate specific features to a category, which is supposed to be known. Once the formalism is found, it can be applied to other patterns to rapidly assign their category. Supervised techniques have been proven to be effective, as even complicated relationships between features and categories can be learnt. However, they can also give misleading results, if applied to patterns belonging to a parent population that differs from the one considered during the learning phase. The problem is known as "overfitting". In geophysical applications, the risk of facing patterns belonging to a new parent population is quite frequent. Indeed, we often deal with time-dependent data, that is, data with pattern characteristics that develop over time, eventually making our classification procedures obsolete. A further issue is the high number of degrees of freedom that nonlinear classification schemes often have. This implies the necessity to consider large training data sets. Defining target values for many examples brings along a considerable effort together with the risk of committing errors (i.e., the choice of partially flawed examples for the learning process).

In unsupervised learning, we leave the whole job to the computer, asking the machine to identify groups or clusters of similar patterns based on a measure of (dis)similarity. Ideally, the groups can be clearly distinguished from each other. In some circumstances, we can use a few patterns—even only one single pattern—as a prototype for a whole group. As the patterns of a cluster are supposed to have similar characteristics, we do not lose much information neglecting the internal variability among the patterns of the same cluster. If the clusters are well distinguished, the variability of the whole ensemble maintains a fair fidelity by considering only the prototypes. We may also limit ourselves to label the prototype patterns, i.e., all members of a cluster are automatically assigned to the same target as the prototype pattern. We can set up a semisupervised classification scheme as follows: (i) carry out the unsupervised classification, identify clusters and prototypes, (ii) assign targets to all patterns according to the one of the corresponding prototype, and (iii) carry out supervised classification on prototypes. At first glance, this looks like nonsense, but turns out as a useful scheme, for instance in the presence of very large data sets. As shown below, we may create the so-called "Self-Organizing Maps", where we first

Advantages and Pitfalls of Pattern Recognition. https://doi.org/10.1016/B978-0-12-811842-9.00003-0

form microclusters using an unsupervised clustering scheme and then use the prototypes in further classification steps.

3.1.1 Metrics of (dis)similarity

A key point here is that we must establish a measure of (dis)similarity or a metric. Recall the four basic properties of a metric mentioned in Chapter 1. Given the feature vectors **a**, **b**, and **c**, with **d**(,) being the distance between two vectors, these properties are

$$\text{Nonnegativity:}\quad \mathbf{d}(\mathbf{a},\mathbf{b}) \geq \mathbf{0}$$
$$\text{Reflexivity:}\quad \mathbf{d}(\mathbf{a},\mathbf{b}) = \mathbf{0} \text{ if and only if } \mathbf{a} = \mathbf{b}$$
$$\text{Symmetry:}\; \mathbf{d}(\mathbf{a},\mathbf{b}) = \mathbf{d}(\mathbf{b},\mathbf{a})$$
$$\text{Triangle Inequality:}\quad \mathbf{d}(\mathbf{a},\mathbf{b}) + \mathbf{d}(\mathbf{b},\mathbf{c}) \geq \mathbf{d}(\mathbf{a},\mathbf{c})$$

The well-known Euclidean distance

$$\mathrm{d}(\mathbf{a},\mathbf{b}) = \left(\sum_{i=1}^{d}(a_i - b_i x)^2\right)^{1/2}$$

is certainly a metric. It can be seen as a specific case of the more general Minkowski metric

$$\mathrm{d}(\mathbf{a},\mathbf{b}) = \left(\sum_{i=1}^{l}|a_i - b_i x|^k\right)^{1/k} \tag{3.1}$$

often referred to as the L_k norm of an l-dimensional distance vector **d**. The Euclidean distance is a Minkowski metric with $k = 2$. Setting $k = 1$, we obtain the "Manhattan" or "city block" distance. We can better understand the implications of choosing different values for k by considering the shapes of equidistance functions. For the Euclidean distance $k = 2$, all points with d(**a**,**b**) = constant lie on a sphere. Setting $k = 1$, the equidistance curves are described by linear elements, such as lines, planes, and hyperplanes; from that, we can easily understand the name of this metric as the "city block" distance. An interesting case arises for $k \gg 1$. Indeed, for $k \rightarrow \infty$, the distance is controlled by only one component i, for which

$$|a_i - b_i| = \max$$

Thus, the elements of equidistance are given by the set of planes perpendicular to the axes of components (see Fig. 3.1).

The "Tanimoto Distance" (T_D) is based on the number of elements two patterns have in common and the number of elements where they differ. It compares two vectors, **a** and **b**,

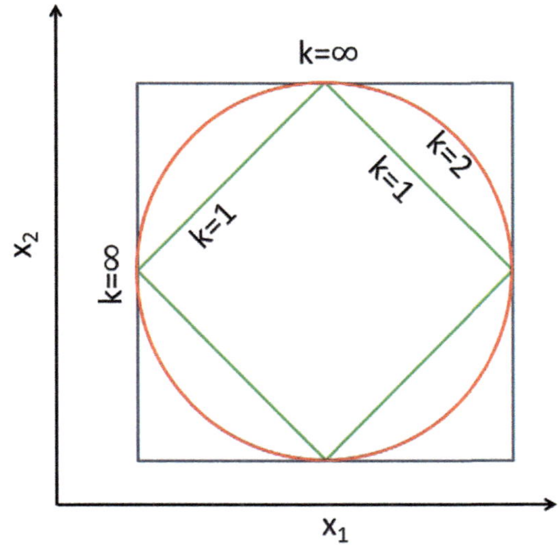

Figure 3.1
Minkowski distance for $k = 1, 2, \infty$. The green, red, and blue figures represent elements of equidistance depending on the exponent k.

having n_1 and n_2 elements, respectively. The number of common elements is given by n_{12}. The distance d(**a**,**b**) is then the Tanimoto Distance

$$T_D = \frac{n_1 + n_2 - 2n_{12}}{n_1 + n_2 - n_{12}} = 1 - \frac{n_{12}}{n_1 + n_2 - n_{12}} \tag{3.2}$$

(see, e.g., Duda et al., 2001). It is used when we deal with categorical data. Note that here we distinguish elements in Boolean form, in which no graded similarity is allowed. For real-valued vectors, we can use the term

$$T_D = 1 - \frac{\boldsymbol{a}^T\boldsymbol{b}}{||\boldsymbol{a}||^2 + ||\boldsymbol{b}||^2 - \boldsymbol{a}^T\boldsymbol{b}} \tag{3.3}$$

The "Jaccard index" J is a general measure of (dis)similarity. It is defined as

$$J(\mathbb{A}, \mathbb{B}) = \frac{\mathbb{A} \cap \mathbb{B}}{\mathbb{A} \cup \mathbb{B}} \tag{3.4}$$

where $\mathbb{A}$ and $\mathbb{B}$ are two sets of elements (see Fig. 3.2).

A further measure for (dis)similarity of real-valued feature vectors is the cosine similarity measure, given by the term

$$r = \frac{\boldsymbol{a}^T\boldsymbol{b}}{||\boldsymbol{a}||\,||\boldsymbol{b}||} \tag{3.5}$$

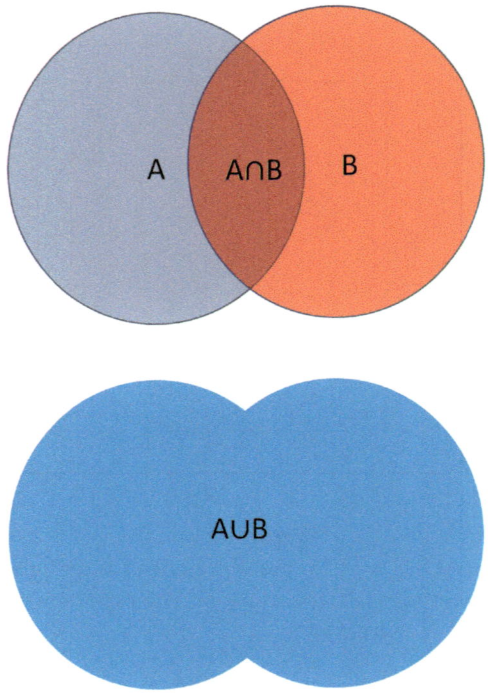

Figure 3.2
The Jaccard index can be seen as a generalization of the approach considered in the Tanimoto Distance.

This measure is of particular importance, when normalized feature vectors are considered. A similar measure is the "Pearson correlation coefficient"

$$r' = \frac{\boldsymbol{a}'^T \boldsymbol{b}'}{||\boldsymbol{a}'|| \, ||\boldsymbol{b}'||} \tag{3.6}$$

where we consider vectors $\boldsymbol{a}' = [a_1 - \overline{a} \ldots a_l - \overline{a}], \boldsymbol{b}' = [b_1 - \overline{b} \ldots b_l - \overline{b}]$, with

$\overline{a}, \overline{b} = \frac{1}{l} \sum_{i=1}^{l} a_i, \ b_i$

3.1.2 Clustering

In cluster analysis, we try to identify groups or segments of objects as subsets of an ensemble to achieve an easier interpretation of the information stored in our whole data set. We are often able to define the so-called prototypes, i.e., objects with features that represent a large number of patterns in our data. A critical issue is the choice of the measure of (dis)similarity or metrics, for which we mentioned some examples above. A further issue is the choice of the clustering strategy (see Anderberg, 1973). This choice depends on the ideas we may have on the underlying structure of data, for example, if they have a hierarchy. For instance, in taxonomy it is useful to follow a strategy accounting for such a structure. Otherwise, we may assume that data form clusters independent of each other. In the latter case, we shall adopt methods of "partitioning clustering".

3.1.2.1 Partitioning clustering

In partitioning clustering, we have to fix the number of clusters a priori. Often we start with a small number of clusters and augment this number in the following steps, until we find a solution—a "partition"—feasible[1] to our goals. The "K-means" algorithm is perhaps the most popular clustering method for its straightforward concept and the computational simplicity.

Consider the dispersion S in our ensemble consisting of n feature vectors, $\boldsymbol{x}_i$ having the global mean $\overline{\boldsymbol{x}}$,

$$S = \sum_{i=1}^{n} \|\boldsymbol{x}_i - \overline{\boldsymbol{x}}\|^2; \boldsymbol{S}_{Tot} = \sum_{i=1}^{n} (\boldsymbol{x}_i - \overline{\boldsymbol{x}})(\boldsymbol{x}_i - \overline{\boldsymbol{x}})^T \tag{3.7}$$

$\boldsymbol{S}_{Tot}$ denotes the dispersion matrix of the whole data set. Having a partition such that each $\boldsymbol{x}_i$ belongs to some cluster j, we measure the dispersion within each cluster by

$$\mathbf{S}_j = \sum_{i=1}^{m} (\boldsymbol{x}_i - \overline{\boldsymbol{x}}_j)(\boldsymbol{x}_i - \overline{\boldsymbol{x}}_j)^T \tag{3.8}$$

The dispersion between the k clusters is

$$\mathbf{S}_c = \sum_{j=1}^{k} m_j (\overline{\boldsymbol{x}}_j - \overline{\boldsymbol{x}})(\boldsymbol{x}_j - \overline{\boldsymbol{x}})^T \tag{3.9}$$

where we assume that each cluster is composed of m_j samples. It can be shown that

$$\mathbf{S}_{Tot} = \sum_{j=1}^{k} \mathbf{S}_j + \mathbf{S}_c \tag{3.10}$$

[1] This implies some subjectivity, which is however an intrinsic characteristic of unsupervised learning techniques. Literature proposes criteria that may guide our decision whether or not to accept a clustering; nevertheless, the user eventually decides. Unsupervised learning is therefore said to be "data-driven and user-defined".

i.e., the total dispersion is given by the sum of the dispersions measured within the clusters and the dispersion measured between the cluster centroids $\bar{\boldsymbol{x}}_j$. In the framework of ANOVA ("ANalysis Of VAriance", Appendix 3.1, also see Davis, 1986; Borradaile, 2003), $\boldsymbol{S}_C$ is termed "between" dispersion and $\sum_{j=1}^{k} \boldsymbol{S}_j$ is referred to as "within" dispersion (or variance). All three matrices are quadratic and symmetric. For those matrices, we may find the total amount of dispersion from the sum of their eigenvalues $\sum_{i=1}^{l} \lambda_i$ or their trace, i.e., the sum of the diagonal elements. For our purposes here, we note that once we find a partition that minimizes the "within" dispersion (trace of $\sum_{j=1}^{k} \boldsymbol{S}_j$), the "between" dispersion (trace of $\boldsymbol{S}_C$) s maximum. This corresponds exactly to our goal "find clusters with the highest degree of compactness (the highest degree of homogeneity), which are the most distant from each other (with the highest degree of heterogeneity)".

In practice, we may follow the scheme outlined in Box 3.1.

As a practical example, consider Table 3.1, which reports the concentration of important chemical components found in volcanic rock samples (see Corsaro et al., 2013). The numbers have quite different physical meanings: SiO_2 and K_2O are components given as the percentage of total weight; Ca/Al and Rb/Nb are ratios between components; others (so-called "trace elements", e.g., Th, La) are present only in very small concentrations expressed as parts per million (ppm). The direct application of the K-means clustering to a set of 103 samples yields results like those shown in Fig. 3.3.

In K-means clustering, we just take numbers as they are, without any a priori assumption about the meaning they may have. For instance, in our example of Table 3.1, "Cr" ranges from 22.15 to 39.88 and "K_2O" from 2.02 to 2.19, by far less than "Ni" or "Sr". Accordingly, "Sr" has a stronger influence than "K_2O", which is not necessarily justified from the viewpoint of a petrologist. To neutralize the effects of the various units of measure—in Table 3.1, percentage, ratios, and ppm—we normalize our data as already proposed in Chapter 1; here the normalization is with respect to the range encountered in each component of the feature vector. Without normalizing the data, the clustering would be controlled by one component (Sr, see Fig. 3.3A). A suitable standardization overcomes this effect. In Fig. 3.3B, we see that more components contribute to clustering, as clusters in the 2D marginal distributions overlap—which is a common effect of low-dimensional representations of high-dimensional data (see Chapter 2, Box 2.1).

Applying normalization to the feature vectors, the K-means clustering tends to create clusters having circular, spherical, or hyper-spherical convex hulls. This is clearly recognized from Fig. 3.1. As K-means uses the Euclidean distance as a metric, the

Box 3.1

Start at step $t = 0$.

Choose the maximum number of iterations, ITMAX > 0, choose the number of desired clusters.

Define an initial partition, such that each pattern i belongs to some cluster Cj. Avoid partitions where some $Cj = \{\ \}$—empty clusters. A way to do this is to define the initial average vectors ("centroids") for each cluster randomly. An alternative is to assign randomly each pattern to a cluster j by setting j randomly.

Loop:

In a procedure called "interchange technique" (see, e.g., Späth, 1983), select a pattern and calculate its distance to each centroid. Assign the pattern to the cluster with the closest centroid. Update centroids and dispersion encountered within the clusters. Suppose that we shift a pattern originally to a cluster p, which becomes a new cluster q. Then,

$$\boldsymbol{S}_q = \boldsymbol{S}_p + \frac{m_p}{m_p + 1}(\boldsymbol{x}_i - \bar{\boldsymbol{x}}_p)(\boldsymbol{x}_i - \bar{\boldsymbol{x}}_p)^T$$

On the other hand, for a cluster p from which a pattern is taken away, we have

$$\boldsymbol{S}_q = \boldsymbol{S}_p - \frac{m_p}{m_p - 1}(\boldsymbol{x}_i - \bar{\boldsymbol{x}}_p)(\boldsymbol{x}_i - \bar{\boldsymbol{x}}_p)^T$$

In a similar way, we update the centroid vectors

$$\bar{\boldsymbol{x}}_q = \frac{m_p}{m_p + 1}(m_p\bar{\boldsymbol{x}}_p + \boldsymbol{x}_i)$$

- for the cluster receiving a new pattern, and

$$\bar{\boldsymbol{x}}_q = \frac{m_p}{m_p - 1}(m_p\bar{\boldsymbol{x}}_p - \boldsymbol{x}_i)$$

- for the cluster from which a pattern is removed. We may accept or refuse the new partition, verifying whether the sum of dispersions, trace $(\sum_{j=p,q} \boldsymbol{S}_j)$, has decreased. Note that it is sufficient to consider only the two clusters involved in the interchange, as the dispersions and centroids of the remaining clusters do not change. The computational effort is limited as the updating formula is very simple.

Continue with this pattern and compare to other clusters until no further decrease occurs.

Set $t = t + 1$. Take the next pattern and start again the interchange technique.

Continue unless the number of iterations t reaches ITMAX.

Table 3.1: Example of concentration of chemical components in volcanic rock samples (see Corsaro et al., 2013).

Mg#	SiO_2	K_2O	Ca/Al	Th	La	Nb	Nd	Sr	Tb	Cr	Ni	Rb/Nb
0.48	47.84	2.14	0.556	7.97	57.0	42.18	46.87	1153	0.99	22.58	21.04	1.16
0.47	48.12	2.19	0.564	8.07	57.2	41.49	46.49	1133	0.97	22.15	20.54	1.16
0.50	47.25	2.02	0.635	6.74	51.0	37.09	44.38	1079	0.98	36.92	29.16	1.23
0.51	46.86	2.06	0.645	6.72	50.6	36.09	44.06	1067	0.97	39.88	29.45	1.25
0.50	46.89	2.05	0.642	6.60	50.0	35.95	43.34	1056	0.95	39.75	28.59	1.22

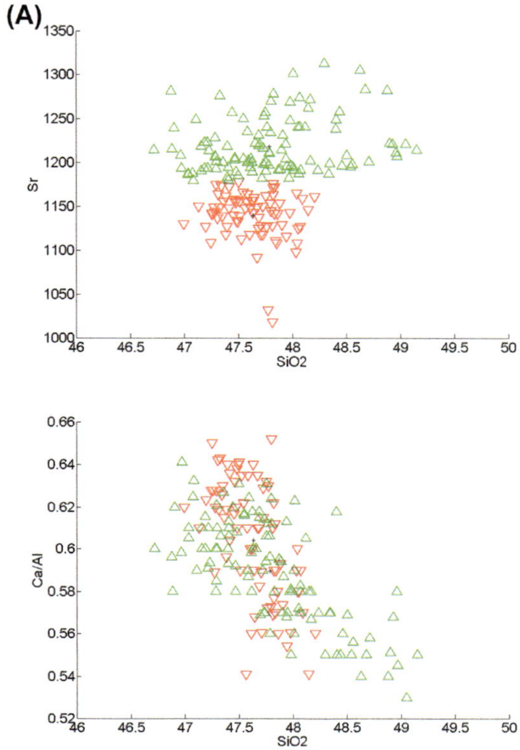

Figure 3.3
Results of K-means applied to petrochemical composition of volcanic rock samples. Marginal distributions of two clusters are shown, selecting two components out of 13. Panels (A) are related to results using data as is, and panels (B) were obtained after a normalization of the components with respect to their units of measure. In panels (A), we clearly recognize that the separation of the two clusters is controlled by one component (Sr).

Minkowski exponent is $k = 2$, and the shape of the equidistance figure is circular or (hyper)spherical. This can be a serious limitation, as such cluster components of the feature vectors are supposed to be statistically independent of each other. This is not necessarily the case even after normalization. For instance, recall the example of the m_b–M_S criterion for the distinction of earthquake and nuclear test seismograms we discussed in Chapter 2; m_b and M_S are correlated—the greater the body wave magnitude m_b, the greater the surface magnitude M_S. Applying the K-means clustering (e.g., script S3_2 accompanying this book) after a normalization, we get the results shown in Fig. 3.4A.

Selecting a partition with two clusters, K-means forms two groups: the first one in the lower left angle of the diagram, where both kinds of magnitudes have low values, and the second group in the part where m_b and M_S have high values. The picture shown in

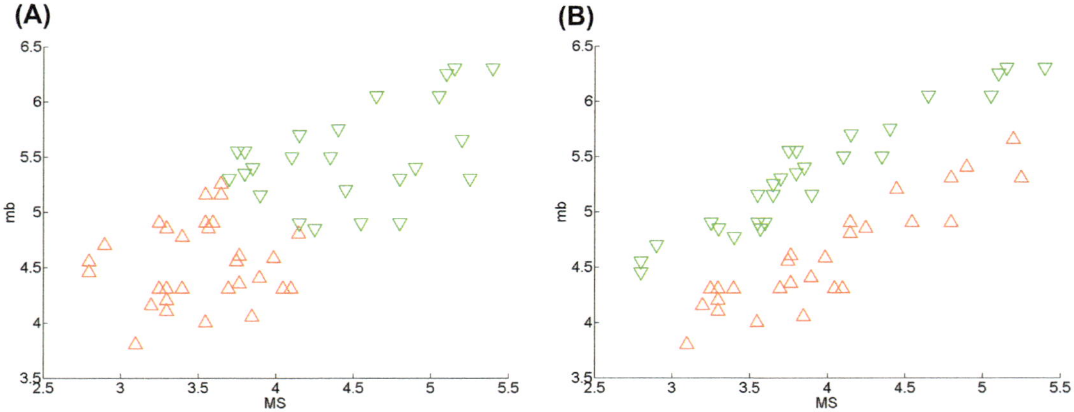

Figure 3.4

Clustering of M_S—m_b data related to earthquakes and nuclear tests using (A) K-means and (B) adaptive distances. Fig. 3.4A may be generated using the MATLAB™script S3.1 and Fig. 3.4B by using script S3_2.

Fig. 3.4A is in contrast with our a priori knowledge we outlined in Chapter 2. Recall that for that discrimination problem, we invoked the Mahalanobis distance, which accounts for groups of feature vectors with correlated components. In fact, accounting for the covariance matrix of our samples, we were able to establish a discrimination criterion.

In clustering, we can exploit the same idea using a metric called "determinant" metric (see Späth, 1983). In K-means, we use the metric for the distance of a sample $\boldsymbol{x}_i$ from a center $\overline{\boldsymbol{x}}_j$

$$d(\boldsymbol{x}_i, \overline{\boldsymbol{x}}_j) = \left[(\boldsymbol{x}_i - \overline{\boldsymbol{x}}_j)^T(\boldsymbol{x}_i - \overline{\boldsymbol{x}}_j)\right]^{\frac{1}{2}} = ||\boldsymbol{x}_i - \overline{\boldsymbol{x}}_j|| \tag{3.11}$$

which describes a sphere if $||\boldsymbol{x}_i - \overline{\boldsymbol{x}}_j|| \leq$ constant for all $\boldsymbol{x}_i$. A more general metric is obtained from

$$d(\boldsymbol{x}_i, \overline{\boldsymbol{x}}_j) = \left[(\boldsymbol{x}_i - \overline{\boldsymbol{x}}_j)^T \boldsymbol{G}(\boldsymbol{x}_i - \overline{\boldsymbol{x}}_j)\right]^{\frac{1}{2}} \tag{3.12}$$

with $\boldsymbol{G}$ being a positive definite matrix.

Similar to K-means, we consider the (squared) sum of the distances measured within the clusters, i.e.,

$$\sum_{j=1}^{k} \sum_{i=1}^{m_j} (\boldsymbol{x}_i - \overline{\boldsymbol{x}}_j)^T \boldsymbol{G}(\boldsymbol{x}_i - \overline{\boldsymbol{x}}_j) \tag{3.13}$$

which we want to be minimum. Keeping the partition fixed for the moment, we find (see Späth, 1983, Appendix 3.2) that the matrix $\boldsymbol{G} = (\det \boldsymbol{S})^{1/l} \boldsymbol{S}^{-1}$, where $\boldsymbol{S}$ is the sum of the

l-dimensional dispersion matrices encountered within the clusters, matches this request. From a geometrical point of view, the metric is very similar to the Mahalanobis distance. Note that the elements of the covariance matrix are obtained from the dispersions by dividing the latter by the number of samples in the ensemble. In other words, the same covariance matrix may correspond to a variety of dispersion matrices obtained from groups of different sizes. In the metrics proposed here, clusters with a large number of members will have a higher degree of homogeneity than those where the number of members is small.

So far, we have applied a concept already followed by Fisher's discriminant analysis described in Chapter 2, where we measured the distance by applying a normalization with respect to the inverse of the pooled covariance matrix (see Appendix 2.1, Eqs. A2.1.9, A2.1.10). In our new context, we also envisaged the case in which the pooled covariance matrix does not truly represent the covariance measured in the single groups. In fact, we had the log-likelihood L

$$lnL = 1/2\ln\left(\frac{|\boldsymbol{C}_1|}{|\boldsymbol{C}_2|}\right) - 1/2(\boldsymbol{x}-\boldsymbol{\mu}_1)^T\boldsymbol{C}_1^{-1}(\boldsymbol{x}-\boldsymbol{\mu}_1) + 1/2(\boldsymbol{x}-\boldsymbol{\mu}_2)^T\boldsymbol{C}_2^{-1}(\boldsymbol{x}-\boldsymbol{\mu}_2)$$

for $\boldsymbol{C}_1 \neq \boldsymbol{C}_2$. In this case, the separation function is not given by a linear element, but has a quadratic form. In our clustering scheme, we can allow $\boldsymbol{W}_j \neq \boldsymbol{W}_k$ by a criterion called "adaptive distance criterion" (see Späth, 1983). We simply reuse Eq. (3.13), but instead of summing over all dispersion matrices, we consider the individual ones obtained for each cluster. Consequently, we minimize the sum

$$\sum_{j=1}^{k}\sum_{i=1}^{m_j}(\boldsymbol{x}_i - \overline{\boldsymbol{x}}_j)^T\boldsymbol{G}_j(\boldsymbol{x}_i - \overline{\boldsymbol{x}}_j) \tag{3.14}$$

where $\boldsymbol{G}_j = (\det\boldsymbol{S}_j)^{1/l}\boldsymbol{S}_j^{-1}$.

We can apply the "interchange method" outlined above in order to find an optimum partition in the sense of Eq. (3.14). At each step, we verify whether the sum

$$\det(\boldsymbol{S}_p)^{1/l} + \det(\boldsymbol{S}_q)^{1/l}$$

with l denoting the dimension of the feature vector, has decreased when transferring a pattern from cluster p to cluster q. The upgrade formula limits the computational burden, as we do not have to go through all data of a cluster to calculate averages and dispersion matrices, considering only the single pattern to be transferred. In Fig. 3.4B, we apply such a kind of clustering to our example of earthquake and nuclear test magnitudes, reproducing fairly well our former results obtained with supervised classification. Use the MATLAB™ script S3_2 for reproducing Fig. 3.4B.

3.1.2.1.1 Fuzzy clustering

In K-means clustering, we have been considering the dispersion

$$\mathrm{S_j} = \sum_{i=1}^{m} \left(\boldsymbol{x}_i - \overline{\boldsymbol{x}}_j\right)\left(\boldsymbol{x}_i - \overline{\boldsymbol{x}}_j\right)^T = \sum_{i=1}^{m} d\left(\boldsymbol{x}_i, \overline{\boldsymbol{x}}_j\right) \tag{3.15}$$

and defined the optimum partition for a configuration where

$$\sum_{j=1}^{M} \boldsymbol{S}_j = min$$

Such a partition is the one where clusters have the highest degree of homogeneity. Up to now, we have assumed that a pattern belongs exclusively to one single cluster. In Box 3.2, we outline a clustering based on probabilistic considerations, in the sense that the distance expresses a probability that a pattern belongs to a certain cluster. Finally, we decided to assign the—crisp—class membership based on the largest probability found among all the clusters. Crisp clustering sometimes comes along with unpleasant threshold effects: a small difference in the distances of a pattern to the clusters may flip a pattern from one class to another. In geophysics, where we often wish to monitor the development of pattern characteristics with time, we may get apparently strongly fluctuating graphs even though the changes—expressed in absolute terms—are minor (Fig. 3.5).

Therefore, we may be interested to keep track of the minor probabilities. For instance, we could do this following the GMDAS strategy outlined in Box 3.2. However, the sum of cluster memberships of a pattern inferred from this method is not 1. At the same time, we use the a priori assumption that clusters follow multivariate Gaussian distributions, for which we can achieve a stable estimation of the parameters. This is not always guaranteed.

A way out of these problems is the minimization of a redefined cost function for the optimum partition (see, e.g., Zadeh, 1965; Bezdek, 1981), i.e.,

$$\sum_{I=1}^{N} \sum_{j=1}^{M} u_{ij}^{q}\, d\left(\boldsymbol{x}_i, \overline{\boldsymbol{x}}_j\right) = J(\boldsymbol{c}_j\,, \boldsymbol{U}) = min \tag{3.16}$$

where $\boldsymbol{U}$ is composed of the fuzzifiers u_{ij} expressing the membership degree of the i-th pattern to the j-th cluster, and $\boldsymbol{c}_j$ is the representative vector for the j-th cluster, such as its centroid. We also require that

$$\sum_{j=1}^{M} u_{ij} = 1$$

Box 3.2 Distances and probabilities

Note that in both K-means and adaptive distance clustering, we work with dispersions rather than variances. In the adaptive distance clustering, we carried out the normalization by using the inverse of the dispersion matrix rather than considering the classical Mahalanobis distance, which we exploited, for instance, in Fisher's discriminant method. The difference between the two strategies is not marginal. Whereas the Mahalanobis distance is a direct measure of the probability that a sample $\boldsymbol{x}_i$ belongs to a cluster p with a centroid $\bar{\boldsymbol{x}}_p$, the measure $(\boldsymbol{x}_i - \bar{\boldsymbol{x}}_p)^T \boldsymbol{G}_j(\boldsymbol{x}_i - \bar{\boldsymbol{x}}_p)$ cannot be read as a probability, being derived from the dispersions (i.e., we do not normalize it with respect to the number of samples in the cluster). Dispersions may be large if the number of samples is high; however, at the same time, the corresponding variance may be small. Dispersion-based measures tend to favor large clusters (large dispersion in relation to their variance), which can be a desired effect from a practical point of view.

Nonetheless, we can apply clustering focusing on the probabilities that a pattern belongs to a certain cluster. The Expectation–Maximization (EM) Algorithm allows us to tackle this task. The algorithm is based on the a priori assumption that the distribution of our samples is obtained by the sum of multivariate normal distributions, each of which is described by the parameters $\boldsymbol{\mu}_j$ and $\boldsymbol{C}_j$ (i.e., their centroid vectors and covariance matrices). Given cluster j, the probability that a pattern $\boldsymbol{x}_i$ belongs to this cluster follows from

$$p(\boldsymbol{x}_i|j) = \boldsymbol{C}_j^{-1/2}\exp\left(-(\boldsymbol{x}_i - \boldsymbol{\mu}_j)\boldsymbol{C}^{-1}(\boldsymbol{x}_i - \boldsymbol{\mu}_j)^T\right)$$

At the same time, we should be aware that clusters j and k themselves are not equally likely to occur.

Therefore,

$$p(\boldsymbol{x}) = \sum_{j=1}^{M} p(\boldsymbol{x}|j)P_j$$

where P_j is the a priori probability of getting cluster j. In other words, all M distributions contribute in a weighted way to $p(\boldsymbol{x})$.

Typically, we do not know a priori the parameters of the distribution, and our task is now the blind identification of all the parameters, i.e., the centroid vectors and covariance matrices of the single cluster, $\boldsymbol{\mu}_j$ and $\boldsymbol{C}_j$, as well as the a priori probabilities P_j. For this purpose, we can apply the "Generalized Mixture Decomposition Algorithmic Scheme" (GMDAS, for more details see Theodoridis and Koutroumbas, 2009).

The algorithm starts with an initial guess for $\boldsymbol{\mu}_j, \boldsymbol{C}_j$ and the a priori probabilities P_j. At step t, we calculate the probabilities

$$p(\boldsymbol{x}_i|j) = \boldsymbol{C}_j^{-1/2}\exp\left(-(\boldsymbol{x}_i - \boldsymbol{\mu}_j)\boldsymbol{C}_j^{-1}(\boldsymbol{x}_i - \boldsymbol{\mu}_j)^T\right)$$

and the a priori probabilities from

Continued

Box 3.2 Distances and probabilities—cont'd

$$P_j = \sum_{i=1}^{m_j} p(\boldsymbol{x}_i|j)$$

Considering the pattern $\boldsymbol{x}_i$, we are interested to find

$$p(j|\boldsymbol{x}_i)$$

that is the probability that—given the pattern $\boldsymbol{x}_i$—we have a member of the j-th cluster. It is easy to understand that this probability depends on the (Mahalanobis-) distance of the pattern to the centroid, but also on the overall probability to find the cluster j at all. Large clusters will be more likely than those with few members, that is, they have a higher a priori probability than small ones. Consequently,

$$p(j|\boldsymbol{x}_i) = p(\boldsymbol{x}_i|j)P_j \Big/ \sum_{k=1}^{M} p(\boldsymbol{x}_i|k)P_k$$

At step $t+1$, we read just the parameters of the distributions, $\boldsymbol{\mu}_j$, $\boldsymbol{C}_j$, and P_j using

$$\boldsymbol{\mu}_j(t+1) = \sum_{i=1}^{N} p(j|\boldsymbol{x}_i)\boldsymbol{x}_i \Big/ \sum_{i=1}^{N} p(j|\boldsymbol{x}_i)$$

and

$$\boldsymbol{C}_j(t+1) = \sum_{i=1}^{N} p(j|\boldsymbol{x}_i)\left(\boldsymbol{x}_i - \boldsymbol{\mu}_j(t)\right)\left(\boldsymbol{x}_i - \boldsymbol{\mu}_j(t)\right)^T \Big/ \sum_{i=1}^{N} p(j|\boldsymbol{x}_i)$$

In an iteration scheme, these steps are repeated until the change of parameters falls below a threshold. We understand that the algorithm converges from a simple consideration: as soon as we find a set of improved parameters of the distributions, we shall get a better estimation of the corresponding probabilities. During the subsequent steps, having better estimated probabilities, we improve the estimation of parameters of the distributions, entailing a still better estimation of the probabilities, and so forth. Eventually, we assign the cluster membership to $\boldsymbol{x}_i$ considering the maximum of $p(j|\boldsymbol{x}_i)$.

In the three plots shown below, we apply the GMDAS scheme to a data set. Panel (A) shows the distribution of randomly generated data using three values for the centroids and covariance matrix; panel (B) is the result when we try to blindly reconstruct the original groups using the GMDAS approach discussed above, assuming that our samples mirror three underlying multivariate Gaussians. Besides the narrow zone where blue and red samples intersect, the identified clusters correspond well to the original distributions of the randomly generated data. The application of GMDAS to our earthquake/nuclear test magnitude data (panel (C)), however, differs from the results of the adaptive distance clustering and from our a priori knowledge. Possible explanations for this different result are i) the a priori assumption of multivariate Gaussians does not hold for this data and ii) the data set (27 samples each) is too small for a stable estimation of the parameters, in particular centroids and covariance matrices.

Box 3.2 Distances and probabilities—cont'd

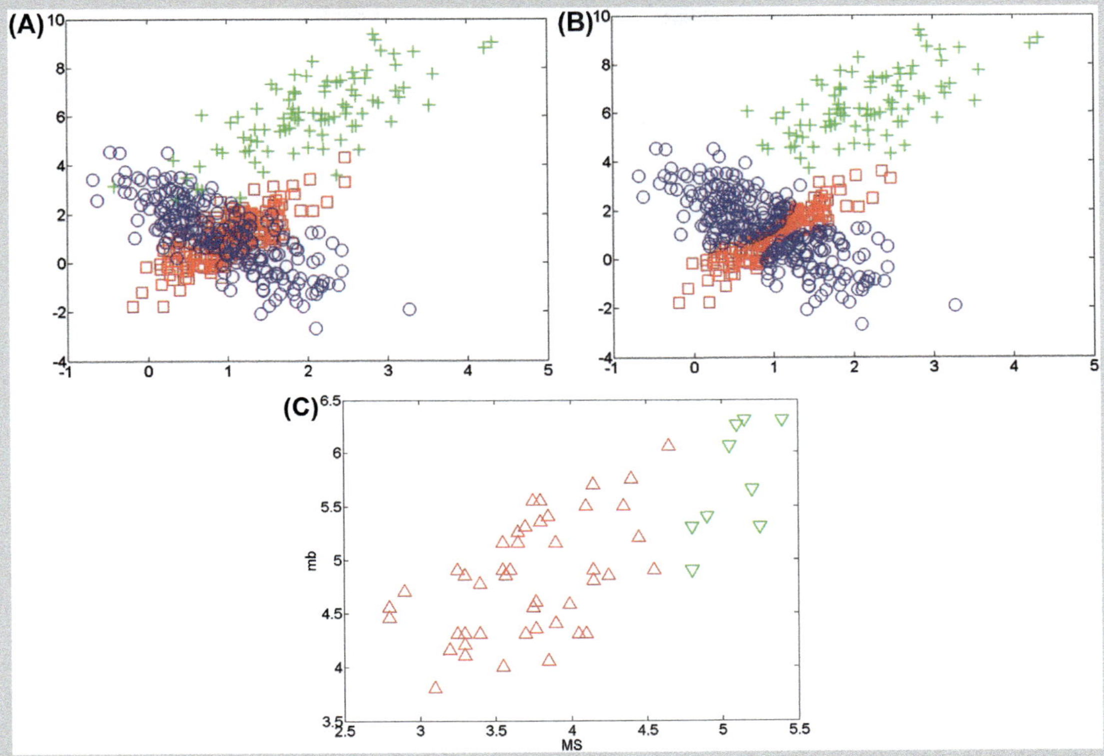

Figure B3.2

Application of GMDAS. (A) Three groups of randomly generated test data. Each group was generated assuming a Gaussian distribution with varying centroids and variance. (B) Blind reconstruction of class membership using GMDAS. (C) Application of GMDAS to the M_S–m_b data (magnitudes of earthquakes and nuclear tests). Note the difference to the results depicted in Fig. 3.4B. Panel (B) was obtained applying the GMDAS method to data shown in panel (A), invoking the MATLAB script S3_3. Panel (B) was created with the MATLAB S3_4a. Panel (C) can be generated with the script S3_4b.

In the fuzzy cost function, the exponent q has a specific role. It determines whether the minimum cost is achieved either by fuzzy or crisp clustering. We demonstrate this with a simple example (see Theodoridis and Koutroumbas, 2009). Suppose $\boldsymbol{X} = [\boldsymbol{x}_1, \boldsymbol{x}_2, \boldsymbol{x}_3, \boldsymbol{x}_4]$, with $\boldsymbol{x}_\mathrm{i}$ being [0,0], [2,0], [0,3], [2,3], and two cluster representatives $\boldsymbol{c}_j$ given by [1,0] and [1,3]. In crisp clustering, the membership vector $\boldsymbol{U} = [u_{11,}\ u_{12}], [u_{21}\ u_{22}], [u_{31}\ u_{32}],$

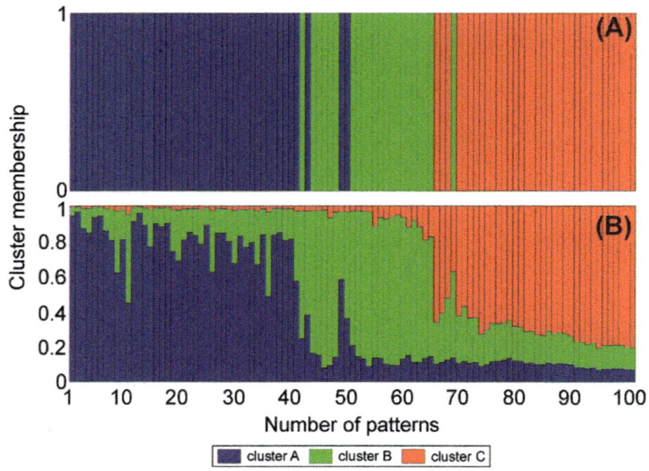

Figure 3.5
Time series of cluster membership: (A) crisp clustering and (B) fuzzy clustering.

$[u_{41}\ u_{42}] = [1, 0], [1, 0], [0, 1], [0, 1]$, and the corresponding cost is 4. Assume now $q = 1$ and the u_{ij} in the range between 0 and 1. Then we get for the cost

$$J(\boldsymbol{c}_j\,, \boldsymbol{U}) = \sum_{i=1}^{2}\left(u_{i1} + u_{i2}\sqrt{10}\right) + \sum_{i=3}^{4}\left(u_{i1}\sqrt{10} + u_{i2}\right)$$

which is greater than 4 for any $0 < u_{ij} < 1$, as $u_{i1} + u_{i2} = 1$. In other words, for $q = 1$, crisp clustering remains the most effective one with respect to the cost. Things change if $q > 1$. Setting $q = 2$, we find $J(\boldsymbol{c}_j,\boldsymbol{U}) < 4$ if u_{i1} falls in the range [0,0.48] for $i = 1,2$ and u_{i2} lies in the interval [0,0.48] for $i = 3,4$. Setting $q = 3$, the ranges where fuzzy clustering gives better cost functions than crisp clustering are $0 < u_{i1} < 0.67$ for $i = 1,2$ and $0 < u_{i2} < 0.67$ for $i = 3,4$. Obviously, these ranges are valid accounting for the condition $u_{i1} + u_{i2} = 1$. In our small example, we learn that the ranges for u_{ij}, for which fuzzy clustering outperforms crisp clustering with respect to the cost function, increase in relation to q. In other words, choosing high q values will increase the fuzziness of the clusters.

In "fuzzy C-means", the most popular method of fuzzy clustering, we face the task of identifying suitable representative vectors $\boldsymbol{c}_j$ —typically the centroids of the clusters—and the cluster membership vector $\boldsymbol{U}$. For this purpose, we use an iterative upgrade scheme, starting with an initial guess of the $\boldsymbol{c}_j$. The elements of $\boldsymbol{U}$ can be calculated from

$$u_{rs} = 1\Big/\sum_{j=1}^{M}\left((d(\boldsymbol{x}_r,\ \boldsymbol{c}_s)/d(\boldsymbol{x}_r,\ \boldsymbol{c}_j)\right)^{1/(q-1)} \tag{3.17}$$

and for the representative vectors, we find

$$c_j = \sum_{i=1}^{N} u_{ij}^q x_i \Bigg/ \sum_{i=1}^{N} u_{ij}^q \tag{3.18}$$

In the subsequent iterations, we use the $\boldsymbol{c}_j$ in order to find improved versions of $\boldsymbol{U}$, which are then used to update the $\boldsymbol{c}_j$. We stop the iteration, once the update rate of the $\boldsymbol{c}_j$ falls below a certain threshold.

In Fig. 3.6, we show the fuzzy cluster membership values obtained for our rock composition data set. In fuzzy-C-means, we have applied three values for q, i.e., $q = 1.05$

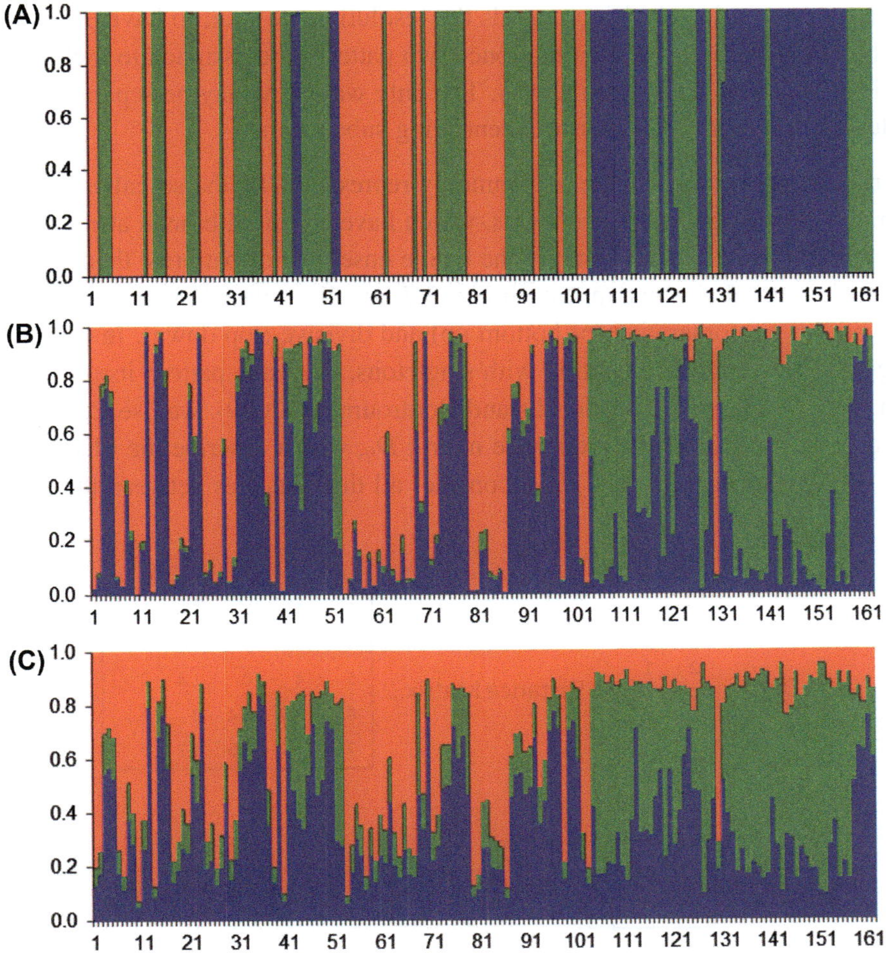

Figure 3.6

Fuzzy clustering of 161 13-dimensional rock composition data, using exponents $q = 1.05$, 2, and 3. Note that with $q = 1.05$ (A), we have an essentially crisp clustering, with increasing q (B and C), the clustering tends to be less crisp.

(quasi crisp), $q = 2$, and $q = 3$, assuming three clusters. As mentioned earlier, the higher the q, the less crisp the clustering. We may set $q = 50$, getting almost equal cluster membership values of $\sim 1/3$. Such a setting turns out as rather useless. In practice q is set somewhere between 1.5 and 3, typical default settings provide $q = 2$. The reader may play with the various options using script S3_5 coming along with this book.

3.1.2.2 Hierarchical clustering

In the methods considered so far we have been using an approach forming disjoint clusters, so to say the data description is "flat" (Duda et al., 2001). However, we may be faced with data structures, where clusters still exhibit heterogeneities, subclusters, etc. In partitioning cluster analysis, we had to choose a priori the number of clusters we wished to form. Here, our a priori decision regards the resolution we want to have. A fundamental of hierarchical clustering is the fact that once two patterns are assigned to a class, they will remain within that cluster all the way. The only way they may be separated is when we introduce a new level of clustering, identifying subclusters.

The hierarchical clustering scheme is commonly represented by the so-called "dendrograms", looking like a net of roots, which have an initial branch at the highest level of the hierarchy—which is simply the whole ensemble of patterns, then split into thinner branches as subclusters appear. In Fig. 3.7, we illustrate this scheme. We have two principal options: agglomerative (or bottom up) and divisive (top down). In Fig. 3.7, we use the agglomerative method starting with singletons, i.e., each pattern itself forms a cluster. In our example (see Theodoridis and Koutroumbas, 2009), we use Euclidean distances as a metric and derive a distance matrix D_{ij}, which describe the (dis)similarities of pattern i with respect to pattern j. We consider all the distances between the samples

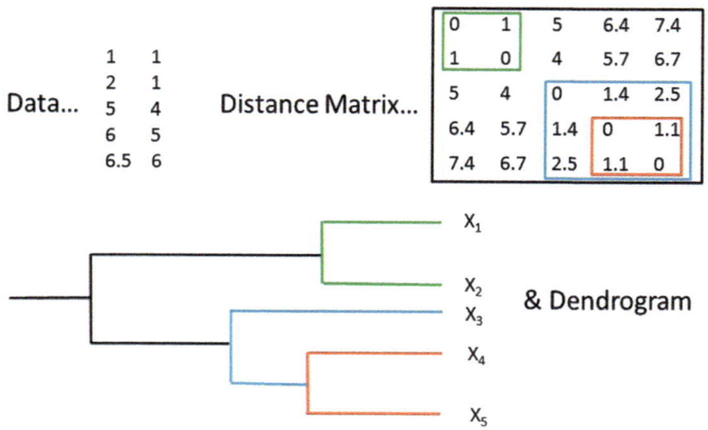

Figure 3.7
Construction of a dendrogram.

and join the two for which the smallest distance is encountered, in Fig. 3.7 this is the case for samples #1 and #2. Then we find that #4 and #5 are close to each other, so we joined the two, forming a cluster. From the distance matrix, we recognize that the sample #3 is close to #4 and #5, but distant from #1 and #2., i.e., we add it to the cluster formed by samples #4 and #5.

During the clustering procedure, we may meet configurations of singletons and clusters containing two or more samples. The choice that controls the next step regards the definition of the distance. In our example, we had to decide what to do with sample #3. As it was closer both to #4 and #5 than to #1 and #2, our decision was clear.

In general, however, we may use various criteria for measuring the distance of singletons or clusters to another cluster. In the "single linkage" method, we decide on the basis of the smallest distance between a sample and the members of a cluster (see Fig. 3.8). In our case, sample #3 is closest to sample #4 ($D_{34} = 1.4$), whereas its distance to the samples #1 and #2 is larger. In the "complete linkage" method, we consider the distance from our pattern to the cluster, for which the largest value is encountered. For our sample #3 and the cluster consisting of #4 and #5, the distance to the latter is relevant ($D_{45} = 2.5$). Considering the cluster consisting of #1 and #2, we must consider D_{13} which is 5, i.e., greater than D_{45}. In our example, complete linkage and single linkage lead to the same dendrogram. In general, when comparing clusters, we base our decision on

$$D(\mathbb{C}_q, \mathbb{C}_s) = min(D(\mathbb{C}_i, \mathbb{C}_s), D(\mathbb{C}_j, \mathbb{C}_s)) \tag{3.19}$$

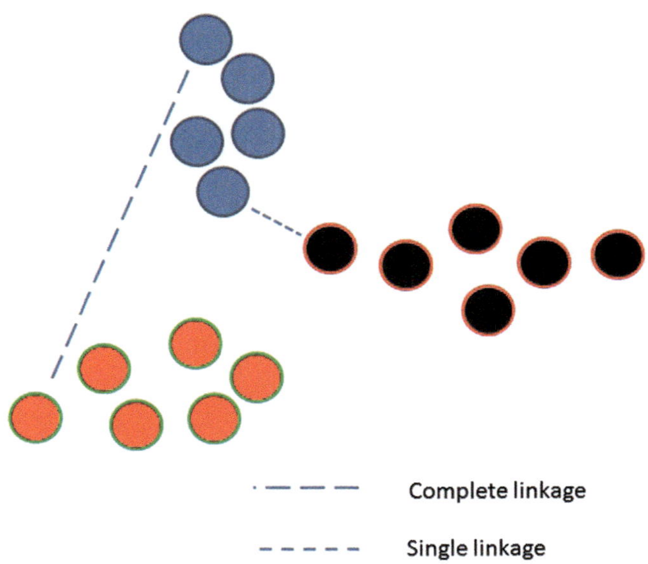

Figure 3.8
Complete and single linkage.

where $\mathbb{C}_i$, $\mathbb{C}_j$, and $\mathbb{C}_s$ are old clusters, and $\mathbb{C}_q$ is the new cluster to be formed. That is we consider the elements of the i-th cluster and those of the j-th cluster and compare them to elements of $\mathbb{C}_s$. We merge on the basis of the minimum distance found, either $\mathbb{C}_i$, $\mathbb{C}_s$ or $\mathbb{C}_j$, $\mathbb{C}_s$. In complete link, we use the distances

$$D(\mathbb{C}_q, \mathbb{C}_s) = max(D(\mathbb{C}_i, \mathbb{C}_s), D(\mathbb{C}_j, \mathbb{C}_s)) \tag{3.20}$$

In other words, the single linkage follows a "min-min" concept—we measure the distance of two clusters by taking the pair of closest patterns (the first "min") and then repeat this comparing all clusters to each other, finally merging the two clusters for which $D(\mathbb{C}_q,\mathbb{C}_s)$ is minimum (the second "min"). The complete link is a "max-min" approach—we measure the distance between two clusters by taking the pair of most distant patterns ("max") and then merging the two clusters for which $D(\mathbb{C}_q,\mathbb{C}_s)$ is minimum (which is the "min").

In Fig. 3.9 we recognize clusters of data in a 2D data space. From intuition, we would start with two groups, having $x_1 > 5$ and the other one $x_1 < 5$. Besides, we notice a somewhat less sharp separation with respect to $x_2 > 5$. It means that we can further split the two groups identifying subclusters.

Applying standard MATLAB™ procedures for agglomerative clustering using the single linkage method, we obtain a dendrogram shown in Fig. 3.10. As outlined above, at the lowest level, we merge singletons to clusters on the basis of their distance to each other, subsequently we merge more and more clusters forming larger groups. At the beginning, we merge patterns and groups being very close to each other. Toward the higher levels, we have to merge groups being more distant to each other. In the dendrogram in Fig. 3.10, we keep trace of the degrees of dissimilarity where a merge of clusters takes place. Jumps in

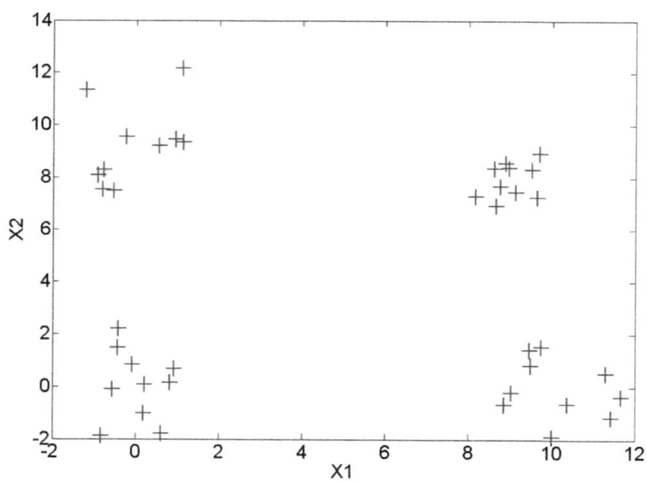

Figure 3.9
Demonstration of a data set for hierarchical clustering.

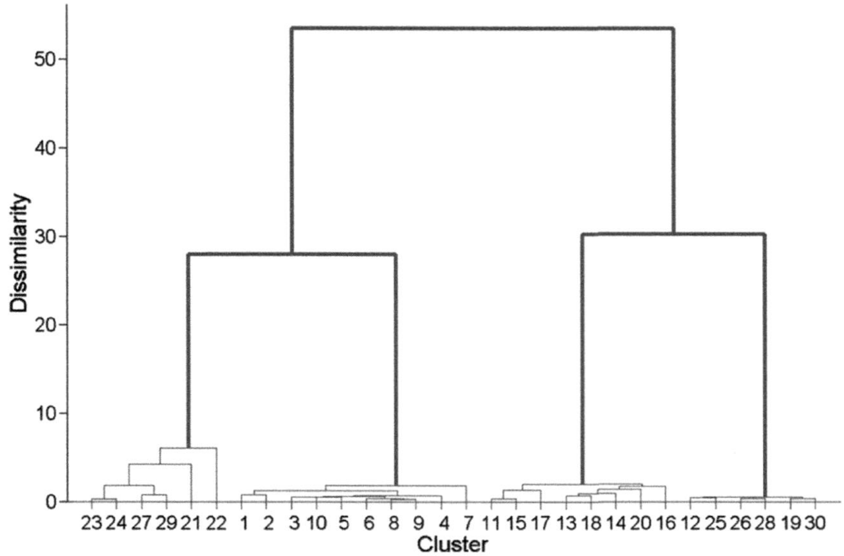

Figure 3.10
Dendrogram for clustering of data shown in Fig. 3.8, using the agglomerative method and single linkage distance metric. Note that the lowest levels of the dendrogram are suppressed ("cut") for the graphical representation.

the dissimilarities are an important feature in the dendrograms. When they appear, we have to bridge large distance for a merge. In Fig. 3.10, we notice only small jumps at the early stage, which means that we are merging at quite close distances. At some moment, we do not find groups having a distance, let's say less than 10. Defining a threshold of 10 as a measure of critical dissimilarity, we end up with four clusters, as shown by the bold lines in Fig. 3.10. Another jump of the dissimilarity is found at a distance slightly below a value of 30. That is if we want to merge our four subgroups to two major clusters we have to bridge this distance.

In Fig. 3.11, we applied the "complete linkage" metric; however, the dendrogram has essentially the same feature as the one obtained with single linkage. Obviously, the numerical values of dissimilarity are higher, but the structure is essentially the same as in Fig. 3.10.[2] The reader may play with some aspects of agglomerative clustering using the MATLAB™ script S3_6 coming along with this book.

Besides the single and complete linkage metrics, one finds other definitions in the literature. For instance, we may use the distances between the centroids of two clusters or

[2] Single and complete linkage metrics may differ considerably when elongated clusters are involved. One end of such a cluster may lie close to a certain group, but the other may be very distant. Therefore, what looks close in single linkage gets far distant with complete linkage. This situation is shown in Fig. 3.8.

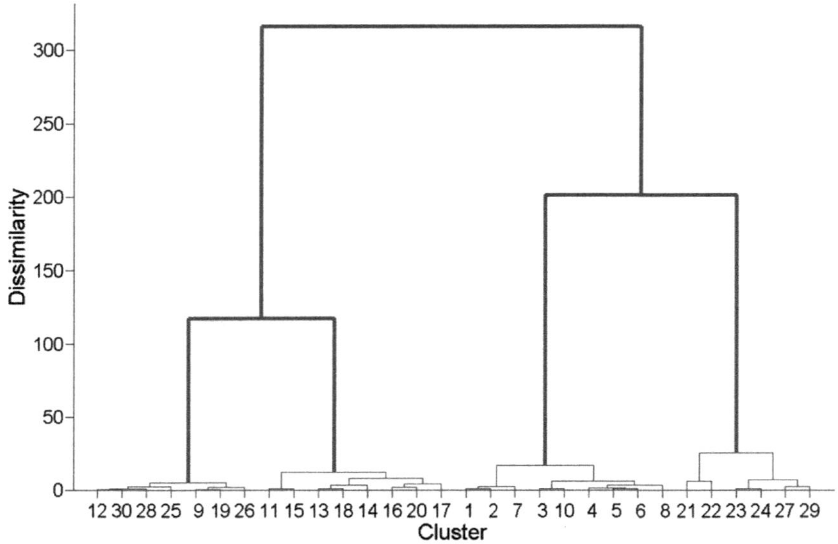

Figure 3.11

Dendrogram for clustering of data shown in Fig. 3.8, using the agglomerative method and complete linkage distance metric. Note that the lowest levels of the dendrogram are suppressed ("cut") for the graphical representation.

their median vectors. A further method is Ward's linkage. It considers the sum of the variances encountered within each cluster and merges the two clusters for which this sum increases less. However, there is a caveat related to the request of monotonicity, i.e., the measure of dissimilarity increases for higher cluster levels as observed in our dendrograms in Figs. 3.10 and 3.11. Single linkage, complete linkages, and the Ward's algorithm comply with this request. The "centroid" and "median" methods, however, can produce a cluster tree that is not monotonic. This occurs when the distance from the union of two clusters, $\mathbb{C}_r$, $\mathbb{C}_s$, to a third cluster $\mathbb{C}_q$ is less than the distance from either $\mathbb{C}_r$ or $\mathbb{C}_s$ to $\mathbb{C}_q$. In this case, sections of the dendrogram change direction. This is an undesired effect and one should use another metric.

Alternative to the agglomerative scheme, we could follow the "top-down" or "divisive" strategy. In divisive schemes, we start with an initial division of the whole data set into two clusters, searching for the partition, which maximizes the distance between the two groups. In the next step, we go through the new generation of clusters. We try to divide them with the scope to establish a new partition of the clusters where we achieved a maximum distance between two clusters. The procedure is repeated unless we end up at the lowest level, with clusters consisting only of singletons. As before, we may adopt various metrics, such as single or complete linkage, Ward's linkage, etc. Again, we may design dendrograms, monitoring jumps in the dissimilarity measures in order to identify suitable partitions, i.e., configurations where the number of clusters is limited, meanwhile their degree of homogeneity is acceptable. The divisive schemes come with a rather high computational burden, which renders them less popular than agglomerative schemes.

3.1.2.3 Density-based clustering

In partitioning clustering, we followed variance and dispersion-based metrics, which represent the scatter of patterns from centroids. Often they have convex hulls,[3] with a (hyper)spherical or (hyper)ellipsoidal aspect. The number of clusters has to be specified by the user. Even though there are some parameters which may give a hint on the suitable number of clusters, this choice depends on the judgment of the analyst. Hierarchical algorithms create a hierarchical decomposition of the data set. In contrast to partitioning algorithms, hierarchical algorithms do not need the number of desired clusters as an input. We can decide having the results in our hands, where to cut. However, we have to provide a criterion—a level of dissimilarity—indicating when the merge or division process should be terminated.

In density based (DB) Clustering, we focus on the local structure of a data set. We consider a unit volume in our data space and derive the density of samples within this volume. Moving toward neighboring volumes, we verify whether the number of samples has dropped below a threshold. If this is the case, we identify a heterogeneity in our data set. In the absence of such heterogeneity, we declare that the neighboring volumes belong to the same cluster.

In a more formal way we define:

A point $\boldsymbol{y}$ (i.e., a feature vector) is directly density reachable from $\boldsymbol{x}$ (another feature vector), if

$$\boldsymbol{y} \in V_\varepsilon(\boldsymbol{x}) \tag{3.21}$$

and

$$N_\varepsilon(\boldsymbol{x}) \geq q \tag{3.22}$$

where $V_\varepsilon(\boldsymbol{x})$ is the unit (hyper)volume around a pattern and $N_\varepsilon(\boldsymbol{x})$ is the number of pattern in the unit volume around that pattern. Consequently, we use the definition:

A point $\boldsymbol{y}$ (i.e., *a feature vector) is density reachable from* $\boldsymbol{x}$ *(another feature vector), if there is a sequence of points* $\boldsymbol{x}_1, \boldsymbol{x}_2 \ldots \boldsymbol{x}_p$, *with* $\boldsymbol{x}_1 = \boldsymbol{x}$, $\boldsymbol{x}_p = \boldsymbol{y}$, *such that* $\boldsymbol{x}_{i+1}$ *is directly reachable from* $\boldsymbol{x}_1$.

In other words, the $\boldsymbol{x}_1, \boldsymbol{x}_2 \ldots \boldsymbol{x}_p$ form a chain or coherent volume where the density does not fall below a certain limit. A pattern $\boldsymbol{x}$ is a "core point" (see Fig. 3.12), if it has at least

[3] Nonetheless, clusters may intersect each other, as we have seen in the Gaussian mixture model in Box 3.2. The adaptive distance clustering allows intersecting clusters, too.

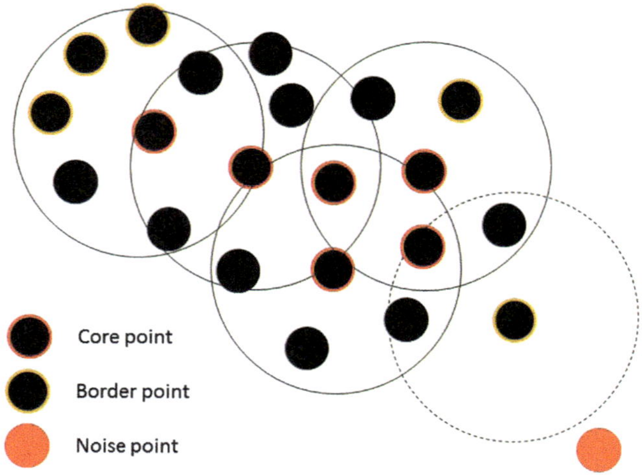

Figure 3.12

Basic elements in density-based clustering. The $V_\varepsilon(\boldsymbol{x})$ are indicated as circles, we further require a minimum of $q = 6$ patterns being present in the elementary volume. Border points are density reachable from the core points, but core points cannot be density reached from border points. Noise points are not density reachable, and no other point can be reached from noise points. They are outside all clusters.

q neighbors. Noncore points can be those situated at the margins of a cluster (these are density reachable from a core point) and are called "border points". Other points, not being density reachable from others, are referred to as "noise points".

A cluster is a set of points, density reachable from a core point. Clearly, a cluster can have many core points and is thus uniquely determined by any of them. There are also noise points, which cannot be density reached and from which no other point can be reached. As the method accounts also for patterns not belonging to a cluster, it is also named

Box 3.3 DBSCAN algorithm

Define the set of unclassified patterns U. At the beginning, all patterns belong to this set.

While $\mathbb{U} \neq \{\}$, select $\boldsymbol{x} \varepsilon \mathbb{U}$. Set number of clusters m = 0.

Check whether $\boldsymbol{x}$ is a noncore point. If TRUE then $\mathbb{U} = \mathbb{U} \backslash \{\boldsymbol{x}\}$, and $\boldsymbol{x}$ is noise.

If $\boldsymbol{x}$ is core, m = m + 1. Search all density reachable points, and form the cluster $\mathbb{C}_m$.

Border points, previously marked as noise, will be added to the cluster. Set $\mathbb{U} = \mathbb{U} \backslash \{\mathbb{C}_m\}$.

End.

"DBSCAN", "Density-Based Spatial Clustering of Applications with Noise" (see Hochspringen et al., 1996, Sander et al., 1998).

In Fig. 3.13, we show an example application of DBSCAN clustering to a test data set. The most interesting aspect is the possibility to form clusters of both convex or concave shapes. In the figure, those clusters are indicated by the blue and red crosses. At the same time, we notice the noise points. These are separated from the other points and do not form a cluster on their own as they are not density reachable from either point of the data set.

Unfortunately, the clustering results depend critically on the choice of the two parameters, ε (for the definition of $V_\varepsilon(\boldsymbol{x})$), and q. Clusters having varying densities may therefore not be separated appropriately. Various alternatives have been proposed in order to bypass the problem, among those is a technique known as OPTICS (*O*rdering *P*oints *T*o *I*dentify the *C*lustering *S*tructure, see Ankerst et al., 1999).

In OPTICS we have two important terms:

core distance of an object $\boldsymbol{x}$: the smallest value ε such that the ε-neighborhood of $\boldsymbol{x}$ has at least q objects and

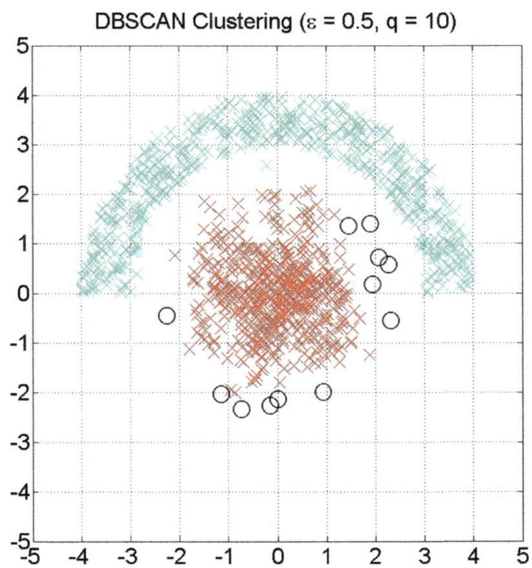

Figure 3.13
Application of DBSCAN to clusters with irregular shapes. Black circles correspond to patterns identified as "noise". See the link http://yarpiz.com/255/ypml110-dbscan-clustering. The tool is also provided in this book (script S3_7).

reachability distance of an object $\boldsymbol{x}$ from the core object $\boldsymbol{y}$: this corresponds to the min radius value that makes $\boldsymbol{x}$ density reachable from $\boldsymbol{y}$.

OPTICS orders the samples, starting from an object not considered so far, and determines its nearest neighbors. All samples being identified as nearest neighbors are temporarily stored applying a ranking with respect to the reachability distance. In the following steps, we take the object situated nearest to the previous one and repeat the search. In this way, one terminates a whole cluster before passing to the next one. Finally, there will be only unprocessed samples available, which have a high reachability distance. It means that such a sample will be positioned at the ultimate rank in the cluster. However, if such an object is a border or core point of another cluster, it may have close neighbors. Such a switch from high to low values in the reachability distance is a clear sign for a new cluster. For demo version, see https://github.com/alexgkendall/OPTICS_Clustering.

DENCLUE (*DENS*ity based *CLU*st*E*ing algorithm, see Hinneburg and Keim, 1998) starts with the identification of "local maxima" of the density of patterns applying a predefined Kernel function, typically having a Gaussian shape. The local density maxima form "density attractors" that are treated as cluster centers. A proper DENCLUE cluster $\mathbb{C}$ is a set of density attractors together with the set of objects. Each density attractor in $\mathbb{C}$ must be reachable along a path with a predefined finite density. For more details, see e.g., Han et al. (2011).

DBSCAN-STRATA—algorithms with the so-called "stratification"—include an analysis of the distribution of distances encountered within a data set. Here stratification is a preprocessing step to divide data into layers where the objects have similar global characteristics (Cassisi et al., 2013). In particular, it sorts subsets in which objects have similar distances between each other. Once those subsets are identified, we can add the information with respect to the distances to the original data, i.e., augment the dimensionality and carry out a density clustering on the augmented data set. As the authors point out, such a procedure has the advantage of being efficient for clusters with varying densities, avoiding the tedious research on the choice of the parameters, ε and q (see Eq. 3.21, 3.22) in DBSCAN, which in general has some difficulties in identifying clusters having differing densities (see Huang et al., 2017). A certain drawback resides in the fact that uniformly distributed noise may blur the results.

Besides the stratification option, Cassisi et al. (2013) proposed a modified definition of neighborhood in density-based clustering. Recall that in classical density-based clustering, we add a pattern $\boldsymbol{y}$ to a cluster, if it is recognized as falling in a range no larger than ε_0 from a core or neighbor pattern $\boldsymbol{x}$ of the cluster. In the modification, the pattern $\boldsymbol{y}$ to be added can "reject" the new membership if it has at least q nearest neighbors, which are closer than $\boldsymbol{x}$. That way we can form clusters with varying densities (see Fig. 3.14).

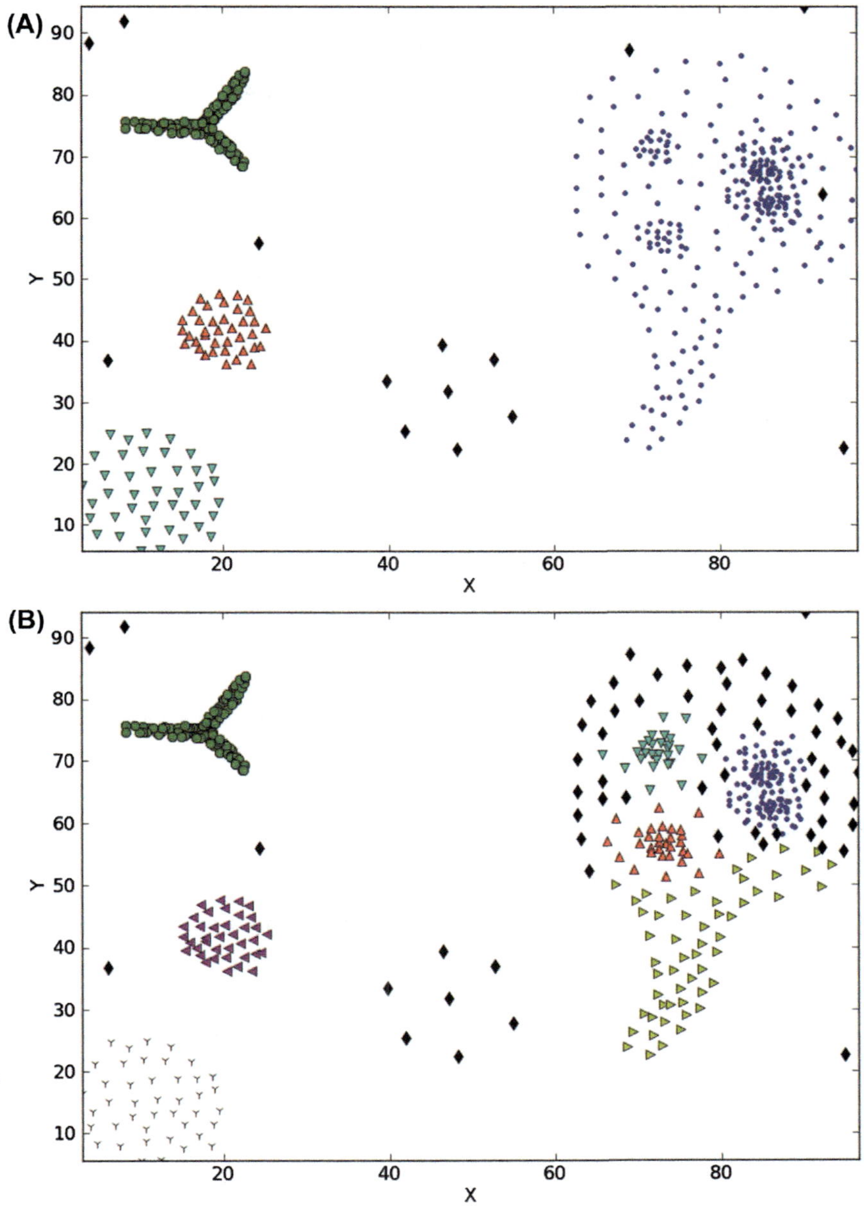

Figure 3.14

(A) Test data, clustered with DBSCAN-STRATA, $\varepsilon = 30$, $q = 10$. (B) Clustered with modified DBSCAN, $q = 16$. Note that in (B) only q has to be specified a priori. Patterns belonging to the "noise" are indicated with black symbols in both plots. *(B) Redrawn after Cassisi et al., 2013.*

For more details, we refer the reader to Cassisi et al. (2013) and references cited therein. A beta version of the DBSCAN_STRATA program can be downloaded from the site http://www.dmi.unict.it/~cassisi/DBStrata/ (see also Aliotta et al., 2011).

3.2 Self-Organizing Maps

The Self-Organizing Maps (SOMs) were invented by Teuvo Kohonen (1984, 2001) and are often referred to as Kohonen maps. They are based on two key ideas: reduction of the number of objects and reduction of the dimensionality of the problem. The first goal is achieved by identifying prototypes of our patterns, each of which represents a number of samples with a reasonable degree of similarity. These prototypes form clusters and are referred to as "nodes" in the terminology of SOMs. In clustering the original set of samples, we achieve a considerable reduction of the number of objects to deal with. The second aspect regards the representation of multidimensional data in much lower dimensional spaces than the original data set. We shall refer to these low dimensional spaces as "representation spaces". For the ease of visualization, the representation spaces are often two-dimensional, and therefore, we use the term "map". The process of reducing the dimensionality of vectors corresponds to a data compression technique known as vector quantization. An intriguing aspect of Kohonen's method is to project the prototypes onto a mesh where the information is stored in such a way that the topological relationships within the patterns of the data set are conserved. In other words, patterns and nodes being close to each other in the original feature space are also close to each other in the representation space.

The usual arrangement of nodes follows a 2D regular spacing along a grid. As shown in Fig. 3.15, we may use different surfaces for the mapping. The most frequently used geometry is the "sheet", i.e., a classical flat and rectangular map (Fig. 3.15A). Alternatives are, for instance, the projection onto a cylinder surface or a torus. The choice of the projection surface may depend on the characteristics of the features. Typically, we define the distance between two patterns exploiting the Euclidean distance between the

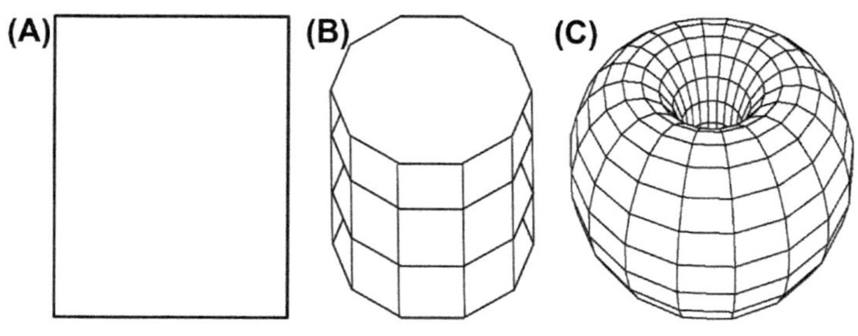

Figure 3.15
2D mapping geometries: (A) sheet, (B) cylinder, and (C) torus (see Vesanto et al., 2000).

corresponding feature vectors. In those cases, the projection onto a sheet is likely the most suitable one. On the other hand, we may define (dis)similarity on some correlation bases metric, i.e., we look at the shape or aspect of an object rather than its overall size. In those applications, a projection onto a cylinder or torus surface may be preferable, as the correlation coefficient corresponds to the angle formed by the two feature vectors rather than the Euclidean distance.

As an alternative to the torus topology, Ritter (1999) proposed a spherical surface for the projection. Using a spherical topology, the results of the SOM can be visualized in a straightforward and easy-to-read way by standard map-projections well known in the field of geodesy. One major shortcoming of a spherical model is that options of perfectly regular placement of neurons on the sphere are rather limited. These options are known as platonic polyhedra. A compromise bypassing the problem has been proposed by Wu and Takatsuka (2006), using so-called geodesic domes (see Fig. 3.16).

The concept starts with the creation of an icosahedron, which has 24 vertices that can be distributed on a sphere with equal distances from each other. The icosahedron has a property comparable to a hexagon in two dimensions, which represents the polygon optimizing the ratio between area and length of edges, and at the same time, its vertices have equal distances to each other. The icosahedron is the platonic polyhedron most similar to a sphere. Unfortunately, an SOM with only 24 units is often insufficient for large data sets. As pointed out by Wu and Takatsuka (2006), this limitation can be solved by carrying out an appropriated tessellation, which essentially consists in splitting the triangle elements making up the icosahedron surface into smaller triangular elements with equal side-lengths. Carrying out tessellation for subsequent steps, we can obtain a large number of points distributed almost equally on the sphere. Most of the points have six neighbors, only 12 of them—the vertices of the original icosahedron—have five. An alternative method approaching a homogeneous mesh consists in deploying the nodes

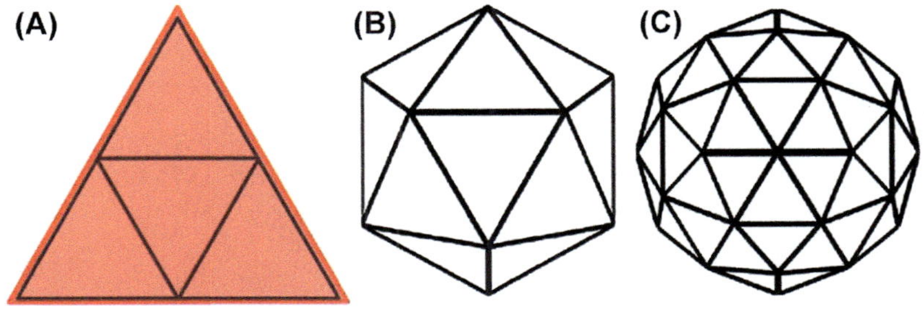

Figure 3.16
Approximation of a sphere by tessellation of icosahedron surfaces: (A) first-order triangle tessellation, (B) icosahedron, and (C) its first-order tessellation. The higher the degree of tessellation, the closer the approach to a sphere. *(C) redrawn from Wu and Takatsuka, 2006.*

along a spiral or in a helix which winds up over the sphere (Nishio, 2005; see also Jagric and Zunk, 2013). The helix structure offers a better degree of homogeneity in the distribution of nodes over the sphere, especially when the number of nodes is large.

Here we focus on the classical sheet topology, which is at the base of one of the most famous applications: the so-called "World Poverty Map" (see Fig. 3.17). It is based on economical parameters provided by the World Bank for the year 1992. In total, there were 39 variables, representing the economic status of a country. These variables include income, education, health, nutrition, etc. Because of the high amount of variables,

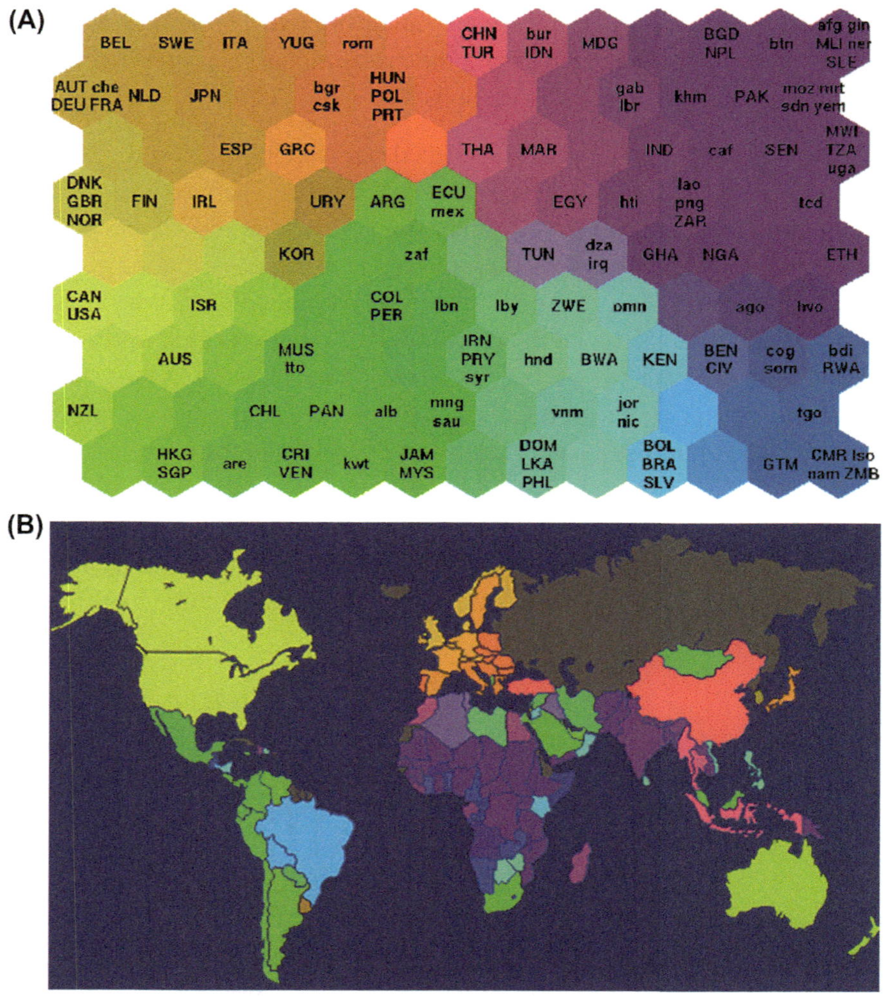

Figure 3.17

World Poverty Map (Neural Networks Research Centre (1997), see http://www.cis.hut.fi/research/som-research/worldmap.html, see also http://www.ai-junkie.com/ann/som/som5.html). (A) The position of the countries on a Kohonen sheet and (B) their geographical position on the globe.

visualization of homogeneous groups is a problem. The use of an SOM is an attractive solution to give a summarized view of which of the patterns may belong to a certain category and which of them show clear differences. In our World Poverty Map, we notice in the lower right corner, countries with the highest poverty, while the countries with a higher economic status are found on the left border of the map.

Besides, we also recognize differences between rich countries, i.e., those having a high procapita gross domestic product. European countries, such as Germany ("DEU" on the map), France, Austria ("FRA", "CHE"), etc., have close positions on the sheet, at some distance to the USA or Canada ("USA", "CDN"). The latter countries resemble to geographically distant states like Australia ("AUS") or New Zealand ("NZL"). Applying a color code to each country, depending on its position on the Kohonen sheet, helps to evidence the distribution of economic features on the world. Thus, we find a block of orange formed by the countries of the Western European Union (Fig. 3.17B). Eastern European countries, which made part of the communist block before 1990, are clearly identifiable as patterns with red colors—mirroring their common history after World War II. On the other hand, the USA, CDN, and AUS resemble each other. They are countries of immigration starting some 100 years ago and have a considerable geographical extension. Compared to the European countries, they have a lower density of population and a different welfare system.

3.2.1 Training of an SOM

For the sake of simplicity, we outline the construction of an SOM with nodes being placed on a map with a sheet geometry. At the beginning of the learning process, each node of the SOM is assigned an initial weight vector $\boldsymbol{W}_i$ (Fig. 3.18A), for instance by assigning random values. The $\boldsymbol{W}_i$ have the same dimensionality as the original feature vectors. The construction of the SOM follows an iterative procedure, considering the differences

$$D_{ij} = \sqrt{(\boldsymbol{W}_i - \boldsymbol{V}_j)^T(\boldsymbol{W}_i - \boldsymbol{V}_j)}$$

between the normalized input vector $\boldsymbol{V}_j$ and the weights $\boldsymbol{W}_i$ of the nodes. A core step is the identification of the closest node to the actual input vector, i.e., the best matching unit (BMU) for the j-th pattern. Throughout this identification process, neighboring nodes lying within a certain radius of influence are considered as well. This is an important characteristic of SOM training, as it makes sure that weights of nodes being situated closely to each other on the sheet will be updated in a similar way.[4] Once BMU and the nodes falling within the area of influence are identified, the weights are gradually adjusted

[4] Ideally, at the end of the training process, we obtain an SOM with "topological fidelity". Nodes being situated closely to each other on the map are close to each other also in the original feature space.

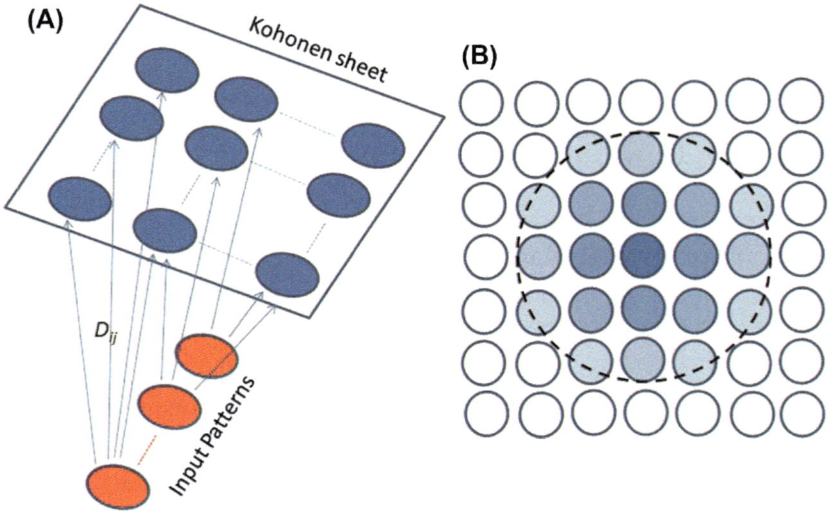

Figure 3.18
Constructing an SOM placing the nodes on a sheet. (A) For each pattern, we search the nearest node on the sheet, accounting for the D_{ij}. Some of the D_{ij} are indicated by the arrows. The nearest node is identified as BMU ("Best Matching Unit"). (B) We upgrade the weights—not only the ones of the BMU (the node in dark blue at the center), but also those of the neighboring nodes, which fall inside a certain radius around the BMU (shown by the *black dashed circle*). The rate of upgrade—the "learning" rate—shrinks with time; besides, it also decreases inversely to the distance of a node from the BMU. We indicate those nodes by blue circles, which get increasingly pale for nodes distant from the BMU.

according to the so-called learning rate. The rate, to which the weight of the nodes are adjusted, decreases with the distance Δ between each node and the BMU. The upgrade of weights follows the relationship

$$\boldsymbol{W}_i(t+1) = \boldsymbol{W}_i(t) + \phi(\Delta, t) \cdot \lambda(t) \cdot D_{ij}(t)$$

where φ describes the distance dependence of the upgrade of a node. $\varphi(\Delta,t)$ can have various shapes, such as Gaussian, inverse parabolic, boxcar (see Messina and Langer, 2011, "KKAnalysis, Extended documentation" downloaded from http://earthref.org/ERDA/974/); it has its maximum for the BMU, whereas nodes outside the radius of influence are not upgraded at all (Fig. 3.18B). During a cycle of training, this procedure is repeated for all input vectors. To achieve the stabilization of the map, both the learning rate and radius of influence decrease during each iterative cycle. Eventually, the map will depict the input vectors, the BMU, and the weight vectors calculated.

As we have seen in the World Poverty Map, visualization becomes extremely effective when pattern characteristics are visualized using color codes. In this context, instead of considering the original feature vector of the single patterns, we use the weights of the

BMU to which a pattern belongs. For instance, countries "AUT", "DEU", and "FRA" belong to the same BMU, and we shall consider the weight vector of this node as being representative for all these countries. The color code of the nodes is determined in a way that it represents its position on the map. We determine the position of the nodes by carrying out a PCA (see Chapter 1) on the covariance matrix of the weight vectors and keep the two biggest eigenvalues and eigenvectors. Say that the first eigenvalue is z_1 and the second is z_2. We now normalize both z_1 and z_2 with respect to a range [0 … 1]. In an RGB (red-green-blue) scheme, we assign the z_i to the degree of saturation of red and green. In order to improve the graphics, we define an auxiliary variable z_3 controlling the saturation of blue tones. We can obtain z_3 for instance as $z_3 = 1 - z_2$. For more technical details for the construction of SOM with a sheet geometry, see Appendix 3.3. There we also give some hints on the assessment of the quality of an SOM.

The color code of the SOM provides a synopsis of the classification results, allowing an immediate comparison between various groups. In the presence of time-dependent pattern characteristics, we may follow their development by plotting sequences of colored symbols exploiting the RGB-coding derived from the SOM. Being time-dependent data frequent in geophysics, SOMs have proven as a successful tool for the monitoring of geophysical observations. We shall report more on these applications in Chapter 5.

Appendix 3

Appendix 3.1. Analysis of variance (ANOVA)

Analysis of variance (ANOVA) is used to analyze the differences among group means and their associated procedures (such as "variation" among and between groups). In its simplest form, ANOVA provides a statistical test of whether or not the means of several groups are equal, and therefore generalizes the t-test to more than two groups.

The term "variance" is a bit sloppy, as ANOVA is based on the dispersion rather than the variance.

The fundamental technique is a partitioning of the total sum of squares into components related to the effects used in the model. A fundamental concept is the split of the total dispersion in terms regarding the variation between groups and the one encountered within (Table A3.1).

Table A3.1: ANOVA table (see Davis, 1986).

Source of variation	Sum of squares	Degrees of freedom	Mean squares	F-test
Between groups	S_B	M-1	MS_B	MS_B/MS_W
Within groups	S_W	N-M	MS_W	
Total variation	S_{Tot}	N-1	MS_{Tot}	

We have

$$S_{Tot} = \sum_{i=1}^{N} (x_i - \overline{x})^2 \tag{A3.1}$$

$$S_w = \sum_{i=1}^{m_j} \left(x_i - \overline{x}_j\right)^2 \tag{A3.2}$$

where $\overline{x}$ is the global mean, $\overline{x}_j$ is the mean of the j-th group, and m_j is the number of samples in that group. The two relations are nothing else than the Eqs. (3.7) and (3.8) for the univariate case. The term

$$S_B = \sum_{j=1}^{k} m_j \left(\overline{x}_j - \overline{x}\right)^2 \tag{A3.3}$$

which is the univariate version of Eq. (3.9). The total dispersion corresponds to the sum

$$S_{Tot} = S_B + S_W \tag{A3.4}$$

we can apply an F-test on MS_B/MS_W
where $MS_B = S_B/(M - 1)$, $MS_W = S_W/(N - M)$, checking whether at least one of the groups differs from the total population with respect to its mean. ANOVA requests that group members are randomly sampled, all groups have the same variance and follow a normal distribution. More details can be found in the referenced textbooks.

Appendix 3.2 Minimum distance property for the determinant criterion

In Chapter 3, we have discussed a normalization of dispersion by introducing a matrix $\boldsymbol{G}$

$$\sum_{j=1}^{k} \sum_{i=1}^{m_j} \left(\boldsymbol{x}_i - \overline{\boldsymbol{x}}_j\right)^T \boldsymbol{G} \left(\boldsymbol{x}_i - \overline{\boldsymbol{x}}_j\right) \tag{3.13}$$

and required a form of $\boldsymbol{G}$ such that the Eq. (3.13) is minimum. First, we require that

$$\det(\boldsymbol{G})^{1/l} = 1$$

and search the minimum with constraints using the Lagrange multipliers, i.e.,

$$F(G, \lambda) = \sum_{j=1}^{k} \sum_{i=1}^{m_j} \left(\boldsymbol{x}_i - \overline{\boldsymbol{x}}_j\right)^T \boldsymbol{G} \left(\boldsymbol{x}_i - \overline{\boldsymbol{x}}_j\right) - \lambda \left(\det(\boldsymbol{G})^{\frac{1}{l}} - 1\right) \tag{A3.5}$$

Differentiating $F(G, \lambda)$ with respect to $\boldsymbol{G}$ and λ and equating to zero yields

$$\left(\boldsymbol{x}_i - \overline{\boldsymbol{x}}_j\right)^T \left(\boldsymbol{x}_i - \overline{\boldsymbol{x}}_j\right) = \boldsymbol{S} = \frac{1}{l} \lambda G^{-1} (\det \boldsymbol{G})^{1/l} \quad \text{(A3.6)}$$

Therefore,

$$\lambda = l(\det \boldsymbol{S})^{1/l}$$

and

$$\boldsymbol{G} = (\det \boldsymbol{S})^{1/l} \boldsymbol{S}^{-1} \quad \text{(A3.7)}$$

(see Späth, 1983).

Appendix 3.3. SOM quality

The issue of SOM quality is a complicated one. Typically, two evaluation criteria are used: resolution and topology preservation. If the dimension of the data set is higher than the dimension of the map grid, these usually become contradictory goals. The first value returned concerns the resolution and is addressed to as the quantization error *QE*. In the Introduction section, we learned that SOMs are created on the basis of the Euclidean distance between map nodes and the feature vectors, i.e.,

$$D_{ij} = \sqrt{(\boldsymbol{W}_i - \boldsymbol{V}_j)^T (\boldsymbol{W}_i - \boldsymbol{V}_j)} \quad \text{(A3.8)}$$

Once concluded the training phase, we define a generalized measure summing the distances between the feature vectors and the BMUs they belong to:

$$QE = \sum_i \sum_j D_{ij} \quad \text{(A3.9)}$$

with j running from 1 to the number of BMUs and j from 1 to the number of patterns belonging to the j-th BMU. *QE* is the quantization error of the BMU and decreases with the number of BMUs.

Topological error

A fundamental clue of SOMs resides in the fact that units having the smallest distances between each other are ideally placed side by side. In terms of map with sheet geometry, we may express the neighborhood relation on the map as some kind of "topological distance" (TD). These distances are calculated along the map grid. Consider, for example, the case of a 4 × 3 map. The unit ("1" to "c") positions for "rectangular" and "hexagonal" lattice (and "sheet" shape) are depicted below (Fig. A3.1):

Most neighboring nodes ideally have a topological distance of 1, such as the elements labeled with 2, 5, a, and 7 with respect to element 6, whereas the elements 1, 9, b, 8, and

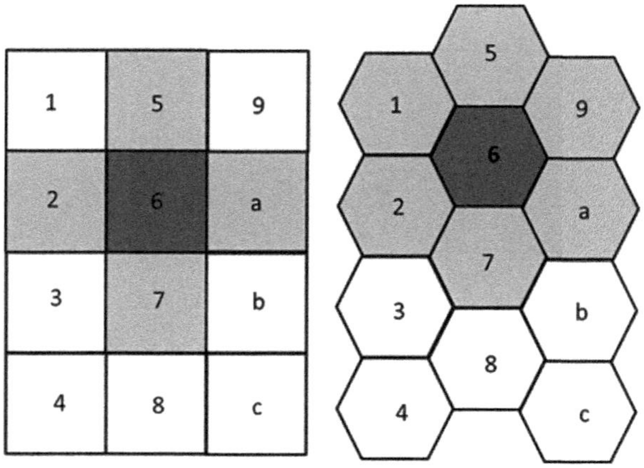

Figure A3.1
Rectangular (left) and hexagonal (right) configuration of nodes on a sheet. Nodes in light gray have a topological distance TD of 1 with respect to the central node ("6", dark gray). There are four nodes with TD = 1 in the rectangular configuration, but six in the hexagonal.

3 have the TD = 2. Suppose now that during training, a node labeled "9" is found to be closer to the node labeled "6" than one of the nodes 2, 5, 7, or a. Then, in the rectangular lattice, there will be a node closer to another, whose "topological" distance is 2 instead of 1. This is undesired and consequently considered as a topological error. On the other hand, in the hexagonal lattice, the element 9 has a TD = 1, so no topological error is encountered in this case.

The global topological error reports the number of cases where a TD > 1 occurs, even though two elements are less distant from each other. A high global topological error is a diagnostics of an unsuitable choice of the map geometry. For instance, the map representations proposed here (i.e., "sheet" shape) may fail for patterns distributed on a circle or a sphere.

Designing the map

A further issue regards the overall aspect of the map and the number of nodes. As we see, for instance, in Fig. A3.2, our nodes are placed on the sheet along rows and columns, i.e., 9 rows and 52 columns. The choice of this aspect ratio, which has to be taken before starting the training, is guided by the 2D PCA carried out on the covariance matrix of the data set, where we keep the two largest eigenvalues together with their corresponding eigenvectors. That way the overall shape of the sheet mirrors the distribution of the data along the two major principal axes. Our nodes will represent our original data with a reasonable homogeneity.

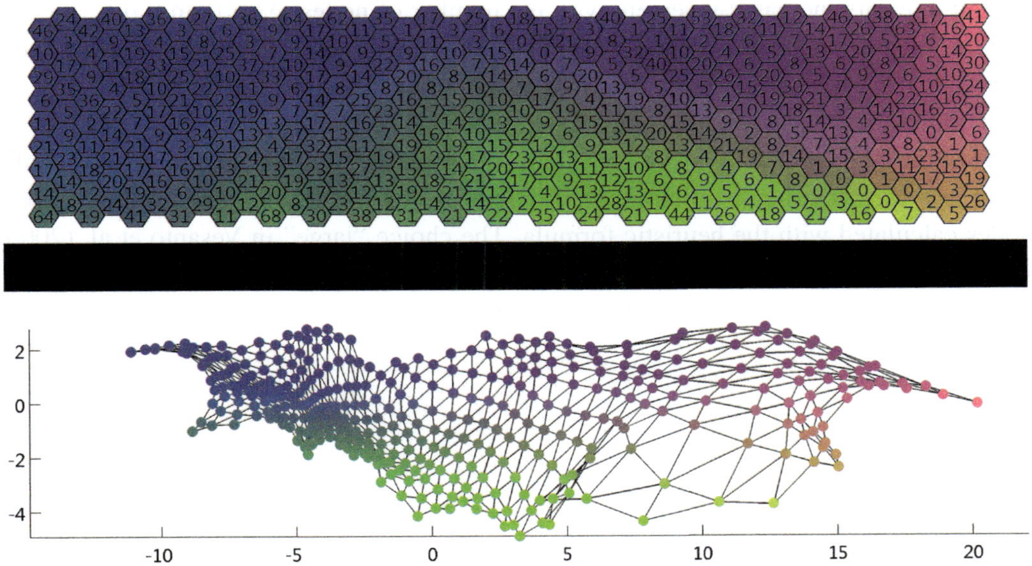

Figure A3.2

A lattice of 9 × 52 nodes. The upper panel shows the distribution of the nodes using the hexagonal configuration. Small numbers in the hexagons give the number of patterns represented by the node. Some nodes carry a "0", they are "loser" nodes, not being BMU for any of the patterns. The lower panel represents the distribution of the nodes in the 2D representation space and gives an idea about the homogeneity of pattern representation by the nodes. "Loser" nodes are more likely in less densely covered areas. Color-coding follows the position of nodes in the lower panel. See also Section 7.2.7 (Toolbox KKAnalysis).

In our specific case shown in Fig. A3.2, the ratio of the two eigenvalues was ca. 5.8. As we decided to use 468 nodes, we choose to distribute nodes on a sheet with 9 rows and 52 columns. Small numbers in the hexagons (upper panel in Fig. A3.2) give the number for which node is a BMU. For instance, the node in the upper left corner of the sheet is a BMU of 41 patterns; the node in the lower left corner represents 64 patterns.

Whereas the upper panel in Fig. A3.2 gives some generic idea concerning the configuration of the nodes on our sheet, their real position in the 2D representation space is shown in the lower panel. It allows a visual inspection of the homogeneity of the node distribution on the sheet. In our example, the homogeneity is quite reasonable for most parts, some areas are poorly covered (such as the range [5 … 15] on the abscissa and [-4..0] on the ordinate). Note that the position of the nodes in the lower panels governs the colors assigned to the nodes.

The choice of the total map size, i.e., the number of nodes, depends highly on the user's needs. Smaller maps be at lower risk of topological error, at the same time the quantization errors increases inversely with the number of nodes. As a thumb rule, we may follow the suggestions in the SOM toolbox 2 (Vesanto et al., 2000, see also Messina and Langer, 2011). Here, the number of nodes for a "normal" sized map is obtained heuristically according to n_nodes = 5 * n_patterns$^{0.5.}$. Then, accounting for the aforementioned ratio of the two largest eigenvalues, we determine the length and the width of the map in a way that the resulting number of nodes is as close as possible to the number of nodes calculated with the heuristic formula. The choice "large" in Vesanto et al. (2000) gives a map with doubled side lengths (i.e., 4 times the number of nodes of "normal), conversely "small" produces a map where the side lengths are only 50% of a "normal" size map.

PART II

Example applications

CHAPTER 4

Applications of supervised learning

4.1 Introduction

In Chapter 2, we understood "learning with supervision" as a way to establish a relation between observations and a target, that is, information that is supposed to be known. In our earthquake-nuclear test example, supervised learning methods were applied to establish a formalism whose predictions had to reproduce with a reasonable accuracy our target, that is, our a priori knowledge. Knowing the two magnitude values—M_S and m_b—we should be able to assign the right category to an event. In a more general understanding, the essence of learning with supervision resides in establishing a function that allows to make predictions from observations in a way that they match our a priori information at best. Applying advanced methods of learning, such as the multilayer perceptron or the support vector machines, the prediction function can be of arbitrary complexity, allowing to handle highly nonlinear problems.

The models we derive from this kind of learning are essentially "black box," as a priori knowledge about the physical relation between observation and the target is absent or very limited—basically helping to choose features or carrying out some data transformation. As our prediction functions can be of arbitrary complexity, we are able—in the extreme case—to fit even noise. Not having any clue whether our black-box models make sense from a physical point of view, specific procedures have to be applied to verify the validity of our models. These procedures imply testing we mentioned already in Chapter 2, and that will be discussed in more details here.

In this chapter, we present practical applications of supervised learning in various geophysical problems, such as the classification of time series (waveforms of explosion quakes and infrasound events recorded at Stromboli volcano and at Mt Etna, respectively). A further example is the classification of rocks. Here, we examine how conventional classification schemes based on the mineralogical composition can be replaced by geochemical analyses.

The application of methods of supervised learning to problems of inversion and regression is also straightforward. Such problems are frequently met in Geophysics. In this chapter, we present the inversion of model parameters from seismic waveforms. The integrated inversion using data of static ground deformation along with magnetic and gravity fields

Advantages and Pitfalls of Pattern Recognition. https://doi.org/10.1016/B978-0-12-811842-9.00004-2

originated from an opening crack is discussed in detail, as it offers specific clues on preprocessing problems as well as the importance of model parameters.

Inversion using supervised learning is based on synthetic data in input and the parameters of the physical model obtained as output. The input/output pairs form the examples from which the inverse relation between data and model parameters is derived. A further straightforward application of supervised learning schemes is given by regression and interpolation, where we aim at making predictions of values based on a set of input data. As in classical regression, we pick up the coefficients of our regression function from a set of examples, in which we know both the input data as well as the target output. In this chapter, we shall outline the concept in a problem of establishing cross-relations between seismic velocities and electric resistivity; a further example regards ground motion prediction equations, which are of paramount importance in seismic hazard analysis.

Successful learning strictly depends on appropriate feature selection and data preprocessing, as we discussed in Chapter 1. In our applications, we will come back to these issues in more detail. In the classification of waveforms (seismic, infrasound, or other data), the direct use of the raw data is troublesome, which makes some transformation necessary. In this case, spectral representations can be envisaged. Another simple transformation is the autocorrelation function of the waveform. Geophysical data often have a large dynamic range, that is, there is a high number of small values mixed with a few large ones. As the latter are not outliers, we cannot just neglect them. Specific normalization techniques can fix the problem.

An alternative approach to black-box learning applications is given by generative models such as hidden Markov models (HMMs) or Bayesian networks, which allow an appraisal of the model dynamics. Consequently, the evaluation of the physical relations each model provides can lead the user to its validation or rejection. In particular, HMM are used to build a classification system that is extremely valuable in case of task-force action. Most automatic classification approaches need a large preclassified dataset for training. However, in case of a volcanic crisis, observatories are often confronted with limitations in the training dataset due to insufficient prior observations. We describe in this chapter a learning-while-recording approach to construct a seismic-event spotting technique that is less dependent on previously acquired databases and classification schemes. Based on a single waveform example, the classification process can be started as early as events have been identified.

4.2 Classification of seismic waveforms recorded on volcanoes

An earthquake source is generally the result of a sudden displacement along a tectonic fault, a manifestation of the activity of our restless planet. Further manifestations of Earth's activity are volcanoes. Similar to earthquakes, volcanoes are mainly located along the margins of tectonic plates continuously remodeling the planet with their eruptions. Nevertheless, volcanic areas offer indispensable natural resources, such as fertile ground and abundance of water, which render their environment attractive for human settlements. This explains why volcanic areas are often densely populated regions.

The paroxysmal activity of volcanoes poses a threat to human life and structural facilities, entailing the necessity of continuous surveillance and monitoring. Among the geophysical data measured on and around volcanoes, seismic data play a key role in the detection of a volcano unrest. Indeed, there is a well-documented evidence of links between enhanced ground motion and volcanic activity. Numerous felt earthquakes heralded the cataclysmic eruption of Vesuvius in 79 AD; however, the population was not particularly concerned, as it had become accustomed to ground shaking in previous years. Moderate to strong earthquakes before eruptions are also documented at Arenal volcano, Mt S. Helens, Mt Usu, Pinatubo, Unzen only to cite a few examples (McNutt, 2000). Volcanic activity is also often heralded (from hours to days) by seismic swarms with hundreds to thousands of small earthquakes. More than 2600 earthquakes with magnitude ≥ 1 were recorded during the 4 days preceding the onset of the July 2001 eruption of Mt Etna (Patanè et al., 2003). Similarly, more than 800 shallow tectonic earthquakes occurred a few hours before and at the onset of the following eruption of Mt Etna, which started on October 27, 2002 and lasted until February 2003.

Earthquakes tied to volcanic activity are the response to deformation of rocks. Indeed, the intrusion of magma and fluids leads to rapid accumulation of stress, with the consequent failure in the form of shear fracturing of the rocks surrounding the rising path. The physics behind such volcano-tectonic (VT) earthquakes resembles that of tectonic earthquakes outside the volcanic environment. Having a broad spectral content with high-frequency peaks, VT earthquakes are also addressed to as "high-frequency" (HF) events (Fig. 4.1A). Well-known parameters in earthquake seismology—such as magnitude, seismic moment, fault mechanism, and stress drop—are also used for their description. On the other hand, there are also seismic transients the characteristics of which differ from tectonic earthquakes. As the parametrization of their sources is still a matter of debate, the type of seismic signals recorded on volcanoes is identified on the base of some morphological aspects of the waveforms rather than addressing directly to source parameters. In a certain sense, the classification of the signals based on their waveforms forms a propaedeutic step in the identification of the physics of the sources.

McNutt and Roman (2015) give a summary of the types of seismic events recorded in volcanic environment:

Long period events ("LP" events; Fig. 4.1B): They are transients having an emergent onset, a wave train without easily recognizable phases, and a clearly dominant period over the whole signal. Their spectra appear narrow-banded, and dominant frequencies (between 1 and 5 Hz) are lower than those of VT earthquakes. The mechanism at the origin of this

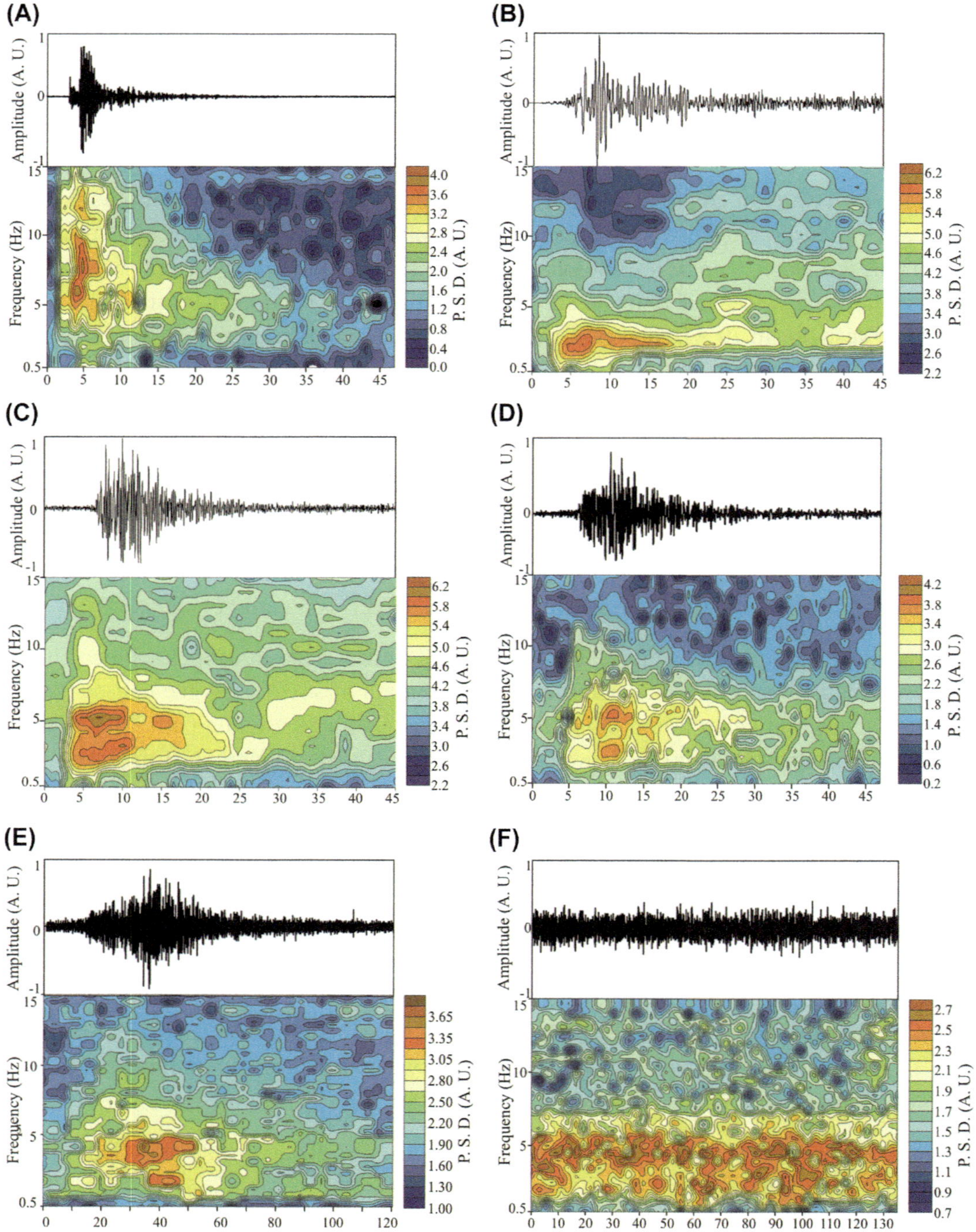

Figure 4.1
Example waveforms recorded at Stromboli (A, D, E, F) and Montserrat (B, C): Volcano-Tectonic (A), Long Period (B), hybrid event (C), explosion quake (D), rockfall (E), and volcanic tremor (F).

event type is a matter of intense research. Most of them are understood as being caused by fluid pressurization, such as bubble formation and collapse. In this light, LP events have attracted considerable attention as potential precursors of an impending volcano unrest. With the advent of broadband seismometers, a subgroup—called very long period ("VLP") events—was identified, being their dominant period well above 1 s.

Hybrid events (Fig. 4.1C): Morphologically, their seismic signature is a wave train with dominant high frequencies at the beginning and lower frequencies in the coda. In some way, they appear as a mixture of HF and LP events. Many authors suppose that the high-frequency part is truly generated by an HF, which triggers oscillation of fluid-filled cavities. Note, however, that volcanoes are particularly complex with respect to their geological structure. Wave propagation effects and attenuation may strongly affect the waveforms of seismic signals, hindering the identification of the real source process.

Explosion quakes (Fig. 4.1D): They can be explained by a mechanism similar to nuclear tests mentioned in Chapter 2, that is, their source is a sudden release of energy caused, for instance, by the burst of bubbles within a magma conduit. When the source is close to the surface, their seismic signature is associated with an air shock, as part of the energy propagates through the air as acoustic wave.

Superficial events (Fig. 4.1F): They originate at the very surface of the volcano edifice and include nonvolcanic processes. Glacial events, shore-ice movements, lahars, landslides, and rockfalls are "secondary" signals recorded on a volcano. Although not directly linked to the dynamics inside the volcano, these signals may be a hint of ongoing processes. For example, the growth of lava domes in andesitic volcanoes with highly viscous magma is often accompanied by an increase in rockfall activity. On the other hand, ground shaking associated with seismic events may act as a trigger for those superficial events.

Volcanic Tremor: McNutt (1992) and McNutt and Roman (2015) define volcanic tremor as a signal with duration ranging from minutes to days. It can have origin from the coalescence of transients, such as Hybrid or LP events. On basaltic volcanoes with permanent open conduits, like Mt Etna or Stromboli, tremor is recorded at any time, forming a persistent signal in which the energy radiation varies smoothly with time. Its amplitude as well as its spectral content changes with the state of the volcano activity. The spectral content is narrow-banded, sometimes with the presence of one dominant peak together with overtones. The source mechanism is supposed to be similar to that of LP events, that is, oscillations of fluid-filled cavities (see Aki et al., 1977; Chouet, 1985). Other authors like Ferrick et al. (1982) and Schick (1988) underscore the role of nonsteady fluid flow causing pressure fluctuations that would generate the seismic signal. An example of volcanic tremor recorded at Stromboli is depicted in Fig. 4.1F.

4.2.1 Signal classification of explosion quakes at Stromboli

Known as the lighthouse of the Mediterranean Sea, for its persistent activity over more than 2000 years, Stromboli volcano belongs to a chain of volcanic islands forming the Aeolian Archipelago in southern Italy (Fig. 4.2). Stromboli's activity consists of continuous degassing as well as so-called Strombolian explosions. These explosions are frequent (10–20 min) and cause the ejection of ash, lapilli, and lava bombs. Periodically eruptive crises occur, with occasional lava effusions and/or paroxysmal activity. In the 21st century, the major reasons of concern for the local population were in December 2002 and March 2003, during the same episode of lava effusion. A landslide—over both land and sea bottom—associated with a tsunami happened 2 days after the onset of the effusive activity on December 28, 2002; 3 months later, these phenomena were followed by a paroxysmal sequence of explosions, which remodeled the summit part of the volcano. Such events had a strong social impact and fostered the setting of a cutting-edge monitoring system.

The majority of seismic signals recorded at Stromboli has a strong link with Strombolian explosions and is known as explosion quakes. The count of the daily occurrence unveils hundreds of such events, the manual analysis of which is a daunting task. Nevertheless, a rapid and reliable analysis of these events is important, for their distribution in classes (according to their seismic signature) changed before a few well-documented paroxysmal activities heralding their occurrence (Falsaperla et al., 1989). For these reasons, Falsaperla et al. (1996) proposed the automatic classification of Stromboli's explosion quakes by using neural networks.

Generally speaking, the motivations for the application of an automatic classification system are inherent to the necessity to quickly process a large amount of signals as well as the lack of easily identifiable and physically well-constrained parameters for their characterization—such as the aforementioned parameters mb/MS for tectonic earthquakes and nuclear test explosions. An answer to this need can be found in supervised classification with neural

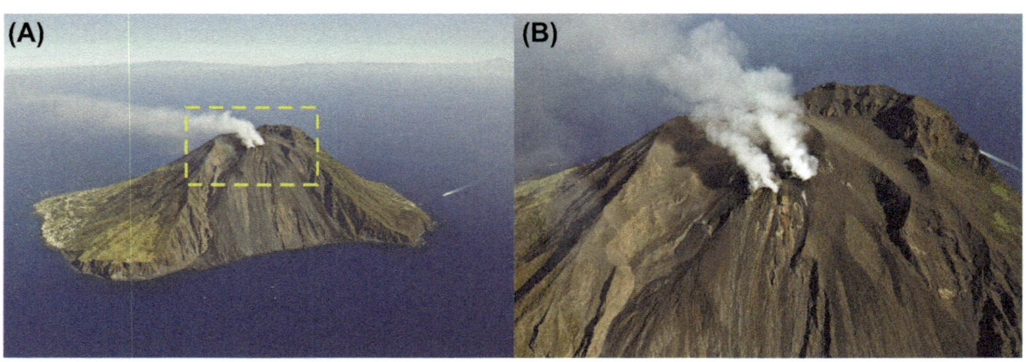

Figure 4.2
The Stromboli island (A); detail of the Sciara del Fuoco with the active summit craters (B). *Reprinted with permission of the author Alfio Amantia.*

networks, for instance using multilayer perceptron (MLP). As we have learned in Chapter 2, they overcome the drawbacks of methods based on statistical considerations like discrimination analysis, which are limited to problems where classes can be separated by linear or quadratic discrimination functions. On the other hand, methods based on cluster analysis, which can deal with nonlinear processes, require an "a priori" definition of metric for the distance of the classes. This is not necessary with MLP networks, as they can generalize the main features of the discrimination functions from examples (see Chapter 2).

The dataset of explosion quakes analyzed by Falsaperla et al. (1996) was recorded by means of a 1 Hz vertical-component seismometer 1.8 km distant from the summit craters. A 12 bit A/D converter digitized the original analogue signal at a sample rate of 33 Hz. Based on a previous study (Falsaperla et al., 1989), the 75 explosion quakes of the dataset analyzed were then divided into four classes labeled K, L, M, and N. Examples for the four event classes are shown in Fig. 4.3.

The first class, "K," depicts a low-frequency phase (1–2 Hz) that lasts ~3 s, followed by a phase with higher frequency content (2–4 Hz) and a much smaller amplitude with respect to the first part The second class, "L," is monochromatic (~2 Hz), generally with brief duration.

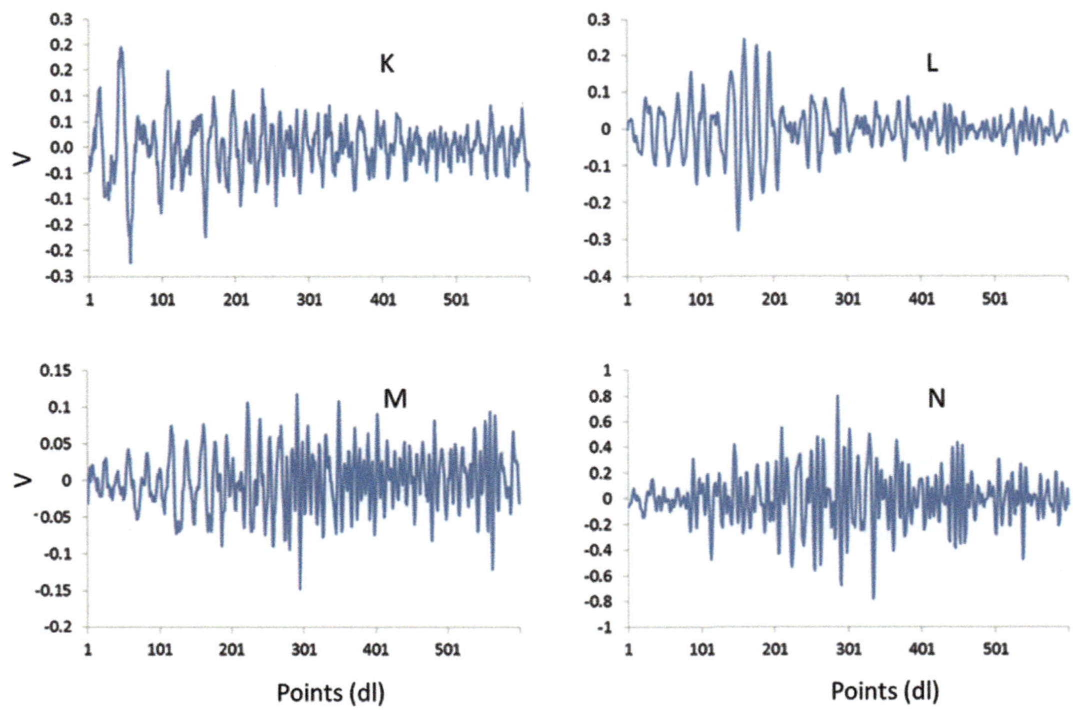

Figure 4.3

Example waveforms of ground velocity for the four classes of explosion quakes recorded at Stromboli (Falsaperla et al., 1996). Abscissa reports the number of points.

The third class, "M," is characterized by a low-frequency phase (1.3–2.5 Hz) followed by another one with higher frequency (4–6 Hz) but comparable amplitude. Finally, the fourth class, "N," shows no clear phase and is dominated by frequencies in an interval from 4 to 6 Hz. As a fixed duration was necessary for the application of the MLP, each explosion quake had assigned a time series containing only 600 samples ($\sim$18 s), independently of the original duration of the signal. This length was a good compromise for representing each individual seismogram without loss of significant information.

For their application, Falsaperla et al. (1996) used an MLP with topology U–V-4, where U is the length of the input feature vector (600 nodes), V is the number of nodes in the hidden layer, and four are the nodes in the output, representing the classes "K," "L," "M," "N." For the input feature vector, the classification task posed the problem of data coding, that is, the choice of the best way to present the features of the input patterns for an efficient separation of the different classes. This is a key aspect, as the performance of a neural network depends on the chosen feature representation. Being poorly guided by theory, this aspect of pattern classification requires a heuristic approach based on trials to reach an appropriate data coding. The authors considered a variety of options for the generation of the feature vectors, such as time series, spectra, envelopes, and autocorrelation functions (ACF). Here, we focus on the latter, which turned out as a good choice in terms of MLP's performance (Falsaperla et al., 1996). Typical ACFs for events of the four classes are depicted in Fig. 4.4. ACF has a zero phase, that is, the phase information of the traces is lost. This overcomes the problem of a proper phase alignment, which may be critical when using time series.

We reproduce here some of the results obtained by Falsaperla et al. (1996), considering the same dataset encompassing 75 patterns. Given the length of the ACF input vectors (600 nodes) and the number of classes (4), the only free parameter in our network is the number of hidden nodes. As we mentioned in Chapter 1 (Section 1.2.7), the choice is guided by two goals: (i) find the mapping function which gives the highest accuracy, and (ii) avoid overfitting effects. Overfitting occurs when a model has too many degrees of freedom. Such a model can fit any kind of data—in theory even noise—but once applied to a set not used during the estimation of the model parameters ("training"), the number of mismatch may be high (see Appendix 4.3). We therefore prefer functions that give a reasonable accuracy with the lowest number of parameters. The identification of a suitable number of model parameters—here the number of hidden nodes—is achieved by a strategy called cross-validation. It consists in setting aside a part of the data, excluding it from the training. In our specific case, we divide the 75 patterns into two sets, namely a training set of 38 patterns and a test set of 37 patterns. During each cycle of training, we apply the current mapping function to both sets, obtaining error histories such as the ones shown in Fig. 4.5. We leave to the readers their own experiments—varying the number of hidden nodes and analyzing the corresponding error histories—using the code and data

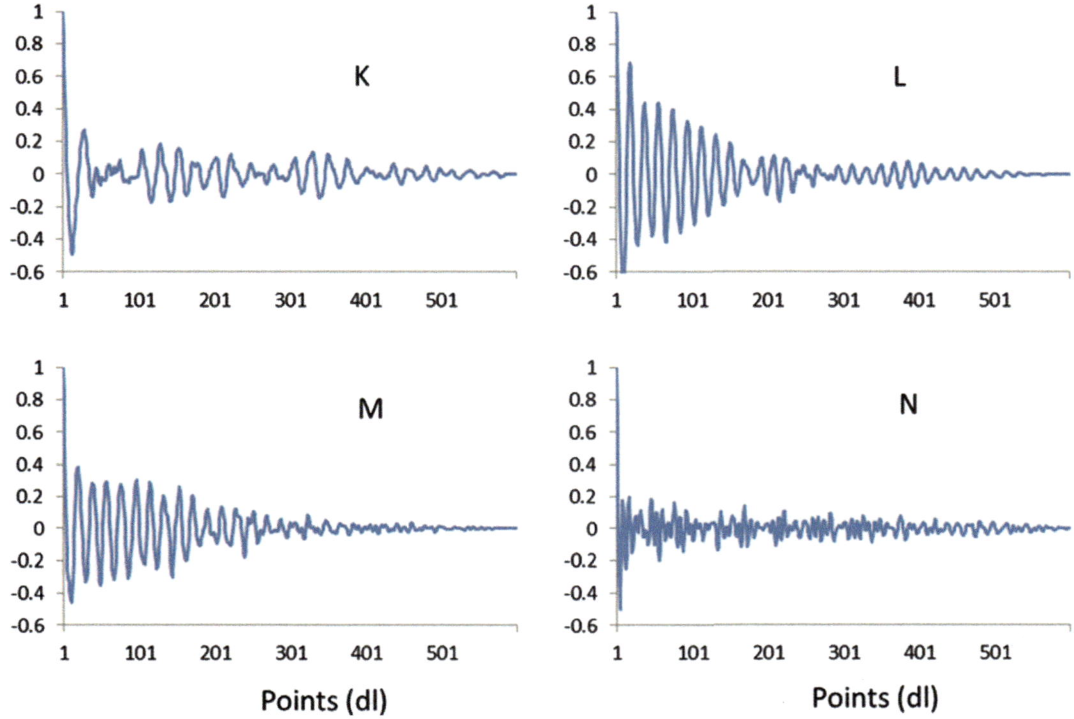

Figure 4.4
Typical autocorrelation functions for the event classes "K," "L," "M," and "N." Abscissa reports the number of points.

accompanying the book (see Chapter 7.3.1 code BPNN, training, and test data). In the following, we shall discuss the results obtained with eight nodes in the hidden layer, that is applying a topology 600-8-4. Note that topologies with only one hidden layer are generally sufficient to solve the same class of problems as more complex networks (Freeman and Skapura, 1992).

We assign an a priori class membership to each pattern using a classification vector of length 4. As targets, we define the class "K" as the vector (1, 0, 0, 0), "L" (0, 1, 0, 0), "M" (0, 0, 1, 0), and "N" (0, 0, 0, 1). During the training phase, the MLP tries to establish a function from the input vector (here the ACF), optimizing the difference between calculated output and target vectors. Compared to the a priori target vectors, where there are only "**1**"s and "**0**"s, the calculated output vector is a floating-point value. Let us consider results and targets for the following example pattern

$$\{1.02998,\ \mathbf{1};\ -0.0116104,\ \mathbf{0};\ -0.00754332,\ \mathbf{0};\ -0.0113994,\ \mathbf{0}\}$$

in which the floating-point values represent the calculated output (the target values are in bold). The target class is "K" ("**1**" in the first position). The calculated output with respect to

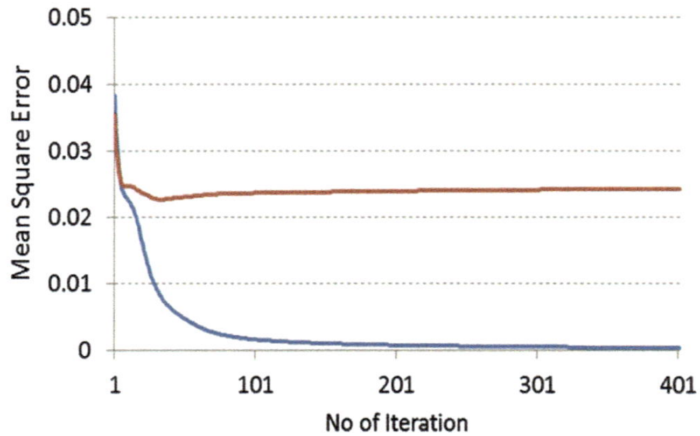

Figure 4.5
Training set (blue) and test set errors (red) throughout the training of an MLP with eight hidden units.

the first element of the class vector is 1.02 ..., whereas the next calculated values are close to zero. This is a typical "success" of the classifier in the sense that target- and calculated-class vector are similar. Now, consider another pattern with the following results:

$$\{0.429063,\ \mathbf{0};-0.0689474,\ \mathbf{0};0.433669,\ \mathbf{1};0.205585,\ \mathbf{0}\}$$

These are still a "success" if we look for the largest calculated value for the identification of the class membership. Indeed, in this case, "**1**" is in the third position (see class "M") and the corresponding calculated output 0.433 ... is the largest among the calculated values.

An example of misfit is

$$\{0.447428,\ \mathbf{0};-0.0216334,\ \mathbf{0};0.237053,\ \mathbf{0};0.337737,\ \mathbf{1}\}$$

with the largest calculated output 0.44 ... in the first position, whereas the target "**1**" is in the fourth position (class "N").[1]

[1] It is worth examining in hindsight the calculated output values, in particular, when they are far from 0 or 1. In our second example, we face with features in between class "K" and "M." Eventually, "M" is the assigned class membership, but the classification is uncertain, as the second highest score is close to the first one. Similar cases merit an in-depth investigation of the a priori input classification. Note that we shall use the gradualness in the MLP output for nonlinear regression and inversion purposes. Here, we should mention that the network training is controlled by the mean square error (MSE) between calculated output and target, whereas in the confusion matrix we simply count the patterns with matching/mismatching classification. The squared errors for the second and the third example are ~0.55 and 0.69, respectively. Even though they share a similar square error, the classification is "successful" only for the second pattern. This is a typical threshold effect and must be kept in mind in the a posteriori analysis of the classification results, especially when the dataset is small.

Based on the number of successes—here achieved considering the maximum of the calculated values—we summarize the results in so-called confusion matrix (Table 4.1).

In the confusion matrix, columns and rows report the a priori classification and the classes identified by the MLP, respectively (Table 4.1). Table 4.1a has no nonzero entries at the off-diagonal elements of the confusion matrix for the training set, documenting a success of 100%. The a priori classification encompasses 11 patterns in class "K" and "L," and eight in class "M" and "N." Using the "BPNN"-code for the MLP (with autotuning of the learning parameters) we get somewhat better results compared to Falsaperla et al. (1996). In fact, we find (see Table 4.1a) that all patterns in the training set were correctly identified. For comparison, the matrix in Table 4.1b reports the results obtained for the test set containing a few nonzero off-diagonal elements, which are misclassifications. For instance, MLP assigned the class "L" and "N" at two of the a priori labeled "K" patterns. Problems of misclassification also rose for patterns of class "M" and "N" (Table 4.1b).

4.2.2 Cross-validation issues

Table 4.1: Confusion matrix for training (a) and test set (b).

(a)				
	A priori K	**A priori L**	**A priori M**	**A priori N**
Calculated K	11	0	0	0
Calculated L	0	11	0	0
Calculated M	0	0	8	0
Calculated N	0	0	0	8
(b)				
	A priori K	**A priori L**	**A priori M**	**A priori N**
Calculated K	10	1	1	1
Calculated L	0	10	0	0
Calculated M	1	0	6	2
Calculated N	0	0	1	4

The results for the training set are slightly better than those reported in Falsaperla et al. (1996), as we used an improved algorithm for network training.

The results of the test set (Table 4.1b) show more misclassifications than the ones for the training set (Table 4.1a). This comes from intuition and corresponds to the error history shown in Fig. 4.5. At the beginning of the training, both errors decrease at each iteration; later, however, the error for the test increases, whereas the one for training continues to decrease. Note that the results reported in Table 4.1 covers the whole training process (400 cycles). From the error history, however, we might suspect that we were better off stopping the training after only ~30 cycles, as the error related to the test set increases, reaching a stable level after ~70 cycles. Although being a tempting solution, it requires caution. If we use the test set to decide when iterations should stop, then the test set is no more a true test! In theory, one could restart the training repeatedly, just changing the initial conditions and taking the best results out of all these attempts. From a statistical point of view, the test error is a random variable itself. Let p being the true but unknown error rate of our classifier, $\tilde{p}$ the estimated value, and k the number of misclassified samples in a test set of size $\tilde{n}$; then k follows a binomial distribution (see Duda et al., 2001)

$$P(k) = \binom{\tilde{n}}{k} p^k \tag{4.1}$$

for which the maximum likelihood estimation of p is

$$\tilde{p} = k/\tilde{n} \tag{4.2}$$

In our example in Table 4.1b, we have 37 patterns in the test set and an estimated rate of misclassification of 7/37 = 0.19. The 95% confidence intervals of true error for the corresponding binomial distribution (Eq. 4.1) is $\sim 0.08 < p < 0.38$ (see, e.g., Kreyszig, 1982). In other words, if we use the test set to stop the training, there is the risk of being biased toward the lower bounds of the confidence interval. Therefore, we must not use the test set for the decision when to stop the learning. In general, we must warrant the principle that the test has to be independent from the training process in any regard. At the same time, we learn that with a single test set of limited size the performance of our classifier is uncertain. There are several ways to overcome these drawbacks:

(i) Validation set: Besides the training and test set, we can create a third ensemble, called validation set. In this case, we can stop the iterations when the error decreases for the training set and increases for the test set; the third ensemble will allow us the final validation of the performance of the classifier, avoiding the problem of being biased toward the lower limits of the confidence interval (see, e.g., Hastie et al., 2002).

(ii) N-fold cross-validation: Having a limited quantity of data, we may find difficult the application of the earlier-mentioned solution to achieve an unbiased and, at the same time, statistically robust estimation of the classifier's performance. An alternative is the random division of the whole ensemble into N roughly equal-sized subsets. After carrying out training and test N-times, we can create statistics on the performance of the classifier with the N-resulting errors obtained.

(iii) Bootstrap and jack-knife methods: Other ways out in case of limited quantities of data are techniques that exploit bootstrap and jack-knife methods. Using bootstrap, we resample the whole dataset, by drawing random patterns from the original dataset. We set aside the patterns that are not drawn during the bootstrap and use them in the test. That way we have on average about 63% of the data in the training set and 37% in the test set (see Hastie et al., 2002). Performing the bootstrap several times, we can get statistics on the performance of the classifier. Jack-knife cross-validation is also known as the "leave-one-out" approach, as one of the patterns is set aside for testing, whereas the rest remains in the training set. In this case, the training goes on *N*-1 times, with *n* being the number of patterns in our dataset. This method is unaffordable if *N* is large.

4.3 Infrasound classification

Seismic monitoring is functional if the sources are buried in the crust, and energy remains trapped in the surrounding material. However, it may lose its efficiency for sources located close to the surface, which release their energy also into the atmosphere. Shallow sources are often accompanied by acoustic and infrasound signals, that is, waves that propagate in the air. For instance, strong superficial earthquakes are reported to be felt by the population as a "bang"; the same holds for explosive events on volcanoes or explosions in the context of human activity (such as quarry blasts). Perhaps the first air waves documented during a volcanic eruption were barometric disturbances recorded during the explosion of the Krakatau volcano in 1883 (see Evers and Haak, 2010). Those disturbances recorded worldwide allowed the precise determination of the origin time of the eruption (see Verbeek, 1885; and Symons, 1888; cited in Evers and Haak, 2010).

With their dominant frequencies well above 20 Hz, acoustic signals propagate only over short distances. On the contrary, infrasound signals have a typical frequency range between 0.01 and 10 Hz and can therefore be recorded over long distances. (Infra)Sounds are longitudinal waves and can stem from numerous sources, such as the aforementioned shallow earthquakes, volcanic eruptions, weather-dependent sources (storms and tornados), and various man-made sources. Infrasound monitoring plays a key role in nuclear test monitoring and is one of the four key technologies exploited in the International Monitoring System of the CTBTO ("Comprehensive Test Ban Treaty Organization" of the UNO). In fact, nuclear testing has been carried out in the crust (with buried sources causing strong seismic signals), in the sea (where sources are revealed by hydro-acoustic monitoring), and in the atmosphere with a strong radiation of infrasound signals.

Following Cook (1962), molecular attenuation is unimportant to infrasonic propagation (5×10^{-8} dB per km). In comparison, a sound at 2 kHz will attenuate at 5 dB per km. Therefore, an infrasound below 1 Hz is virtually unattenuated by atmospheric absorption, and remains detectable at distances of thousands of kilometers from the source. At long distances, several phenomena may complicate the propagation of infrasound signals. Temperature changes in the atmosphere affect the signal in the same way that light waves are refracted by lenses. Wind speed changes contribute to sound refraction, guiding sound waves as they travel for long distances. Some waves escape and travel upwards to great heights in the ionosphere where they dissipate, while others are trapped in the upper atmosphere as they travel horizontally (see Georges and Beasley, 1977).

Although the atmosphere is nonhomogeneous and dynamic, it is relatively uncomplicated compared to the solid medium through which seismic waves propagate (Johnson and Ripepe, 2011). In general, the volcanic edifice is a medium with large impedance contrasts, therefore with high scattering and attenuation of seismic waves. Compared to the solid earth, the atmosphere is relatively homogeneous and isotropic at short propagation distances, such that infrasonic pressure records can be fairly directly related to source processes occurring at a volcano. Therefore, on local scale, infrasound data have achieved considerable attention in the framework of volcano monitoring, as they allow a quantitative description of the eruptive behavior. At local recording distances, amplitude and power, coda duration, signal envelope, and frequency spectra are easily quantified. This enables a comparison of infrasound signals for suites of eruptions either at a single volcano or a group of volcanoes. Similar to seismograms, infrasound signals are commonly processed as time series or as spectrograms, which depict the frequency content in a signal over time (e.g., see Fig. 4.1). Qualitative insight into infrasound signals may also be realized by speeding up infrasound into the audio band. Although the human ear does not have a flat frequency response, it is sensitive to subtle variations in tone.

Compact volumetric sources are the most efficient sources of infrasound signals in a volcanic region (Johnson and Ripepe, 2011). For these sources, the dimension of the volcanic vent (or acoustic radiator) is small compared to the wavelength of the radiated sound. A simple monopole source approximation may then be made, and the resultant sound field will be proportional to the source strength or change in rate of mass injection (Lighthill, 1978)

$$P_M(r,t) = \frac{r}{\Omega} Q\left(t - \frac{r}{c}\right) \tag{4.3}$$

with t being the time variable, r the source-to-receiver distance, and c the sound velocity. Q is the mass acceleration at the source, which is the effective volumetric acceleration of the atmosphere at the source. Under the assumption of a monopole source mechanism, the mass flux history and cumulative explosive flux may be recovered from single and double time integration of Q.

Infrasound waveforms from volcanic eruptions often begin with an abrupt compressional phase followed by a rarefaction phase of similar amplitude (e.g., Morrissey and Chouet, 1997). These bipolar pulses have an N-like shape similar to a chemical explosion shock wave after it decays to acoustic wave. N-shaped waveforms are quite a common feature of the onset of many explosive eruptions. Infrasound signal may acquire a more complex waveform when followed by a broadband or harmonic "tremor"—a signal being persistent for some time, similar to its equivalent in seismology. Such a waveform has been interpreted either as a sequence of pulses or as resonance modes of fluid-filled conduits (Johnson and Ripepe, 2011, and references therein).

Air displacement caused by rockfall associated with volcanic activity has been modeled by Moran et al. (2008) using the monopole approximation mentioned earlier.

4.3.1 Infrasound monitoring at Mt Etna—classification with SVM

Mt Etna, situated on the eastern coast of Sicily, is Europe's largest and most active volcano. The volcanic edifice of Mt Etna covers an area of $\sim 1200\ \text{km}^2$. The flanks of the mountain are densely populated. Catania, Sicily's second largest city with ~300,000 inhabitants, is about 15 km south of the summit craters. There are also important infrastructures—like the International Airport Catania-Fontanarossa; the Freeways Catania—Palermo, Catania—Messina, and Catania—Siracusa. To protect this metropolitan area, Mt Etna is continuously monitored for the identification of precursory phenomena heralding a possible volcanic threat.

Mt Etna's summit area has undergone profound modifications even in relatively brief time spans. For example, it has five active summit craters at present, but there was only one central crater at the beginning of the 20th century (Bonaccorso et al., 2004). In an analysis of infrasound signals during a period of unrest in fall 2007, Cannata et al. (2009) located the sources of infrasound signals using the so-called "semblance" approach (Neidel and Tarner, 1971). This is a commonly used technique for determining both seismic and infrasound source locations at volcanoes, especially when signals have no easy-to-pick onset. The method applies a forward grid search in which each unique sensor pair is time-shifted for each grid node, and then cross-correlated to determine which time shift and respective node provide the best fit. During the location of the infrasound sources it turned out that many of them were situated close to three main craters of the volcano, namely "North-East Crater", "Bocca Nuova" and "South-East Crater" (Fig. 4.6).

It turned out that waveforms can be distinguished on the base of the crater that is closer to the source location (see Fig. 4.6). Considering for instance the recordings of the EBEL station, there were clear differences between the waveforms of signals radiated at the South-East Crater and those related to the North-East Crater. In terms of classification with supervision, one can exploit the location of the source as a priori information of the crater of origin. Accordingly, one can form a training set assigning the class membership with respect to the location of the signal source. The learning phase will attain a function that enables to infer which of the craters yielded a certain signal. This can be extremely helpful especially when more than one sensor is out of order, overcoming the necessity to locate each signal (Fig. 4.7).

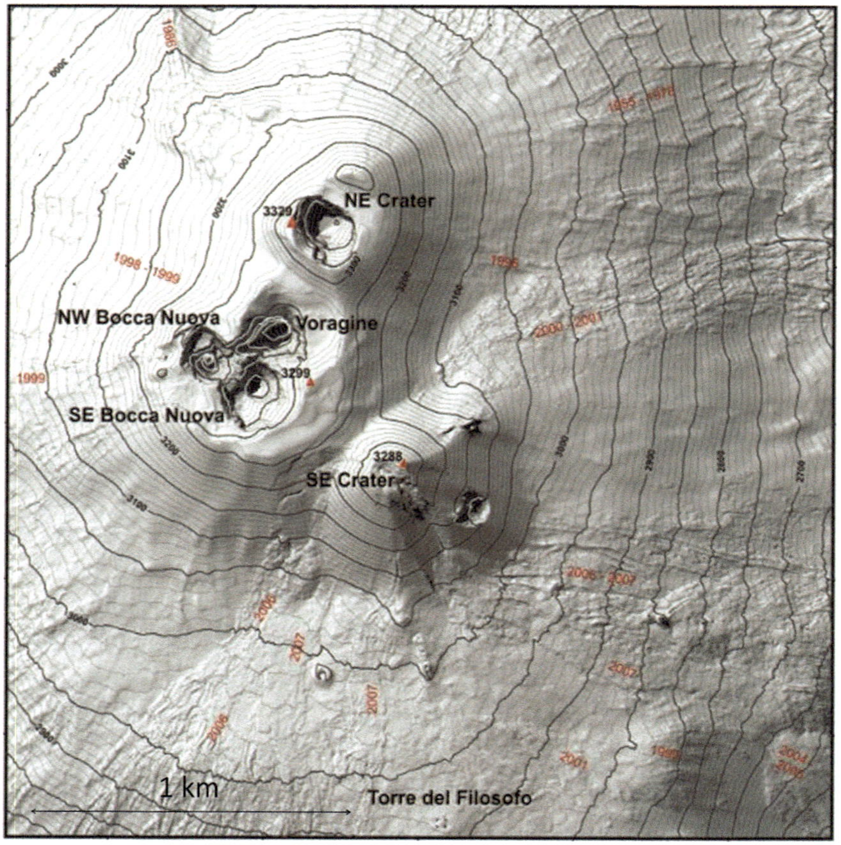

Figure 4.6

Digital elevation model of the summit craters of Mt Etna by Neri et al. (2008) (reprinted by permission of the publisher John Wiley & Sons, Inc.). Red numbers give the year of relevant lava flows observed in the years from 1955 to 2007.

For the feature extraction, Cannata et al. (2011) applied the so-called Sompi method. This accounts for the specific characteristics of infrasonic events, the waveforms of which can be represented as decaying complex exponential functions (Kumazawa et al., 1990, and references therein). The method is a high-resolution spectral analysis technique based on an autoregressive filter, which describes a given time series as a number of "wave elements" consisting of decaying harmonic components, and additional noise.[2]

[2] Besides classical spectral representations and autocorrelation functions, the features of infrasound signals can be also obtained from Hilbert Transforms, the Hilbert Huang Transform (see Huang and Wu, 2008), or using the Wavelet Transforms.

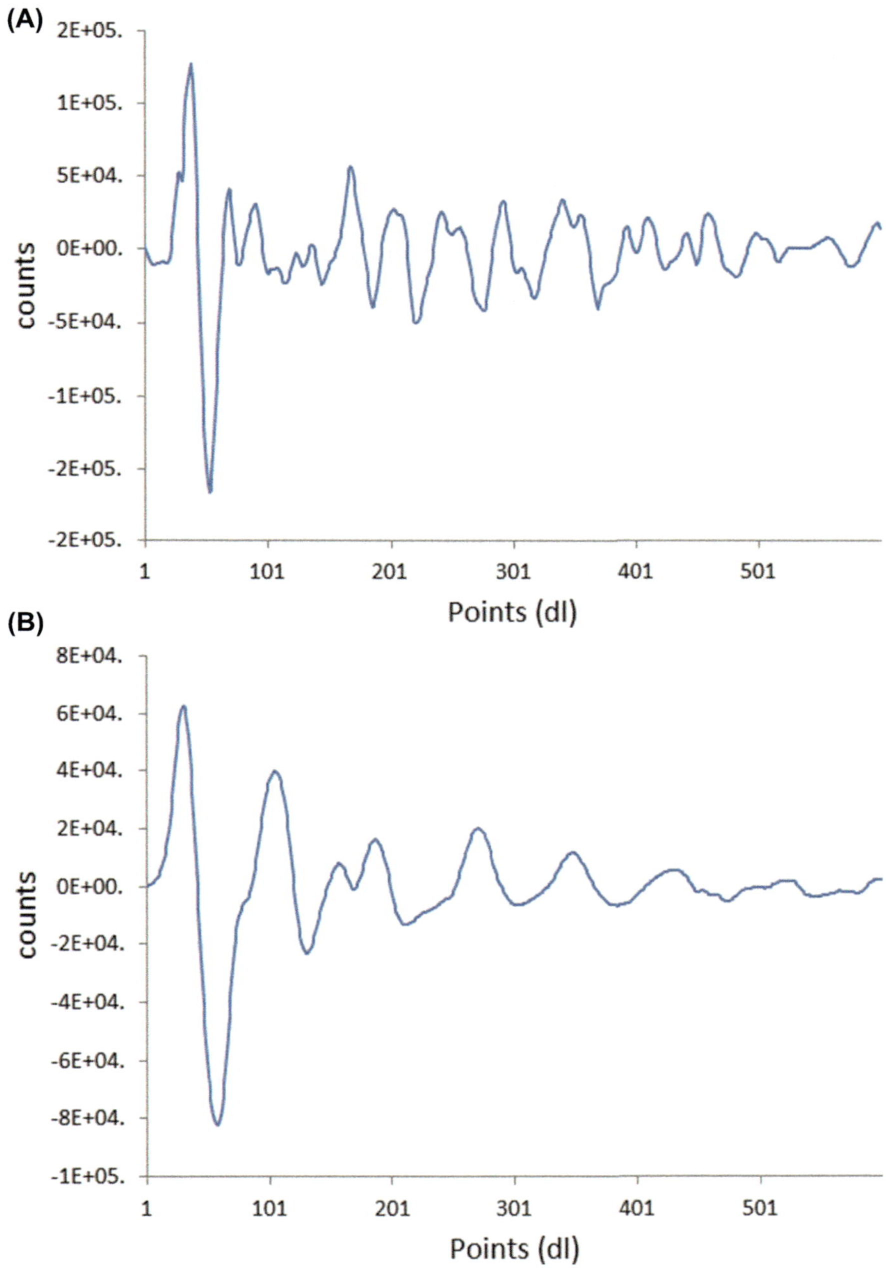

Figure 4.7

Typical infrasound waveforms recorded during fall 2007: (A) infrasound event occurred on September 4, 2007, located at the South-East Crater; (B) infrasound event recorded on October 2, 2007, located close to the North-East Crater (Cannata et al., 2009, 2011).

Cannata et al. (2011) defined the a priori class membership by a DBSCAN-based clustering and applied support vector machine (SVM) classification using this target definition. The testing phase provided the following confusion matrix.

On the whole, the results gave about 12% off-diagonal patterns, which provide the rate of misclassified items. The separation of clusters 2 and 3 is rather poor in the SVM application (Table 4.2). Note, however, that these two clusters represent infrasound events located close to the South-East Crater. Combining the two clusters into one single family representing the location, the mismatch decreases to 5.25%. This example highlights the important role of the confusion matrix, as it can help the user in the choice of the number of classes to handle for the definition of the a priori target.

Mt Etna and, in particular, its summit area undergoes changes at any time due to frequent eruptive activity. We shall discuss later on how such changes in active volcanoes may pose severe problems for the application of supervised classification schemes over long-time spans, as the characteristics of targets may change. For instance, modifications in the geometry of a volcano conduit can alter the waveforms of seismic and infrasonic signals. It is recommendable to repeat the whole learning and testing scheme from time to time (see Cannata et al., 2011). In case of relevant changes in the source characteristics, the definition of the target becomes obsolete and must be replaced accounting for the new situation.

4.4 SVM classification of rocks

Conventional classification of igneous rocks is based on the rock composition considering a few key minerals. The famous Streckeisen diagrams provide a descriptive classification of "rock modes" on triangular diagrams in which the key minerals are Quartz, Orthoclase, and Plagioclase (e.g., Streckeisen, 1974, 1978). In complete diagrams, one also adds the Feldspathoid-branch (Fig. 4.8). As Feldspathoid and Quartz cannot be together in the same rock, they are placed at the opposite vertices of the triangles.

Let us focus on the upper triangle of Fig. 4.8. The three components Quartz, Orthoclase, and Plagioclase are calculated from the mode to sum to 100%. Each component marks the corners of the equilateral triangle, the sides of which have a length divided into 100 equal

Table 4.2: Confusion matrix of infrasound classification at Mt Etna (Cannata et al., 2011).

Actual/Predicted	Cluster 1	Cluster 2	Cluster 3
Cluster 1	**476**	9	6
Cluster 2	9	**15**	8
Cluster 3	8	33	**46**

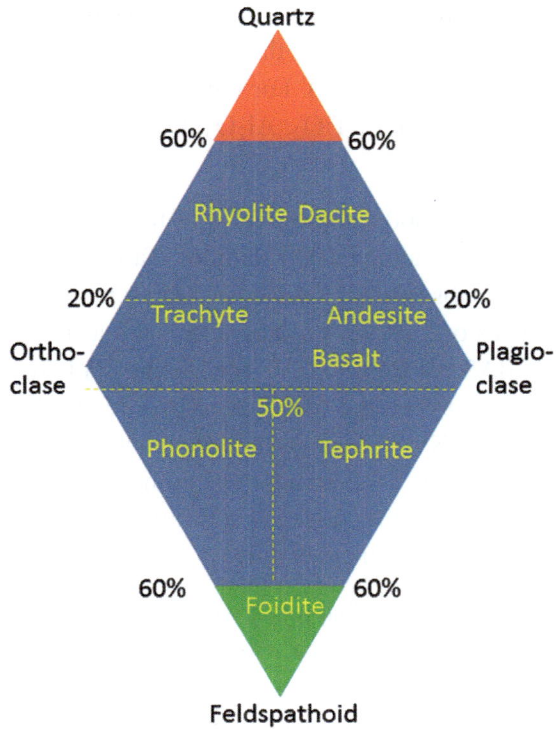

Figure 4.8
Streckeisen diagram for volcanic rock (simplified from Streckeisen, 1978). See also https://web.archive.org/web/20110930102012/http://geology.csupomona.edu/alert/igneous/igclass.htm.

parts. Any composition plotted at a corner has thus a mode of 100% of the corresponding component. Any point on the sides of the triangle represents a mode which is the sum of the two adjacent corner components. For example, a rock with 60% Quartz and 40% Orthoclase lies on the left hand side of the triangle (Quartz—Orthoclase) at a "distance" of 60% from Orthoclase; this composition belongs to the field of Rhyolite (Fig. 4.8). A rock with 60% Quartz and 40% Plagioclase lies on the right hand side of the diagram (Quartz—Plagioclase) at a distance of 60% from Plagioclase. In this case, the composition belongs to the field of Dacite (Fig. 4.8).

In case of aphanitic (fine grained) texture of volcanic rocks, the mode cannot be readily determined; the chemical classification overcomes the problem and is widely used by most petrologists. One popular scheme of such a classification is based on both chemical components and normative mineralogy.

Conventionally, a chemical classification is obtained by plotting the content of SiO_2 versus the content of Alkali minerals ($Na_2O + K_2O$) in the so-called total alkali silica diagram (Le Maitre, 2002). An alternative to these classical schemes is the application of supervised classification. In the following, we examine the application of SVM to geochemical data that were downloaded from the "Geochemical Rock Observatory" (http://georoc.mpch-mainz.gwdg.de/georoc/). For the sake of simplicity, we consider nine chemical components only, that is SiO_2, TiO_2, Al_2O_3, CaO, MgO, MnO, K_2O, Na_2O, and P_2O_5 (see Table 4.3).

In total, our dataset consisted of 8000 samples; 7000 of them were used for training, while the test was carried out on the remaining 1000 samples. Table 4.3 reports eight rock names given in the "Georock Observatory" database, which mark eight classes. The acronyms stand for rock types of Streckeisen diagrams, such as the one shown in Fig. 4.8: AND stands for Andesite; BAS for Basalt; DAC for Dacite; RYD for Rhyo-Dacite, that is, a rock composition falling in between the two fields Rhyolite and Dacite; RYL stands for Rhyolite; TRA for Trachyte; TAN for Trachy-Andesite (rock with composition in between Andesite and Trachyte). Eventually, TRB stands for Trachy-Basalt. In this application of SVM, we use the aforementioned eight classes as a priori targets; the numbers in the corresponding rows of Table 4.3 form their features.

In our previous description of MLP in Chapter 2, we tackled similar multiclass problems, and set as output a vector the length of which was equal to the number of classes. In this vector, the target value was "1" for the class to which a pattern belonged, and "0" for all the other classes. In Chapter 2, we also introduced SVM as a "binary classifier," as SVM establishes whether a pattern belongs to the class A or to a complementary class B. In the light of this binary scheme, the literature proposes two strategies to tackle multiclass problems with SVM. In the strategy called "One-against-All" (it would be better to say "One-against-All others"), we set an SVM classifier for each class. For instance, we take all examples labeled "AND"—which form the class $\mathbb{A}$ and run the training of SVM so that patterns belonging to this class are truly distinguished from the others that are all members of the complementary class $\mathbb{B}$. Repeating this procedure for the eight classes, we finally obtain eight classifiers. During the test, we have to apply all the eight classifiers to the test patterns: ideally, for each test pattern, all classifiers except one will assign "true" to the complementary class $\mathbb{B}$. In reality, we often have more than one classifier giving "true" to the class $\mathbb{A}$, leaving us with an ambiguous classification. Criticisms toward the "One-against-All" strategy also comes from the typical imbalance between the two classes $\mathbb{A}$ ("One") and $\mathbb{B}$ ("All others"). The complementary set $\mathbb{B}$ is almost always larger than A. Having many classes, this imbalance can get indeed substantial, limiting the reliability of the results.

An alternative strategy is the "One-against-One" method. In this case, we take one class, say "AND," and run the training of a classifier able to distinguish its patters from another class, for example, "BAS." In the next step, we set a classifier that distinguishes again "AND" patterns from those of another class, for instance "DAC." Going through all

Table 4.3: Example feature vectors used in rock classification. Numbers report the composition in weight%. Values and corresponding rock names were taken from the "Georock Observatory" available on the site mentioned in the text.

SiO_2	TiO_2	Al_2O_3	CaO	MgO	MnO	K_2O	Na_2O	P_2O_5	Rock name
6.64E+01	4.75E−01	1.44E+01	1.45E+00	3.00E−01	1.1E−01	5.80E+00	3.55E+00	7.00E−02	RYD
4.61E+01	2.01E+00	1.63E+01	1.16E+01	5.98E+00	1.6E−01	1.08E+00	2.58E+00	3.20E−01	BAS
6.84E+01	7.90E−01	1.25E+01	2.13E+00	1.13E+00	1.5E−01	4.37E+00	3.24E+00	1.50E−01	TAN
6.36E+01	6.20E−01	1.49E+01	3.42E+00	2.89E+00	6.0E−02	4.85E+00	2.27E+00	2.10E−01	DAC
6.41E+01	6.10E−01	1.65E+01	5.62E+00	2.34E+00	1.3E−01	1.00E+00	4.03E+00	1.50E−01	DAC
6.88E+01	4.70E−01	1.62E+01	1.29E+00	4.20E−01	7.0E−02	6.41E+00	4.00E+00	1.00E−01	TRA
6.63E+01	8.40E−01	1.54E+01	3.40E−01	2.80E−01	1.1E−01	6.03E+00	7.03E+00	7.00E−02	TRA
6.94E+01	4.80E−01	1.32E+01	3.25E+00	1.32E+00	1.0E−02	2.86E+00	3.73E+00	1.40E−01	TAN
5.29E+01	1.39E+00	1.63E+01	7.93E+00	5.57E+00	1.3E−01	2.72E+00	2.84E+00	4.60E−01	TAN
5.04E+01	2.37E+00	1.84E+01	7.53E+00	2.89E+00	2.1E−01	2.45E+00	4.98E+00	8.10E−01	TAN
6.57E+01	6.30E−01	1.61E+01	4.15E+00	1.73E+00	7.0E−02	2.75E+00	4.32E+00	1.70E−01	DAC
4.96E+01	9.50E−01	1.58E+01	6.96E+00	5.31E+00	1.9E−01	2.44E+00	3.07E+00	3.20E−01	TAN
4.62E+01	7.90E−01	1.35E+01	1.36E+01	9.35E+00	1.7E−01	1.40E−01	4.00E−02	7.00E−02	BAS
7.48E+01	5.50E−01	1.29E+01	2.37E+00	6.70E−01	1.0E−01	1.57E+00	4.31E+00	4.00E−02	RYL
5.86E+01	8.50E−01	1.54E+01	5.02E+00	1.99E+00	1.5E−01	2.05E+00	2.30E+00	3.00E−01	AND
6.56E+01	1.08E+00	1.37E+01	2.70E−01	7.50E−01	3.4E−01	4.99E+00	6.57E+00	7.00E−02	TRA
7.46E+01	1.96E−01	1.35E+01	8.90E−01	1.80E−01	3.6E−02	5.19E+00	3.69E+00	3.13E−02	RYL
4.97E+01	1.60E+00	1.18E+01	8.40E+00	1.00E+01	1.2E−01	3.50E+00	3.30E+00	1.40E+00	TRB

possible configurations, we end up with $m(m-1)/2$ classifiers, where m is the number of classes. Even though the high number of classifiers increases the computational burden, it may be preferred as the imbalance problem between classes A and B is certainly less important than in the "One-against-All" method. Ambiguities in the results can be resolved by assigning the class-membership using a voting scheme. For example, we may assign the membership after counting how many times a "true" was encountered for the classes, and selecting the one with the highest number of hits. However, even in this case, we cannot rule out the occurrence of ambiguities.

Standard routines, such as the ones in the MATLAB toolboxes or in "LIBSVM"-libraries (a GUI version of the SVM library—see Chapter 7.3.2—comes along with this book), solve the problem of ambiguities exploiting the shape of the decision function calculated by the SVM, that is, the scores. In this context, recall Section 2.4 and Fig. 2.14, where we generated a map of scores applying SVM to our earthquake-nuclear test problem. The isoline plot in Fig. 2.14 was obtained by applying the trained SVM to a dataset forming a 2D mesh of m_b-M_S values ranging from 0 to 9 on both axes, and calculating the score for each m_b-M_S couple on the mesh. In our rock classification problem, we shall get a set of scores for each classifier. In case of ambiguities, we may base our decision simply on the absolute value of the scores, that is, adopting the class for which the highest score was achieved.

In Table 4.4, we report the scores obtained for 1000 samples in the test set by the eight classifiers. Class #1 corresponds to "AND" against all others, class#2 corresponds to "BAS" against all others, etc. As we did with MLP, we can evaluate the performance of SVM by considering the confusion matrix, which compares results and target classes (see Table 4.5).

Mismatches are typically concentrated along the borders of the table (see Table 4.5). For instance, in the second row of Table 4.5, there are 25 TRB (Trachy-Basalt) classified as class #2 corresponding to BAS (Basalt). We also notice 15 TRA (Trachyte) classified as class #6 (TAN, Trachy-Andesit). Summing up over the diagonal elements of the confusion matrix, we get an overall performance of 72%. This value can increase to 78% considering only the main fields of the Streckeisen diagram in Fig. 4.8, that is, summing up "RYL + RYD" (Rhyolites in Fig. 4.8) and "TAN + TRA + TRB" (the Trachyte group in Fig. 4.8), which correspond to the boxes highlighted in green in the confusion matrix (Table 4.5).

Beside using directly the scores, it has been proposed to translate them into some probability of a pattern to belong to class $\mathbb{A}A$ (e.g., the "LIBSVM" by Hsu et al., 2016; http://www.csie.ntu.edu.tw/~cjlin or Curilem et al., 2014). In the "LIBSVM" libraries, scores are supposed to follow a La Place distribution given by the formula

$$p(x) = 1/(2b)e^{-(x-\mu)/b} \tag{4.4a}$$

Table 4.4: "One-against-All" SVM using an "RBF" kernel.

ID	Class #1	Class #2	Class #3	Class #4	Class #5	Class #6	Class #7	Class #8
1	-1.026206	-1.041498	-1.017813	-0.656297	**0.635242**	-1.100609	-1.028002	-1.028422
2	-0.882416	-0.819696	-0.920802	-0.989131	-0.920967	-0.858945	**0.260838**	-0.875199
3	-1.009073	-0.969436	-0.618396	-0.676063	-1.062127	**0.171056**	-0.962861	-1.081983
4	**1.193229**	-1.079358	-1.075857	-1.040684	-1.013825	-1.313198	-1.075944	-1.023170
5	-0.989012	-0.994284	-0.989902	-1.002267	**0.911086**	-0.987550	-0.990446	-0.988781
6	-1.011119	-1.017025	-1.007411	-1.006480	**0.894498**	-1.013574	-1.014625	-1.011816
7	-1.215311	-1.022992	**0.857179**	-0.831355	-1.072003	-1.007886	-1.178329	-1.016602
8	-0.361467	-0.875871	-1.027402	-1.008877	-1.021268	**0.169757**	-0.957333	-0.989916
9	-0.752209	**-0.620382**	-0.836054	-0.932597	-0.778129	-0.678760	-0.698196	-0.737281
10	**0.421636**	-0.838195	-0.990682	-0.995358	-0.974988	-0.808162	-0.961391	-0.971999
11	**-0.567711**	-0.719801	-0.673550	-0.854544	-0.838898	-0.764921	-0.781981	-0.807309
12	-0.851733	-0.735595	-0.905792	-0.961258	-0.872379	-0.541764	-0.828286	**-0.277139**
13	**-0.378405**	-0.839352	-0.930913	-0.971455	-0.906271	-0.672945	-0.542975	-0.904009
14	**0.978692**	-1.072714	-1.013912	-1.005605	-1.018832	-1.240773	-1.025870	-0.957699
15	-0.985000	-0.974639	-0.981705	-1.022176	-0.983381	-0.996069	**0.892069**	-0.982918
16	-0.757292	**-0.597234**	-0.839427	-0.933986	-0.782677	-0.675464	-0.705980	-0.742681
17	-1.012813	-1.032598	-1.009627	-1.193033	**1.089109**	-1.015531	-1.016664	-1.014071
18	-1.094160	**1.554906**	-1.042149	-1.017559	-1.057200	-1.078973	-1.076955	-1.213966
19	**0.570700**	-0.777291	-0.995448	-0.998178	-0.993680	-1.044610	-0.991730	-1.267975
20	-1.417787	-0.971778	**1.352412**	-1.169737	-1.034851	-1.018345	-1.002835	-0.977666
21	-0.755587	**-0.623361**	-0.838352	-0.933544	-0.781224	-0.686447	-0.704012	-0.710577
..	..	..	..	..	..	..	..	..

Scores of the eight classifiers for 21 patterns of the test set. On the whole, the classifiers give a unique preference for all the 21 patterns. For instance, the first pattern is assigned class #5 (score in bold). For the patterns highlighted in red, we have only negative scores. Accepting the "least bad" score, we assign class #2 to pattern 21. For this table, we use the class coding: 1 = AND, 2 = BAS, 3 = DAC, 4 = RYD, 5 = RYL, 6 = TAN, 7 = TRA, and 8 = TRB. The results were obtained using the MATLAB routine "fitcsvm" and "predict." In alternative, the user may use the GUI version of the SVM library in the accompanying material. For more details, see Chapter 7.3.2.

Table 4.5: Confusion matrix of the rock classification using SVM trained with the "One-against-All" method.

	AND	BAS	DAC	RYD	RYL	TAN	TRA	TRB
#1	111	6	7	1	0	13	2	0
#2	13	107	13	4	11	14	11	25
#3	9	1	137	16	8	2	3	0
#4	1	0	6	16	7	0	1	0
#5	0	0	6	10	91	1	1	0
#6	12	8	3	1	1	68	15	4
#7	0	0	4	5	3	5	121	1
#8	1	10	0	0	0	11	1	72

We use the class coding: 1 = AND, 2 = BAS, 3 = DAC, 4 = RYD, 5 = RYL, 6 = TAN, 7 = TRA, and 8 = TRB. Summing up "RYL + RYD" (Rhyolites in Fig. 4.8) and "TAN + TRA + TRB" (the Trachyte group in Fig. 4.8), the performance of SVM improves (see text for explanations). This combination would allow to refer the rock composition to the five main fields (here highlighted in green) of the Streckeisen diagram (upper triangle in Fig. 4.8).

and its cumulative form

$$D(x) = 1/2\left[1 + sgn(x-\mu)\left(1 - e^{\frac{-|x-\mu|}{b}}\right)\right] \tag{4.4b}$$

(Fig. 4.9; Abramowitz and Stegun, 1972; Weisstein, 2018). In Eqs. (4.4a) and (4.4b) μ is the mean, and b is related to the variance: $\sigma^2 = 2b^2$.

We may use the scores obtained for the training set to obtain the necessary parameters—the mean μ and the variance σ^2 of the scores—for the design of the La Place distribution. In a multiclass SVM, the use of the La Place probabilities instead of the mere

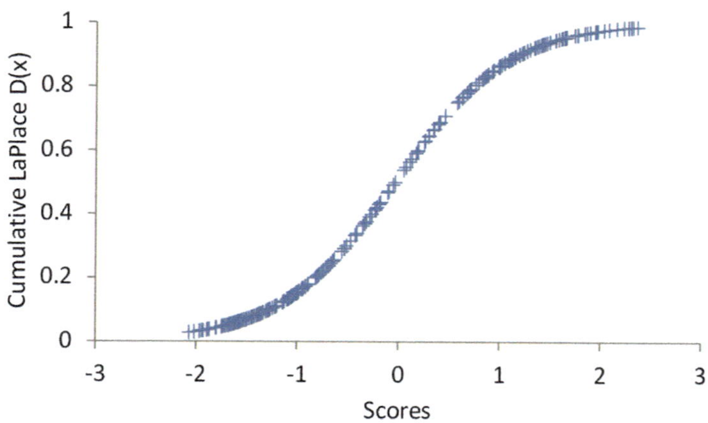

Figure 4.9
Scores and probabilities (Laplace distribution). With a score of 0, we assumed a 50% probability that a pattern belongs to class $\mathbb{A}$. For positive scores, the probability increases. Values were obtained from a test example in "LIBSVM."

scores may be preferred. In fact, having more than one classifier, the meaning of the scores can vary from classifier to classifier.

4.5 Inversion with MLP

4.5.1 Identification of parameters governing seismic waveforms

In Section 2.4 we introduced the MLP that allows to resolve the famous XOR classification problem. Furthermore, exploiting Cybenko's theorem (see Eq. 2.12), we were able to solve classification problems of arbitrary complexity. This was achieved by constructing a function that maps an input vector $\boldsymbol{X}$ to some output vector $\boldsymbol{Y}$. In classification problems the desired output vector $\boldsymbol{Y}$ has a form like (0, 0,..1, 0, 0..), where "1" is in the position of the class to which our object is supposed to belong to. We have seen that the calculated output, however, is a float. In other words, MLP solves the classification task as a problem of nonlinear regression rather than separating in TRUE and FALSE, as SVM does.[3] Being the MLP a type of nonlinear regression scheme, we are able to apply it to target vectors, whose components are not necessarily "1"s or "0"s, but some floating point value. In geophysics, those target vectors may represent model parameters we want to identify from data. Röth and Tarantola (1994) were among the first to apply MLPs in the nonlinear inversion of seismic waveforms for the identification of a one-dimensional velocity structure. Along this line, Langer et al. (1996) investigated the application of ANN (*A*rtificial *N*eural *N*etworks) to the inversion of waveform governing parameters of earthquake seismograms, essentially the geotechnical parameters of a horizontally layered structure and the parameters of the seismic source (see Fig. 4.10). In all cases, the inversion is based on the forward modeling of the training set of $\boldsymbol{X}/\boldsymbol{Y}$ pairs, varying randomly (within given limits) the parameters y_j of the model, and calculating the vector of observations $\boldsymbol{X}$. With the increasing computing capacities and the development of efficient and reliable simulation methods, the forward modeling of many $\boldsymbol{X}/\boldsymbol{Y}$ pairs becomes affordable. On the other hand, little or nothing is known about the inverse relation, that is, how to recover the model vector $\boldsymbol{Y}$ from the observations given by $\boldsymbol{X}$. The lack of knowledge about the inverse relation justifies the treatment of inversion problems with a *black-box* approach, such as MLP. Indeed, no a priori assumptions are made on the mapping function and, at first glance, the obtained results have to be accepted as they are. The application of MLP to unknown data presumes that they can be described by the models used during the forward modeling. In other words, it is assumed that the unknown data belong to the same parent population as the ones of the training set.

[3] We shall see later that SVM can be also exploited in regression. However, that needs a restatement of the optimization problem. MLP does not require this reformulation.

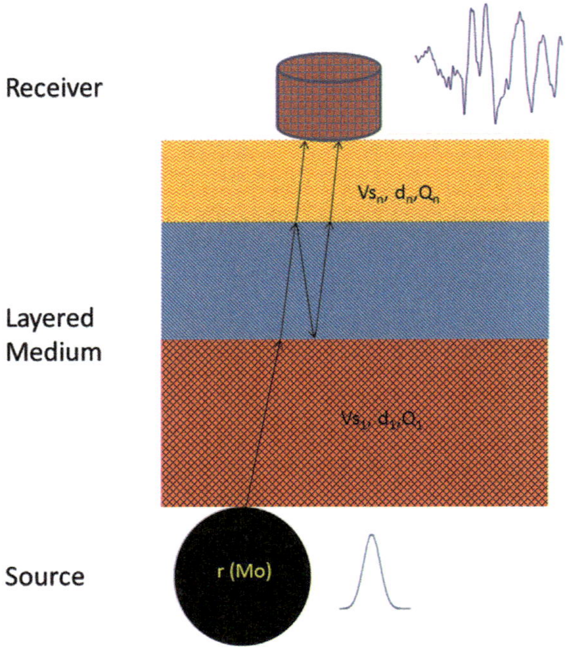

Figure 4.10

Inversion of seismic waveform governing parameters (see Langer et al., 1996). Along its path to the receiver, the signal radiated from the source undergoes changes due to reflection and refraction as well as viscoelastic attenuation. The spectral characteristics of the seismogram recorded at the surface is controlled by the duration of the signal radiated by the source, that is, the source radius r, and reflection and refraction, which essentially depend on the seismic velocities Vs_i. Furthermore, the shape of the seismogram is affected by the thickness d_i of the layer and the quality factor Q_i that controls the viscoelastic attenuation of the signal within a layer. The parameter Mo (seismic moment) governs the amplitude of the signal but has no further impact on its shape. Mo is left out in the inversion process.

4.5.2 Integrated inversion of geophysical data

Earth's dynamics mirrors in a wide variety of geophysical observations, being them seismic signals, but also static ground deformation, changes in the electromagnetic field or gravity. Volcanoes are places where processes of tectonic deformation, rock fracturing, and fluid movements are particularly intense; therefore, they are environments able to provide a huge quantity of observations in relatively short times. In addition we consider the example of Mt Etna, which is the Europe's largest and most active volcano.

Static ground deformation may reflect a possible intrusive process when magma starts to fill cracks, leading to tensile deformation of the ground. Ground deformation can be expressed either as shift vectors measured in a network of sensors (nowadays GPS sensors) or as variations of the distance between two sensors, in terms of strain. The direct use of shift vectors requires the possibility of referring them to some fix baseline, which was not an easy task in a volcanic environment before the advent of high-sensitive GPS sensors (see Nunnari et al., 2001).

Changes in the gravity field occur when the intruding material that forms a dike has a different (typically higher) density than the surrounding rock. During important eruptions at Mt Etna (1989, 1991–93, 2001) there were changes in the gravity field of up to several hundreds of μgal, compared to changes of some tens of μgal, caused by the variations of the ground water level (e.g., Budetta and Carbone, 1995).

During phases of volcanic unrest, the geomagnetic field may be affected by various effects, such as re- and demagnetization due to thermal effects, piezomagnetic phenomena in consequence of ground deformation, and electrokinetic effect due to the movement of electrolyte carrying fluids (see Del Negro et al., 1997). The electrokinetic effect was discussed in the framework of earthquake prediction (see Fitterman, 1979, 1981), but can be expected to have even higher intensities in volcanically active areas.

In the following, we outline the application proposed by Nunnari et al. (2001), who used MLP for the inversion of dike parameters using simultaneously ground deformation, gravity and magnetic field data. This application is particularly interesting for several reasons. It highlights the importance of multidisciplinary data analysis and analyzes the role of noise in training and test. Similar to Langer et al. (1996), the authors compare the results obtained with MLP inversion to those of nonlinear optimization using simulated annealing. Among the electronic material coming along with this book, we provide a software, the "DMGA" package for the generation of the synthetic datasets. The output of this program allows the user a straightforward application of the "bpnE" program, which was already used in the classification example of Strombolian explosions. Using the two programs, the reader may try to apply the software in a similar way as Nunnari et al. (2001).

The model of the volcanogenic source considered here is represented by an opening crack filled with magma. The density of the intruding magma is supposed to differ from that of the surrounding material. Depending on various parameters (see Fig. 4.11, Table 4.6), the dike is assumed to produce a static ground deformation, changes in the gravity field, and an electrokinetic-magnetic effect. The measure of changes in the gravity field is attained after Ehrismann et al. (1966); the static ground deformation for the tensile crack is calculated using the Okada model (1985); and the electrokinetic-magnetic field is obtained following the models by Fittermann (1979, 1981) and Murakami (1989).

In terms of the inversion scheme, these observations form the input vector X. In Fig. 4.12 we show the graphical representation of X as an isoline plot. The output vector Y is given by the searched parameters of the dike, i.e., its length and width, the crack opening, its

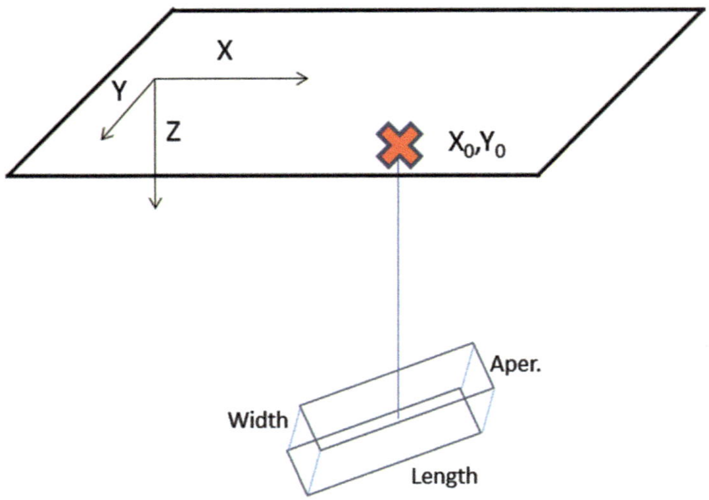

Figure 4.11
Geometry and parameters of the dike.

Table 4.6: Model parameters used for synthetic pattern generation azimuth and dip angles are given in degrees (deg).

Crack length	5 km	±4 km
Crack width	2 km	±2 km
Longitude (UTM)	0 km	±10 km
Latitude (UTM)	0 km	±10 km
Depth	5 km	±4 km
Azimuth	0 deg	±90 deg
Dip	90 deg	±50 deg
Crack opening	2 m	±1 m
Poisson ratio	0.25	±0.0
Density contrast	300 kg/m^3	±100 kg/m^3

coordinates (longitude, latitude, and depth), and its orientation (azimuth, dip angle; see Fig. 4.11). The examples used for the network training are obtained by creating a set of synthetic models in which each parameter has a random fluctuation according to the values given in Table 4.6, and the corresponding expected observations are computed for each model.

To verify the consistency of the inversion, the network has to be tested applying a test set not used during the network training.

In their application, Nunnari et al. (2001) considered 16 stations deployed on a rectangular mesh with 5 km spacing. Ground deformations, originally calculated as absolute three-dimensional shift vectors, were converted to changes of the distances among the stations, which makes 120 values. Moreover, 16 more values represented the changes in the gravity

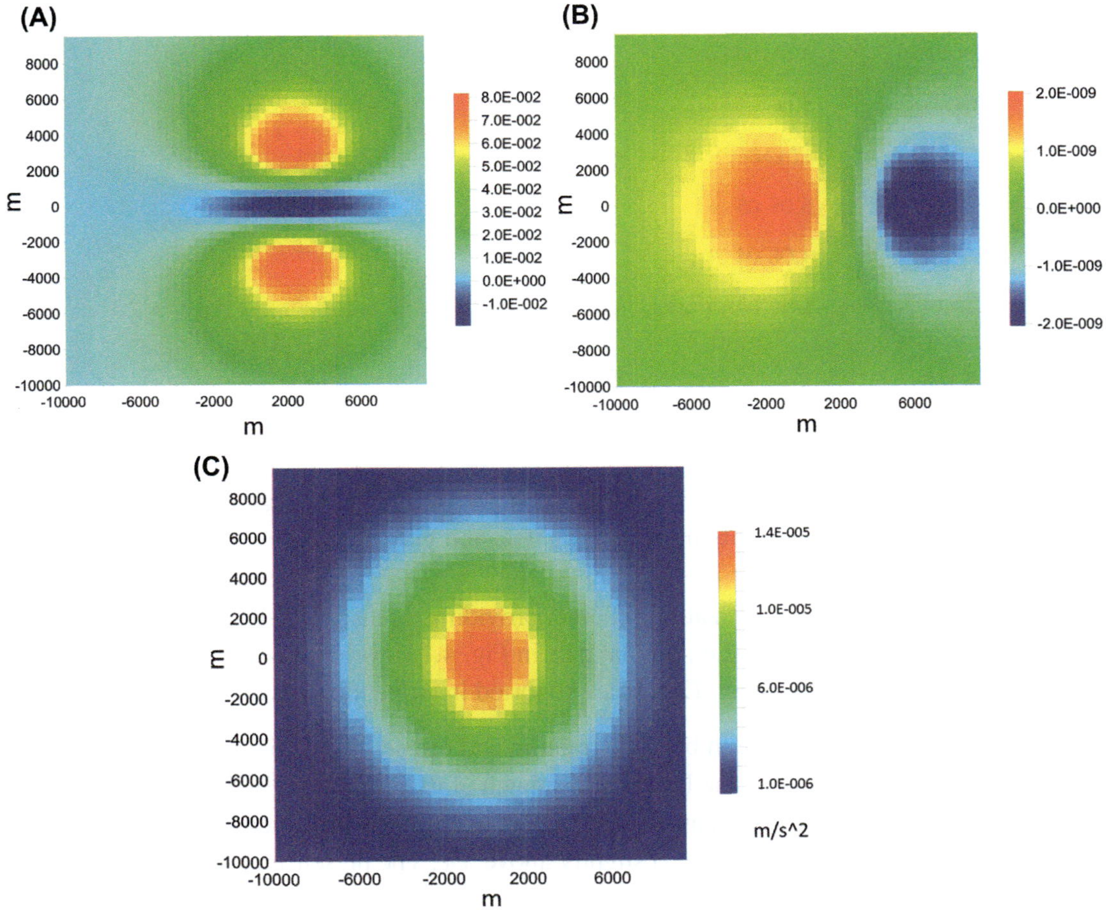

Figure 4.12

Sample fields for (A) vertical ground deformation, (B) magnetic field (vertical component), and (C) gravity changes. All physical dimensions are given in SI units. Axes are longitude and latitude expressed in m. Note that in Nunnari et al. (2001), the three components of the magnetic and ground deformation fields are considered.

field at the stations; eventually, 48 values were obtained for the vertical and horizontal components of the electrokinetic-magnetic effect. The size of the input layer of the neural network was obtained as sum of the considered observations (e.g., $120 + 48 + 16$). The number of neurons in the hidden layer had to be found using a total of 45,000 patterns in the training set. A good accuracy of the parameter identification was achieved with 130 nodes in the hidden layer.

Normalization turned out as a critical issue: During the generation of the 45,000 dataset vectors, the absolute values of ground deformation ranged from $\sim 4 \times 10^{-15}$ to 7.3×10^{-3}. In Fig. 4.13 there is $\sim 90\%$ of the values in the range $\sim \pm 10^{-5}$. It is obvious that this type

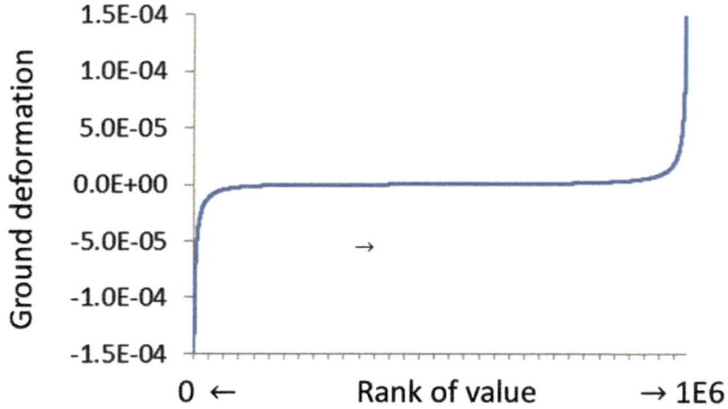

Figure 4.13

Cumulative plot of the ranked ground deformation values (~1 million). For graphical reasons, we omitted the extremes. Nonetheless, we notice a distribution with a large part of values having very low absolute values. This peculiarity requires specific precautions during the normalization. Distributions with a large flat part in the middle of the graph (i.e., where small values are found) are also obtained for magnetic and gravity values.

of distribution required a normalization scheme, providing a reasonable resolution in the range where most of the values (i.e., the flat part in Fig. 4.13) were found. Some of the options discussed in Nunnari et al. (2001) are reported in Appendix 4.1.

Neglecting the presence of noise in both the training and test set, the resolution for the length and width of the dike is ~10%; the location mismatch of its center (longitude, latitude, and depth) is ~5%; the azimuth (orientation with respect to North) is identified with an error <5%; the error for the dip angle is slightly above 5%. The presence of noise only in the test set considerably affects the accuracy of the results. With a noise level of 20% in the test set, the relative mismatch almost doubles for all parameters. On the other hand, the effect of noise is fairly weak if noise is present in both the training and test set. In other words, the inversion scheme is robust under the condition that training and test set belong to the same parent population, that is, they are generated by the same physical process (here a crack undergoing tensile deformation) and are subject to the same random fluctuations.

One of the main aspects of the aforementioned application is the effect of combining various types of geophysical data in an integrated inversion scheme. Using only ground deformation data, the mismatch obtained with the test set is considerably higher than the mismatch reported in Table 4.7. It almost doubles for the parameters concerning the location and orientation of the dike, and the mismatch is ~50% higher for the other parameters. We invite the reader to conduct further experiments using the software "DMGA."

An additional important conclusion of the paper was drawn from the comparison between the inversion with MLP and that with nonlinear optimization carried out with simulated annealing (SA hereafter). For the SA application, first a reference data vector for a set of all

Table 4.7: Results of the parameter inversion with ground deformation, magnetic and gravity data. The table reports the relative mismatch in % with respect of the total range of parameter variation given in Table 4.6.

Crack length	12
Crack width	10
Longitude (UTM)	4.5
Latitude (UTM)	4
Depth	6
Azimuth	2.5
Dip	8
Crack opening	15
Poisson ratio	—
Density contrast	21.5

model parameters was created. Then, SA was applied starting from some initial guess of the model. In an iterative scheme (see Kirkpatrick et al., 1983; Sen and Stoffa, 1991), SA adjusted the model parameters until there was an acceptable fit between the original dataset and the data generated by the inverted model parameters. A solution was accepted if the correlation between the reference data and those obtained with the new model parameters was 0.999. SA was run 85 times, every time using a different initial model. The model parameters found during the optimization differed on average only to a few percent from the original ones. SA was carried out under various conditions, assuming different levels of noise, and using only ground deformation data or ground deformation plus magnetic data. Similar to the inversion with MLP, the mismatch of the inverted parameters with respect to the target increased in a considerable manner when only ground deformation data were used. Besides, the inversion with SA was sensitive to the presence of noise. Similar to the inversion with MLP, the parameters for the location and orientation (longitude, latitude, depth, azimuth, and dip angle) were the ones for which the mismatch was relatively low.

The findings achieved by Nunnari et al. (2001) resemble those reported in the previously cited applications by Röth and Tarantola (1994) and Langer et al. (1996), where model parameters were identified with different accuracy. In particular, Langer et al. (1996) investigated the reasons of these differences by carrying out a number of inversions using the SA global optimization for the identification of the underlying model parameters. Similar to the strategy discussed earlier, synthetic seismograms were calculated varying the model parameters as long as the mismatch between synthetic calculations and some reference data was beyond a given threshold. Langer et al. (1996) run SA 31 times on the same reference seismogram, and accepted a solution for the model parameters only when the cross-correlation between reference and calculated seismogram was 0.999 or higher. The authors found a considerable scatter for a few parameters. Interestingly, the scarcely identified model parameters during the SA corresponded to the ones where the MLP inversion reported major errors. In conclusion, the poor results obtained for those parameters could not be attributed to a failure of the MLP application, but are probably intrinsic to the inversion problem.

4.6 MLP in regression and interpolation

In our previous examples, we established a function able to map a vector of observations to get a vector of parameters associated with a physical model. During the test phase, we have applied this mapping function to predict model parameters for a set of observations not considered before, using the mapping function instead of an inverse physical model. The mapping function can be understood indeed as some kind of regression, where we predict a value y on the base of a number of observations x.

Nonlinear regression is frequently implemented in empirical prediction analysis. In a recent paper, Spichak and Goidina (2016) applied MLP for establishing cross-correlations between seismic velocities and electric resistivity of rock. Seismic and electric resistivity soundings of rock are among the most important noninvasive methods revealing the properties of the ground. Electric resistivity is applied in studies of geological mapping, hydrogeology, void mapping (detection of cavities), environmental studies, and exploration for minerals and oil resources. Seismic measurements provide wave velocity and attenuation properties that can be translated into stiffness and quality factors. Most of the relevant parameters for oil prospection mirror in the electric conductivity and wave velocities. In particular, the presence of voids in a rock and its microstructure are a first-class issue in this context.

Mixture theories have been used to obtain conductivity and velocities of rock in the presence of voids and pores (see, e.g., Carcione et al., 2007). On the base of mixture theories, a number of cross-property relations between electric conductivity and seismic velocity data have been established. Those relations are exploited when one type of information is missing, as both seismic velocities and electric resistivity are highly dependent on the porosity of the material. In Archie's classical model for sands and sandstone (Archie, 1942), we neglect the electric conductivity of the matrix with respect to the conductivity (the inverse of the resistivity) of the pore fluid, that is,

$$\sigma_{Tot} \approx \sigma_f \phi^m \tag{4.5}$$

where σ_{Tot} is the total conductivity, σ_f the conductivity of the fluid, ϕ^{-m} the so-called "formation factor" with m being the "cementation factor." These two factors represent internal structural characteristics of rock and pore volumes. Often, sandstone contains clay minerals, which are good conductors. For such a material, Bussian (1983) proposed a relation

$$\sigma = \left(\frac{1 - \frac{\sigma_c}{\sigma_f}}{1 - \frac{\sigma_c}{\sigma}} \right)^m \sigma_f \phi^m \tag{4.6}$$

where σ_c is the conductivity of clay.

On the other hand, we may consider for example the relation for seismic velocities in consolidated sand (Raymer et al., 1980)

$$v_B = (1 - \phi)^2 v_S + \phi v_f \tag{4.7}$$

where v_B is the bulk P-wave velocity of the porous sandstone, v_S the P-wave velocity of the matrix, and v_f the seismic velocity of the fluid.

Combining Archie's (1942) law and the relation by Raymer et al. (1980), we obtain the cross-property relation

$$v_B = \left[1 - (\sigma/\sigma_f)^{1/m}\right]^2 v_S + (\sigma/\sigma_f)^{1/m} v_f \tag{4.8}$$

For more details and other cross-property relations, we address the reader to Carcione et al. (2007). Although being rather simple, a number of criticalities can be recognized immediately in the cross-property relation. First, Archie's law holds for pure sandstone, whereas the modification proposed by Bussian reveals the importance of clay for the conductivity. Even though limiting ourselves to these two types of sedimentary rock, we face a considerable dependence with respect to the lithological composition of the material—here just the presence of clay minerals. Besides, we have to account for the formation and cementation factors, which are supposed to vary, for instance, as a function of depth. Indeed, the geometrical characteristics of the pores are expected to undergo changes, such as flattening and changes in their connectivity, due to compression. As Spichak and Goidina (2016) point out, porosity is not the only factor controlling seismic velocities and electric resistivity (or conductivity). A way out from the shortcomings of model-based empirical relations is a full black-box strategy, where we use input/output pairs and learn their relation from given examples. In doing so, we follow a classical scheme of MLP application similar to the one discussed earlier about inversion problems.

In their application, Spichak and Goidina applied MLP to a dataset measured in Central Siberia, close to the city of Abakan. The measurements were carried out along a profile within the structural formation of early orogenic molassa Caledonides, which frame siliceous-carbonate massifs of baikalides. Both seismic and electric soundings reached a depth of 10 km, whereas the profile covered a length of ~60 km. Both P- and S-waves were considered together with data inferred from magnetotelluric soundings. The authors started with an estimation of seismic velocities from logarithmic values of electric resistivity. To examine the role of the size of the dataset, they varied the ratio of the number of samples in the training and test set, considering the cases $\kappa = 4:1$, 1:1, and 1:4. In all cases the best results were obtained with a training set four times larger than the test set. Comparing the configurations $\kappa = 1:1$ and 1:4, the authors reported a slightly higher mismatch for the ratio 1:4 (see Table 4.8)

Besides, we note that the mismatch for shear-wave velocities is somewhat higher than that for P-waves. We may speculate about the reasons for this evidence. From observations reported by Hamilton (1978, see also 1979), Schön (1983) concluded that P-wave velocities in dry sediments essentially depend on the matrix, whereas the influence of pore-fluid effects becomes important in case of wet sediments. On the other hand, shear wave velocities strongly depend on the characteristics of the matrix both for dry and wet rock. As electric resistivity is strongly controlled by the pore-fluid properties, we can easily imagine that the link between seismic wave velocity and electric resistivity is weaker for shear waves than for P-waves.

In their study, Spichak and Goidina considered also the inverse case, that is, the estimate of electric resistivity from seismic waves.

On the whole, the mismatch numbers reported in Table 4.9 are higher than in Table 4.8. We should recall, however, that the mismatch in Table 4.9 regards logarithmic values, whereas the ones in Table 4.8 are linear values. Besides, the comparison of the two tables highlights that the assessment in inversion problems of seismic velocities from electric resistivity is more robust than the opposite case.

An interesting aspect of Table 4.8 is the difference between the data in the input layers with and without the location, as including the geographical coordinates of the sites where the data were taken improves the estimation of seismic velocities. From a mere physical viewpoint, this appears strange, but it can be understood as a "tendency of conservation" frequently noticed in geology. This means that data measured at some point tend to be similar to those obtained in the neighborhood. Comparing the mismatches reported in Tables 4.8 and 4.9 we infer that the tendency of conservation is strong using resistivity data in the input layer of the

Table 4.8: Mismatch of estimated seismic velocities using resistivity/resistivity plus location data in the input layer of the MLP. Results are averaged over five cross-validation tests.

K	Output P-wave velocity (%)		Output S-wave velocity (%)	
	Only resistivity	Resistivity + location	Only resistivity	Resistivity + location
4:1	6.0 ± 0.6	1.4 ± 0.3	7.7 ± 1.0	4.0 ± 0.8
1:1	7.9 ± 0.3	2.5 ± 0.6	12.7 ± 1.3	5.0 ± 0.6
1:4	8.4 ± 1.1	3.8 ± 1.5	13.0 ± 1.8	6.0 ± 1.5

Table 4.9: Mismatch of resistivity from seismic velocities/seismic velocities plus location data in the input layer of the MLP. Results are averaged over five cross-validation tests. The mismatch regards logarithmic values of resistivity expressed in %.

κ	Input P-wave velocity		Input S-wave velocity	
	Only seismic velocity	Velocity + location	Only seismic velocity	Velocity + location
4:1	15.5 ± 2.9	14.8 ± 0.8	14.7 ± 0.7	17.5 ± 5.6
1:1	19.6 ± 2.2	14.7 ± 0.2	16.7 ± 0.9	20.0 ± 4.6
1:4	24.4 ± 5.0	16.1 ± 0.6	15.3 ± 0.6	25.6 ± 10.5

MLP, whereas combining geographical information and seismic velocities in the input layer does not improve the accuracy of estimations in Table 4.9.

MLP have been also applied in regression in the context of seismic hazard analysis. In probabilistic seismic hazard assessment we apply so-called "Ground Motion Prediction Equations" (GMPE), in which a parameter of ground shaking, such as the peak ground acceleration, is inferred from earthquake magnitude, focal depth, and source-to-receiver distance. In more sophisticated relations, further parameters are added, such as site correction factors or the focal mechanism of the seismic source. A simple empirical GMPE is given in Eq. (4.9)

$$log(y) = w_0 + w_1 M + w_2 log\sqrt{R^2 + h^2} + w_{3i} S_i \tag{4.9}$$

where y is the ground shaking parameter, M the magnitude, R the epicenter distance, h the focal depth[4] and S_i a site specific parameter, depending from an a priori assigned soil class i. Here, we consider measured data concerning earthquakes for which we know the depth h, the distance to our recording sites R, the magnitude M, and a site specific parameter S_i (see Ambraseys et al., 1996; Sabetta and Pugliese, 1987), along with instrumentally recorded ground shaking parameters (y in Eq. 4.9). It is worth mentioning that the GMPE is not fully a black-box assessment, as it encompasses a physical background. For instance, one assumes a linear relation of log(y) to magnitude and hypocenter distance.

In the empirical prediction of ground motion during an earthquake, traditional models such as the one given in Eq. (4.9) come with severe limits. For instance, theoretical modeling demonstrates that a simple linear relation between log(y) and magnitude is highly questionable (see, e.g., Langer et al., 2016); the same also holds for the distance dependence of ground motion parameters. In the light of these limits, various authors applied MLP for the prediction of peak ground motion during earthquakes (see Derras et al., 2012, and references therein). Similar to the GMPE in Eq. (4.9), they considered magnitude, focal depth, and distance as relevant parameters. In the seismological community, site characteristics are also

[4] The parameter h is often kept fixed, that is, constant for all events of the dataset. Its value can be estimated during the nonlinear regression together with the other coefficients w_j.

claimed to have a relevant impact on ground motion. In particular, the presence of weak material at the surface, having low shear-wave velocities, may lead to considerable amplifications of ground motion during an earthquake. To account for these effects, site-specific parameters, such as the site-resonance frequency and the shear-wave velocity of the uppermost 30 m-thick layer (in seismology known as "V_{s30}"), are also acquired. For this reason, in their MLP application, Derass et al. (2012) considered five nodes (for magnitude, depth, distance, site-resonance frequency, and V_{s30}) in the input layer, and one node in the output layer (for the ground motion parameter of interest, here the peak ground acceleration).

The number of hidden nodes was obtained on the base of a joint analysis of the error and the Akaike criterion (Akaike, 1973). The Akaike criterion *AIC* is given by

$$AIC = L\ln(MSE) + 2m \tag{4.10}$$

where L is the number of patterns, and m the degrees of freedom given by the number of weights of the MLP. An optimum score was achieved with 20 nodes in the hidden layer. From the analysis of the results, the authors also found that the residuals had a normal distribution and, considering skewness and kurtosis (see Appendix Chapter 1) of this distribution, there was no obvious trend with respect to the input parameters. Regarding the misfit, the MLP-based predictions of ground motion parameters outperformed all considered conventional models. The relevance of the parameters with respect to the goodness of the fit was estimated in two ways. First, the authors compared the accuracy achieved using all five parameters in the input layers; then they left out some parameters and recalculated the misfit. In alternative, they looked directly at the weights of the connections between input and hidden layer, considering the term

$$W_i = \frac{\sum_{j=1}^{N_h} w_{ij}^h \vee}{\sum_{i=1}^{N} \sum_{j=1}^{N_h} w_{ij}^h \vee} \tag{4.11}$$

where W_i is the importance of the i-th input parameter, w_{ij}^h the node connections between the i-th node in the input layer and the j-th node of the hidden layer, N is the number of input parameters or nodes in the input layer, and N_h the number of nodes in the hidden layer. The results highlighted the major importance of the epicenter distance (29.5%) and the magnitude (26.1%), whereas V_{s30} (9.1%) was found of minor importance for the final result.[5] Note that such an analysis requires that input parameters are properly normalized, and that they are not correlated. The latter condition is, however, not warranted. For instance, we may suspect that large magnitude earthquakes are more likely to have records at long distances than smaller ones.

[5] The limited importance of this parameter is surprising at first glance, as a low-velocity material is supposed to produce considerable amplification of ground motion. However, these effects are known to be distance dependent, with a tendency to be stronger near the seismic source and having a rapid decay at more distant sites (see Scarfi et al., 2016).

4.7 Regression with SVM

4.7.1 Background

Although achieving a considerable improved fit with respect to conventional regression methods, the MLP is known to come with some drawbacks, such as problems of overfitting or the risk of being trapped in local minima. In Chapter 2, we introduced the SVM as an alternative to MLP in the discrimination problem for classes that cannot be separated by a linear function or a hyperplane. As the discriminating elements in SVM are found by solving a quadratic programming problem, their training is not affected by the issue of local minima.

As we have seen in Chapter 2, classification with SVM is based on the criterion of separating two classes with the highest possible margin. This margin is defined by two hyperplanes. Ideally, no pattern should be found inside this margin (see Fig. 2.13). Using the SVM concept in regression, we have to restate the optimization problem: Instead of searching two hyperplanes separated by a margin where no patterns are found, the hyperplanes should be the smallest possible separating margin where possibly all patterns are present (see Fig. 4.14).

The basic idea can be outlined starting from a linear problem (Smola and Schölkopf, 2004). We search a function f

$$f = \boldsymbol{w}^T \boldsymbol{x} + \boldsymbol{b} \tag{4.12}$$

from a training dataset $\{(x_1, y_1 \ldots \ldots (x_N, y_N)\}$. The principal condition f should obey is the "maximum flatness," which corresponds to minimize $||w||^2 = w^T w$. This can be rewritten as the following optimization problem

$$\text{minimize} \left| \, ||w||^2 \right.$$

$$\text{subject to} \begin{cases} y_i - \boldsymbol{w}^T \boldsymbol{x}_i - b \leq \varepsilon \\ \boldsymbol{w}^T \boldsymbol{x}_i + b - y_i \leq \varepsilon \end{cases}$$

We may prefer to allow some errors, for instance when we have most of our data concentrated in a narrow space and a few ones scattered in a larger space. We are then interested to define a narrow dominated range or "tube" (in case of dimensions larger than 2) at the cost that some patterns remain outside. Formally, we rewrite our optimization problem by introducing the slack variables ξ_i, ξ_i^*

$$\text{minimize} \left| \, ||w||^2 + C \sum_{i=1}^{N} (\xi_i + \xi_i^*) \right.$$

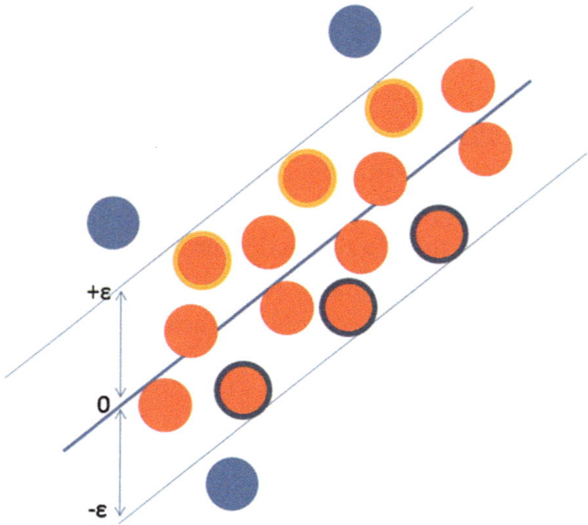

Figure 4.14
The concept of the SVM regression. Red dots represent samples falling within the range $\pm\ \varepsilon$, that is, inside the margin. The blue dots are outliers. Appropriately choosing C, the influence of those samples on the optimization can be limited keeping the width of $\pm$ ε small. Support vectors are represented by circled red dots.

$$\text{subject to} \begin{cases} y_i - \boldsymbol{w}^T\boldsymbol{x}_i - b \leq \varepsilon + \xi_i \\ \boldsymbol{w}^T\boldsymbol{x}_i + b - y_i \leq \varepsilon + \xi_i^* \\ \xi_i, \xi_i^* \geq 0 \end{cases}$$

The constant $C \geq 0$ determines the importance we want to give to errors. For instance, if one sets $C = 0$—no errors permitted—then all patterns have to fall into the "tube." Eventually, we deal with so-called "ε-insensitive" loss function (Smola and Schölkopf, 2004), i.e.,

$$\left| \xi \right|_{\varepsilon} = \begin{cases} 0 & \forall\ |\xi| \leq \varepsilon \\ |\xi| - \varepsilon & \forall\ |\xi| > \varepsilon \end{cases} \tag{4.13}$$

In Fig. 4.14, we show the geometrical ideas behind regression with SVM for the linear 2D case: we aim at finding two linear elements, that is, lines or (hyper)planes that embrace the highest number of samples (represented as dots). For the prediction of y, we can use the element placed at the center of the margin. The concept of "flatness" is graphically explained in Fig. 4.15. In minimizing the flatness, we look for the hyperplane having the lowest possible coefficients. In the 2D case, this corresponds to the search of the line having the smallest slope $w = w_{min}$. As shown in Fig. 4.15, there may be a trade-off between minimizing the flatness and the goal of finding a narrow tube, that is, a small ε.

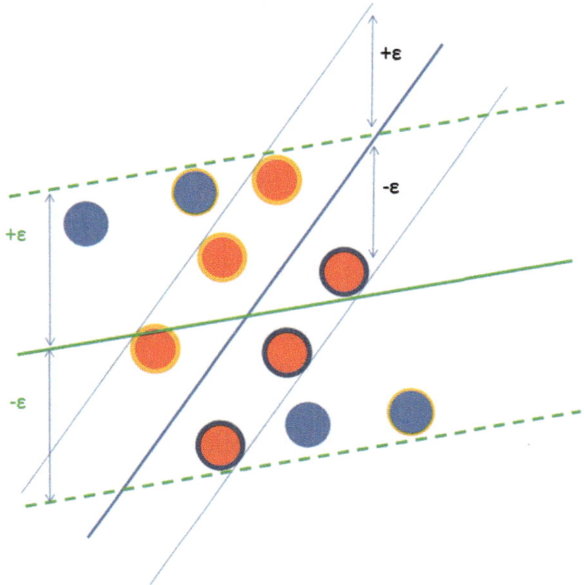

Figure 4.15
Flatness and ε trade-off in regression with SVM. Support vectors are indicated by encircled dots. We can embrace the red dots by a steep flatness but within a narrow tube (blue lines) or a wider tube (green lines) having a better flatness. With such a wider tube we can also embrace some outliers (shown as blue dots). In case we decide to "tolerate" the presence of outliers, this is achieved by an appropriate choice of C.

Allowing larger ε—in other words, a larger tube inside which patterns do not contribute to the error—can help improve the flatness of the regression function f.

In both conventional- and MLP-based regression, we have been identifying the optimum solution considering the difference between target and prediction. As we have seen in the formalism earlier, the optimization in SVM regression accounts only for samples falling outside the margin, whereas samples falling between the two elements do not contribute to the error. Clearly, choosing a narrow width (ε small) comes along with the risk of leaving many samples outside the margin. With the introduction of the slack variables and the factor C, we can adapt the regression to the structure of our problem. Having a rather compact dataset nicely arranged along a regression function, we can keep C small as we are likely to find our samples inside a narrow tube. Suppose now that besides such a dataset we also find a few outliers. Trying to embrace them all by the tube requires that its width has to be large. "An appropriate choice of C" means that we can keep the width of the tube narrow, as the importance of the outliers is downsized by the constant C.

The minimization of the "ε-insensitive" loss function is an optimization problem with constraints, for which we outline more details in Appendix 4.2.

As in classification tasks, we can generalize to the nonlinear problem by introducing the kernel functions mentioned in Chapter 2:

$$\text{Linear}: K(\boldsymbol{x}_i, \boldsymbol{x}_j) = \boldsymbol{x}_i^T \boldsymbol{x}_j$$

$$\text{Polynomial}: \quad K(\boldsymbol{x}_i, \boldsymbol{x}_j) = \left(\boldsymbol{x}_i^T \boldsymbol{x}_j + 1\right)^q, \; q > 0$$

$$\text{Radial Basis Functions (RBF)}: \quad K(\boldsymbol{x}_i, \boldsymbol{x}_j) = exp\left(- \left|\left|\boldsymbol{x}_i - \boldsymbol{x}_j\right|\right|^2 / \sigma^2 \right)^q$$

$$\text{Hyperbolic Tangent}: \quad K(\boldsymbol{x}_i, \boldsymbol{x}_j) = \tanh\left(\beta \boldsymbol{x}_i^T \boldsymbol{x}_j + \gamma\right)$$

The appropriate choice is based on similar considerations as in classification, that is, we shall train the SVM regression using a part of our data as training set and another part as test set, applying the trained SVM to the test dataset and then comparing the mismatch obtained for the training and test dataset. As an applicative example, we present the prediction of ground motion parameters mentioned earlier. Instead of using a model such as the one given in Eq. (4.9), we perform a full black-box regression with SVM.

In Fig. 4.16, we demonstrate the application of SVM regression to a dataset related to shallow earthquakes at Mt Etna. The independent input vector $\boldsymbol{X}$ is given by the local magnitude, epicenter distance, and focal depth. The output vectors used for training are represented by observed peak ground accelerations encountered at ca. 50 stations. Stations were deployed on the volcano and adjacent areas, with epicenter distances ranging from

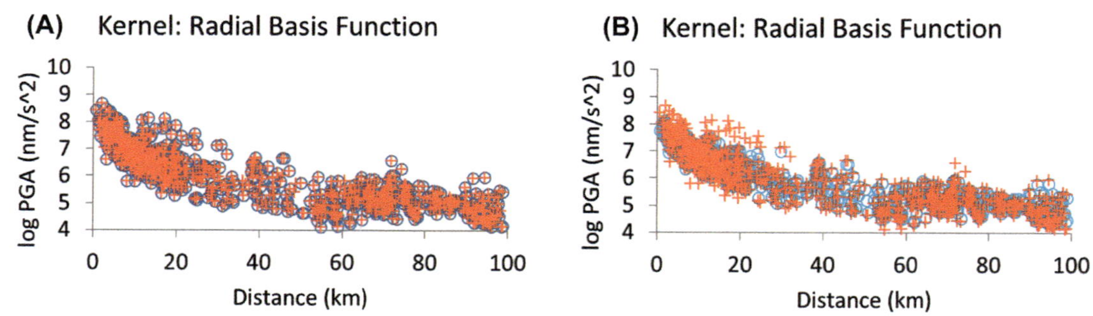

Figure 4.16

Regression of peak ground acceleration (PGA) using an SVM with a Gaussian ("RBF") kernel. Data correspond to 1000 shallow earthquakes recorded at Mt Etna in the years 2006–12 (Tusa and Langer, 2016). Earthquake magnitudes range from 3.0 to 4.3. In Fig. 4.16a we plot the observed PGA (red crosses); the support vectors are shown as blue circles. In Fig. 4.16b we again plot the support vectors as blue circles, whereas the red crosses indicate the predicted PGA. Note the interval between 30 and 70 km, where PGA are rather constant. This behavior is represented fairly well in the SVM predictions and the support vectors.

less than 1 to 100 km. Here we have been applying the MATLAB™ machine learning functions, in particular the "fitrsvm" tools. The "Gaussian kernel" in the toolbox uses an exponent $q = 1$ (and $\sigma^2 = 1$). Besides, we set the "epsilon" parameter to 0.2 keeping the default value for C (here addressed to as "BoxConstraint," which depends on the interquartile range of the response *y*, *iqr(y)*, divided by 1.39). Playing with the settings, we found that the results were rather robust.

The predicted ground accelerations (PGA) are expressed as log(nm/s^2). In Fig. 4.16a, we show only the marginal distributions for PGA and epicenter distance; the scatter in the PGA values is partly due to the fact that we plot the data for all ranges of magnitudes and focal depth. In Fig. 4.16b, we represented the predicted PGA obtained with a trained SVM regression. We notice that the predictions (red crosses) fall nicely inside a range delineated by the blue circles, which mark the position of the support vectors. In the right hand side of the figure we compare the position of the support vectors—which represent the space where our predictions are found—to the truly observed values. The tube delineated by the support vectors embraces a large part of the observations. On the whole, we find an average prediction error of ~0.32, which is below the values ~0.39 obtained by Tusa and Langer (2016) for the conventional GMPE prediction models proposed by Sabetta and Pugliese (1987) and Boore and Atkinson (2008). Note, however, that with the black-box approach (SVM- or MLP-based regression) we have no physical clue about the validity of the model. In a certain sense, with this approach we minimize the aleatoric uncertainty by choosing a sufficiently complex regression function to match the target. The remaining error in this logic is basically epistemic—it regards the model itself. If we tried to fit an error being truly random, we would get model parameters that are falsified during the test phase. Testing is therefore an important issue.[6] Here, we have carried out a fivefold cross-validation of our SVM regression. We obtain a test error of 0.357, which is still below the errors obtained with the conventional GMPE models.

4.7.2 Brief considerations on pros and cons of SVM and MLP in regression problems

Both SVM and MLP methods have their pros and cons when applied in regression problems. In the literature, MLPs are often reported to be more sensitive to overfitting than SVM. On the other hand, their application to problems with a multivariate prediction is rather straightforward. SVMs are based on the solution of binary problems—in classification (*A* or *B*) as well as regression ("in" or "outside" the tube)—which makes their application to problems with multivariate targets questionable. At the same time, the mathematical structure of MLP is somewhat more transparent than that of SVM, allowing

[6] Obviously, cross-validation is also useful in nonlinear regression based on physical models. Unfortunately, it has not become a common practice as yet.

to get a first guess on the importance of the input parameters just from the analysis of the weights (see Eq. 4.11). In the example by Derass et al. (2012), we found that some parameters were of minor importance for the prediction, as their corresponding weights were low. Such an analysis—similar to the one leading to Eq. (4.11)—is not straightforward in SVM regression.

4.8 Classification by hidden Markov models and dynamic Bayesian networks: application to seismic waveforms of tectonic, volcanic and lunar origin

4.8.1 Background

In the applications discussed before, we considered patterns described by single feature vectors, being them spectral components, a combination of various geophysical observations, rock composition, etc. Each of the patterns was assigned to a target, typically a category or a set of target values related to a physical model. The HMMs discussed in Chapter 2 are based on a different approach. Rather than considering the single patterns, they focus on a sequence of observations. In our context here, the sequences are made up of patterns, even though the strategy allows for more general sequences, such as actions and events.

In Chapter 2, we mentioned HMMs being a double stochastic process, where the first process describes the transitions between hidden states given the hidden state sequence $\boldsymbol{q} = (q_1, q_2, .. q_T)$, and the second process determines the output of the HMM. Based on the emission probability of the current state, an observation symbol is issued. Thus, an HMM with N states is described by the triple $(\boldsymbol{\Pi}, \boldsymbol{A}, \boldsymbol{B})$:

initial state probability vector $\boldsymbol{\Pi} = (\pi_1, \pi_2, \pi_3, \ldots, \pi_N)$
transition probability matrix $\boldsymbol{A}_{NxN}$, where $a_{ij} = P(q_t = s_j | q_{t-1} = s_i)$
emission probability matrix $\boldsymbol{B}_{NxM}$, where $b_j(k) = P(o_k \text{ at time } t | q_t = s_j)$
the number of states N.

In waveform classification, a suitable description of an event through an HMM requires the identification of the HMM parameters based on a first training set, i.e., a set for which the targets are known. The optimum models are found following a maximum likelihood strategy. An early version of such a strategy is the Baum-Welch algorithm, which was described in Chapter 2 (Appendix 2.4.3). A more general method is the expectation maximization (EM) algorithm, which we also mentioned in Chapter 3 (Box 3.2). An alternative is the Viterbi algorithm (see Appendix 2.4.2 in Chapter 2).

4.8.2 Signals related to volcanic and tectonic activity

In case of a continuously recorded data stream, each single sequence of time frames can be classified as a specific category, being it simply noise, a rockfall or another event. In contrast to the previous methods illustrated in Chapter 4, the approach allows detection and classification at the same moment. In other words, the classification method is not "input signal present → classify," but immediately assign each frame in the continuous data stream to one of the classes (Hammer et al., 2012). In conventional applications of HMM, some issues remain. First as in all supervised methods mentioned so far, there is a need of having enough examples for training. Besides, the number of states must be defined at the beginning. This may require some a priori knowledge, for instance, the number of relevant sections in the signal (P-wave, S-wave and coda in an earthquake record). On volcanoes, however, signals have a widely varying aspect, with clear phases in "vt," sharp onset and smooth amplitude decay in the "hy," or the cigar-like envelope of the "rf" example in Fig. 4.17.

In their application to seismic signals recorded at Soufrière Hills volcano, Montserrat, Hammer et al. (2012) developed a strategy offering more flexibility, which is described in the following.

Consider the case in which there is a large number of unlabeled data and a researcher is interested in a classification system based on a limited number of waveform samples, which means saving preparation time. In this context the reference event (e. g., Fig. 4.18,

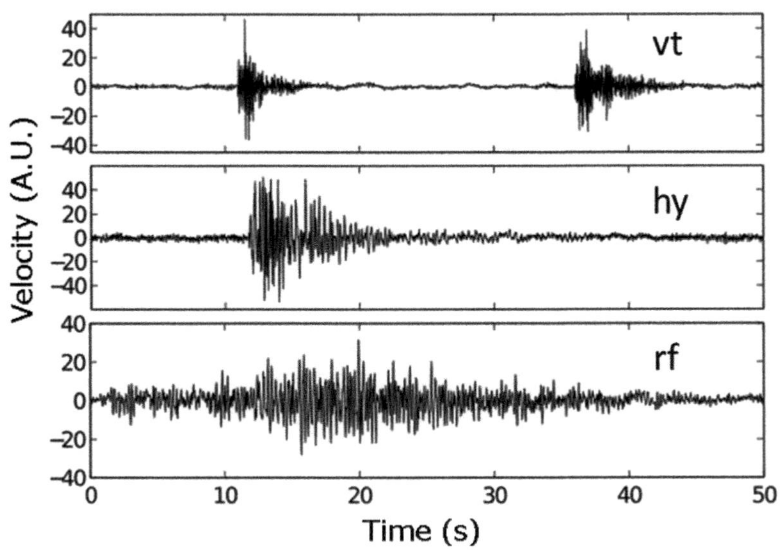

Figure 4.17
Example waveforms of volcano tectonic events (vt), hybrid events (hy), and rockfall events (rf) recorded at Soufrière Hills volcano, Montserrat. *Figure modified from Hammer et al. (2012).*

upper panel) is described by the statistics of the unlabeled data stream. The overall feature distribution *P(**fv**)*, where ***fv*** is the feature vector, is modeled by Gaussian mixture densities (see Chapter 3, Box 3.2).

The overall density function is therefore modeled by the convex combination

$$P(\boldsymbol{x}) = \sum_{m=1}^{M} P(m) \times \mathcal{N}(\boldsymbol{x}|\boldsymbol{\mu}_m, \boldsymbol{C}_m)$$

with $P(m)$ being the mixture weight, and $\mathcal{N}(\boldsymbol{x}|\mu_m, \boldsymbol{C}_m)$ is a normal distribution with mean $\boldsymbol{\mu}_m$ and the covariance matrix $\boldsymbol{C}_m$.

Using the EM method, the authors describe their unlabeled data as a combination of 16 Gaussian distributions and defined the characteristics of each category on the base of the Gaussians. To fix the appropriate number of states for the event models, the statistical description of the overall wavefield (Fig. 4.19, upper panel), expressed as Gaussian mixture density (Fig. 4.19, lower right panel), is used. A reference waveform for each class of interest has to be chosen. Starting from the raw time series—such as the one shown in Fig. 4.18 (upper panel)—features are extracted for gliding windows, here the normalized instantaneous frequency and the largest eigenvalue are calculated from the covariance matrix over the three seismic components (for more details, see Hammer et al., 2012). In a plot of feature #1 versus feature #2, one obtains a trajectory, which represents the

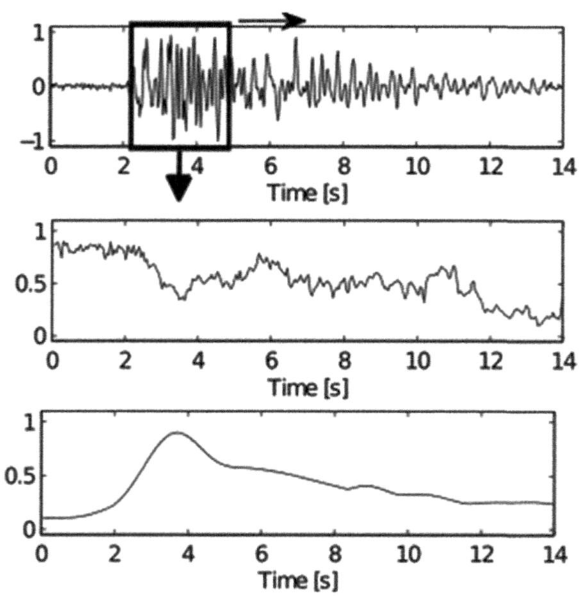

Figure 4.18
Raw waveform of a hybrid event (top trace) replaced by time series of normalized instantaneous frequency (middle trace) and normalized largest eigenvalue (measure of correlation, bottom trace).

development of the time series in the feature space. At the same time, one checks, at each time step, which of the Gaussians is closer to the actual position on the trajectory. Keeping that indication, information on the state of the time series over time is obtained. We summarize the whole procedure of feature extraction and state identification in Fig. 4.19.

From the time series of the unlabeled data (top of Fig. 4.19), the features x_1 and x_2 (Fig. 4.19, bottom left) are extracted and their distribution is modeled by Gaussian mixture densities (Fig. 4.19, bottom right). Reflecting its random structure, the corresponding background model is composed of a parallel connection of M states, one for each Gaussian mixture component (Fig. 4.20, left). The model can change from each state in every other state with a probability >0, that is given by the mixture weights P(m) of the m-th mixture component. Consequently, the noise is a signal without a specific temporal structure.

In the next step, features x_1 and x_2 are extracted from a reference event waveform. However, here a temporal structure is assumed, so a specific path in the feature space is followed as shown by the black trajectory (Fig. 4.19, bottom right). During the way across the feature space, the path approaches one of the Gaussians, and one denotes the closest one in that moment as the current state. In a prewhited feature space, the assignment is based on the minimum Euclidean distance. Following this procedure, one obtains a

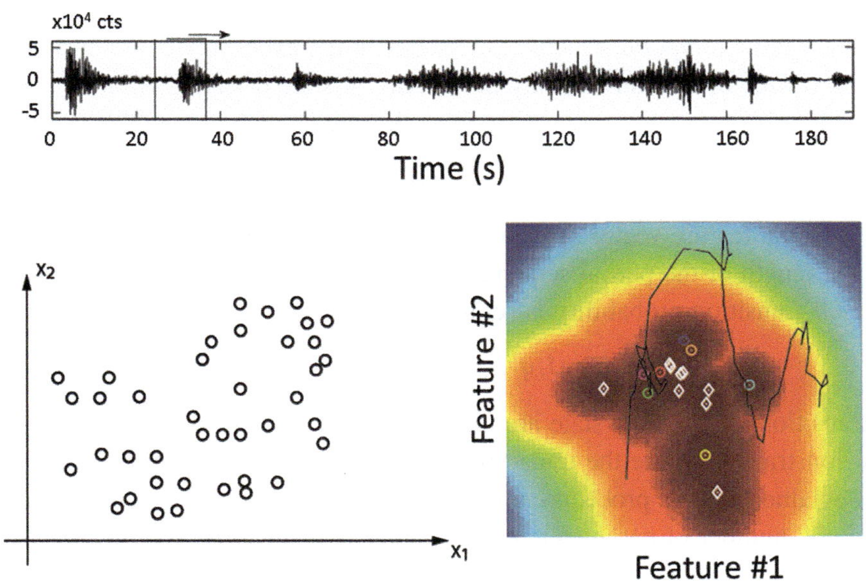

Figure 4.19
Schematic description of feature extraction (modified from Hammer et al., 2012). From an unlabeled data stream (top trace), features ×1 and ×2 are extracted (bottom left). Note that no temporal structure is assumed. Their overall distribution is modeled by Gaussian mixture densities (bottom right). Event trajectory is shown in black. Means of visited and nonvisited mixture components are marked by colored circles and white diamonds, respectively.

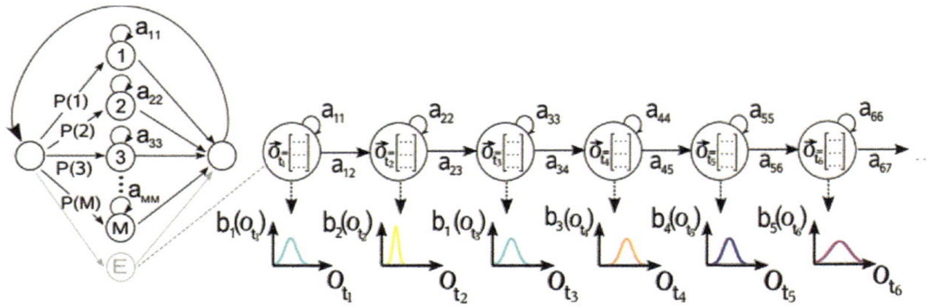

Figure 4.20

The complete model (see Hammer et al., 2012). The noise model (left) is a parallel connection of M states. In the event model, states are arranged in sequences. The number of states is estimated from the sequence of "visited" mixture components in Fig. 4.19. Colors correspond to means in Fig. 4.19 (bottom right).

sequence of "visited" mixture components that gives the number of states appropriate to describe the reference event. Each time the index of the assigned mixture component changes, a new state is created in the corresponding event model. If an already selected Gaussian is "visited" later again, this generates a new state in the model. In this way a data-driven segmentation of the event in parts of similar observations is obtained without any preconceptions of the researcher. Further on, we shall notice for each event class differing trajectories, and differing sequences of states.

Capturing its specific temporal structure, the states in the event model are arranged in sequence. Each time the model can stay in the current state or change in next state only (Fig. 4.20 right). The emission probability of each state is a Gaussian with the provisional mean and covariance of the less distant mixture component. The transition probabilities are estimated as follows. During the procedure described earlier, a partial sequence FV = $(fv_{t1} \ldots fv_{tN})$ of the feature vector sequence is assigned to the closest mixture component. Assuming the length of the partial sequence is equal to the expected state duration d_i of state i, the self-transition probability is set to

$$P(q_t = i, q_{t+1} = i) = a_{ii} = 1 - 1/d_i$$

The state transitions to the next state are then set to a fixed value of $1 - a_{ii}$ as the probability constraints require the transition probabilities to sum to unity. The model parameters obtained by this procedure are reestimated as follows. With a small training sample size, the number of parameters to be estimated needs to be kept as low as possible. For that reason, states corresponding to the same mixture component are tied together, that is, states generated by the same mixture component share the same mean and the same covariance matrix. Thus, the means are reestimated using all time segments assigned to the corresponding mixture component. The covariances as well as the transition probabilities are not updated, as a larger training dataset would be necessary to robustly estimate the parameters.

Hammer et al. (2012) applied their classification system to seismic data recorded at Soufrière Hills volcano, Montserrat. The authors focused on three event classes, that is, rockfalls ("rf"), hybrid events ("hy") and volcanotectonic events ("vt," see Fig. 4.17). These signals were embedded in a "noise" part, which has no temporal structure. The considered dataset was recorded in January and May 1997 by the permanent seismic network run by the Montserrat Volcano Observatory (MVO). This time span was characterized by a high level of volcanic activity that culminated in a major dome collapse on June 25. Hammer et al. (2012) used data from a single station only. The station was equipped with a three component broadband sensor; the sampling frequency was 75 Hz. The dataset was composed of individual recordings with duration from 25 s to several minutes. Similar to the application discussed in Section 4.1, the authors did not use the waveforms directly, but extracted a number of features. In particular, the features regarded:

Spectral attributes (half octave bands, dominant frequency, bandwidth, central frequency, cepstral coefficients).
Complex trace attributes (Instantaneous bandwidth, normalized envelope, instantaneous frequency, centroid time)
Polarization attributes (planarity, rectilinearity, largest eigenvalue)

More details regarding the features are given in Hammer et al. (2012). After computing the features, the authors performed a normalized Karhunen-Loève transform to prewhiten the signal. That way, in the covariance matrices of the multivariate Gaussian distributions only the diagonal elements must be estimated which reduces the computational burden efficiently.

The application allows the recognition of transients within continuous seismic data. That means seismic events were not pretriggered. Training was carried out on an HMM for each event type of interest while an HMM was trained for the background noise. In the classification step, the incoming signal was then assigned to the most similar signal class, which was either an event or noise. The complete dataset consists of 193 hybrids, 197 rockfalls, and 115 volcanotectonic events. Hammer et al. (2012) showed that the classification based on the described draft models provide an appropriate description of individual event classes. More than 84% of the events were classified correctly (Fig. 4.21 left). However, additional information about the event class may improve the recognition accuracy. Thus, Hammer et al. (2012) suggested to use correctly classified events to even better adjust the system to individual class patterns. The manually positively confirmed samples provide a larger training dataset to retrain event models. Given the availability of a sufficiently large training dataset, the retraining is done by using the classical formulas by Baum and Welch (see Chapter 2, Appendix 2.4.2). The draft models are here used as a seed model for the iterative training process, and states generated from the same mixture component are still tied. However, now, the covariances of the emission probabilities as well as transition probabilities are adjusted to the newly added training data. This reestimation increases the classification rate to more than 97% (Fig. 4.21 right).

4.8.3 Classification of icequake and nonterrestrial seismic waveforms as base for further research —HMM

4.8.3.1 Icequakes

The automatic detection and classification of interesting seismic signals, as outlined earlier, has encouraged further research investigations beyond the classical field of earthquake and volcano monitoring. Hammer et al. (2015) applied the system for the automatic detection and classification to icequakes recorded by the Neumayer seismic network in Antarctica. As we have seen, the approach described earlier allows to construct classifiers on the fly while enabling the recognition of highly variable time series. In that way, we do not rely on a large number of manually classified training events, which are not available for the Neumayer seismic network. Hammer et al. (2015) focus on a specific event type occurring close to the grounding line of the Ekström ice shelf. Seismic events cover a very narrow frequency range shifting from 1 Hz in the beginning to 2 Hz at the end of the events (Fig. 4.22A-C).

The waveform is first "translated" into different wavefield parameters (e.g., spectral features) which facilitates the discrimination between different seismic signal types. Then, for each signal class of interest (e.g., calving event and basal event) and the background noise, a generative probabilistic model (here an HMM) is learned from prelabeled training data. Using prototypes for each class of interest, the trained HMM aims to summarize

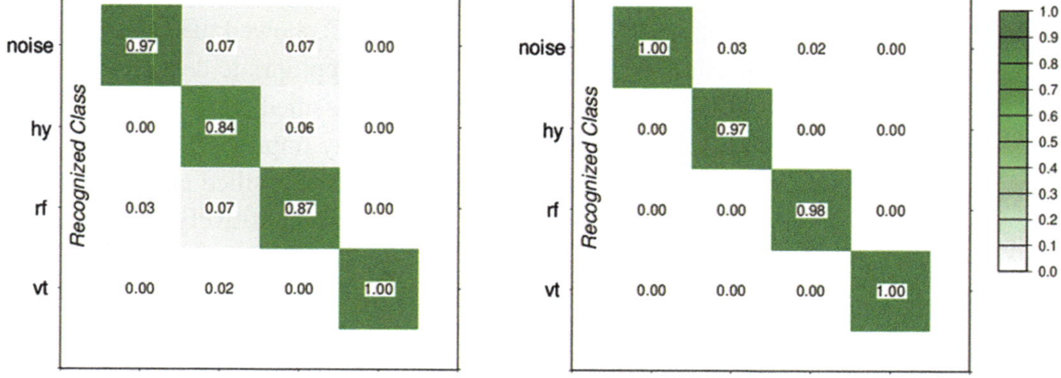

Figure 4.21
Classification results of draft models (left) and after re-training with 30 positively confirmed events. *From Hammer et al. (2012).*

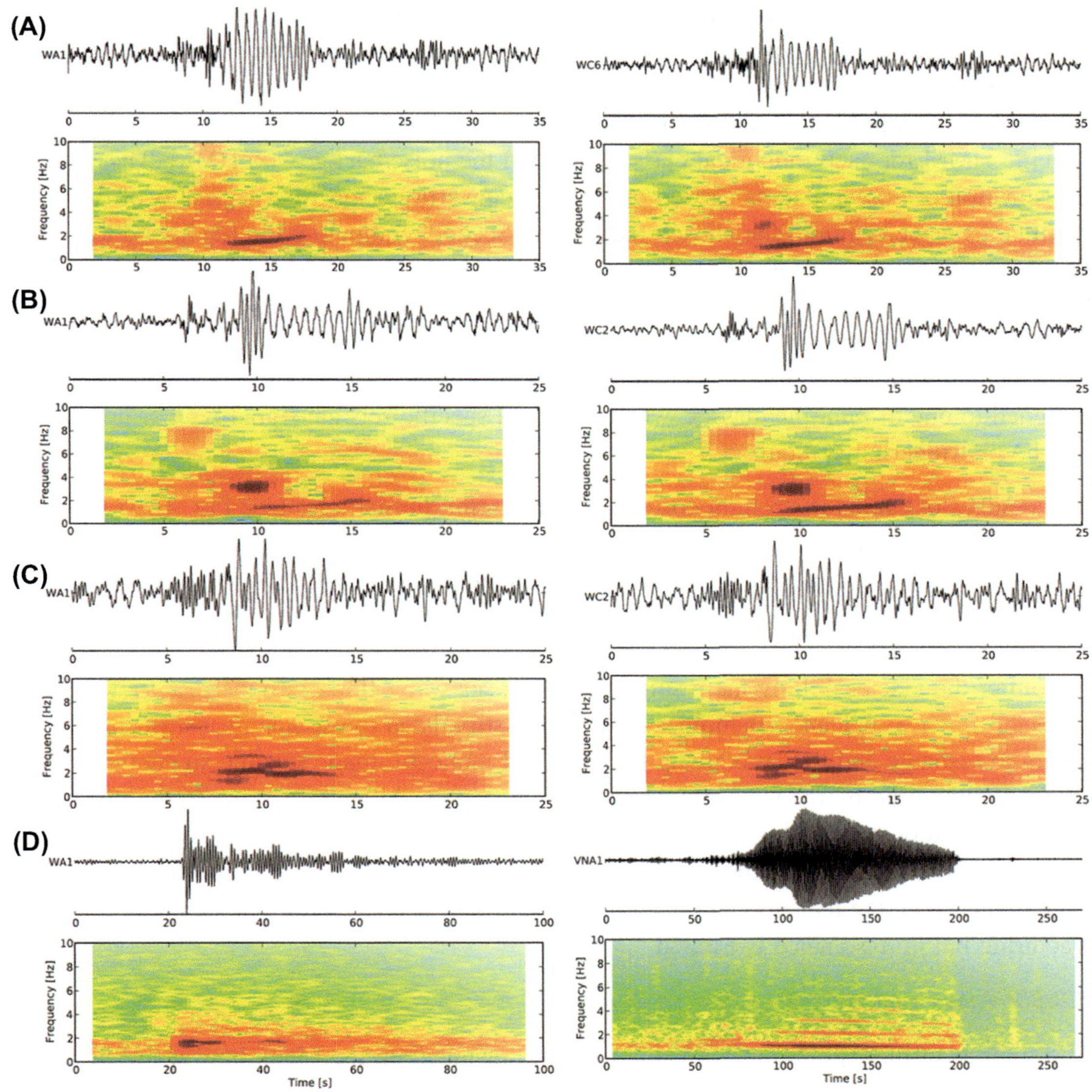

Figure 4.22

Waveforms and spectrograms of manually detected icequakes (A—C) and other events (D). The two panels in a—c show the events recorded at two stations under back-azimuths of 300^0 and 45^0. *Modified from Hammer et al. (2015).*

typical class characteristics, and the number of training samples directly influences the system performance. As by now there is no systematic study of cryoseismic events observed at Neumayer seismic network, a reliable a priori classification scheme as well as a sufficiently large database of preclassified events for learning appropriate prototypes is not available. For that reason, the authors relied on an approach that requires a minimum amount of training data, that is, a single reference waveform (Hammer et al., 2012). After construction of the prototypes the classification of unseen data can start. First, the incoming continuous data stream is converted into the chosen feature set. Second, the likelihood is computed that the observation sequence at hand has been generated by a specific HMM for each individual class HMM and the noise/background HMM. The classificaton of unseen data is carried out by first converting short time segments of the continuous waveform into the chosen feature set. Second, computing the likelihood that the observation sequence at hand has been generated by a specific HMM for each individual class HMM and the noise/background HMM. Third, the HMM that best describes the observed feature sequence (i.e., achieving the highest likelihood) is chosen as the winning model. In this way each time instance in the continuous data stream is assigned to one of multiple classes, which can be a specific seismic signal type or noise, and no pretriggering is required.

The number of events is strongly correlated to the semi-diurnal tides (see Fig. 4.23). More precisely, events exclusively occur during rising tides. In combination with observed waveform characteristics, this observation leads Hammer et al. (2015) to propose the following model. Near the grounding line, ice masses are partly floating. The vertical tidal movement then generates an extensive state of stress causing fracturing at the top or bottom of the glacier. Filling the resulting cavities, sea water flows into the resulting cracks. Similar to volcano seismology, the fluid is stimulated to oscillate, causing the cavity to resonate and thus producing the observed monochromatic seismic signature. This model is supported by the temporal position of the events. Although ice fracturing due to bending is likely to be active at rising and falling tides most of the events are observed during rising tide. Although the sea/ice interface at the bottom provides conditions for event generation, the ice/snow interface does not. Consequently, falling tides induce fracturing but no resonance is excited.

4.8.3.2 Moon quakes

The stations of the Apollo lunar seismic network, installed on the Moon between 1969 and 1972, recorded the only confirmed seismic events on any extraterrestrial body so far (e.g.,

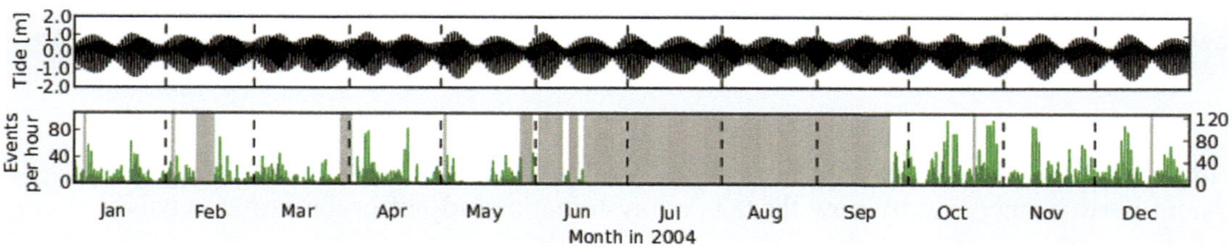

Figure 4.23

Hourly detected events in 2004 together with sea level changes (top panel). Gray periods indicate data gaps. Dashed lines mark individual month. *Modified from Hammer et al. (2015).*

Lognonné, 2005; Lognonné and Johnson, 2007). The large number of events recorded on the Moon, which had been considered as tectonically "dead," came as a big surprise at that time (e.g., Frohlich and Nakamura, 2009). The waveforms recorded on the Moon were readily recognized to be very different from what had been expected based on Earth data (Gold and Soter, 1970; Dainty and Toksöz, 1981). Lunar event seismograms have emergent onsets, a strong body wave coda, and a lack of coherent, dispersive surface-wave trains. In contrast to classical terrestrial signals, the lunar events are characterized by longer event durations of up to 100 min (Fig. 4.24). Source processes of these events probably differ from those on the Earth, where most quakes is related to plate tectonics.

The Apollo seismic recordings are a valuable test case for any future experiment in planetary seismology and algorithms that could be employed in it. The main lunar event types, listed in the long-period event catalog (Nakamura et al., 1981), are given by deep moonquakes, impacts, and shallow moonquakes. Deep moonquakes, occurring in a depth range between 700 and 1200 km (Nakamura, 2005), make up more than 55% of all identified events and form spatial clusters of periodic activity (e.g., Lammlein et al., 1974; Lammlein, 1977; Nakamura, 2003; Bulow et al., 2007). The established periodicities of 1 month, 7 months, and 6 years correlate with the Moon's position with respect to the Sun–Earth–Moon system, indicating variations in tidal stress as cause of deep moonquakes (Lammlein, 1977; Nakamura, 1978; Minshull and Goulty, 1988; Weber et al., 2010). Frohlich and Nakamura (2009) have argued that fluid phases or partial melts in the lunar mantle could additionally play an important role, similar to the case of intermediate-depth subduction-zone seismicity on the Earth, as deep moonquakes occur at depths where brittle failure is impossible in dry rock. The second largest group of lunar seismic events, around 13% of all identified events, are generated by impacts, both of natural meteoroids and of artificial sources. Artificial sources are spent stages of the Saturn-V launch vehicles and lunar modules that were directed at the lunar surface on purpose to create seismic sources of well-defined timing and location. The natural impacts, especially of meteoroids with masses smaller than 1 kg, show temporal clustering correlating with known meteor showers (Nakamura et al., 1982; Oberst and Nakamura, 1991). In contrast to deep moonquakes, shallow moonquakes with focal depths of less than 200 km are rare events, making up less than 1% of all identified events. They are significantly larger than the deep moonquakes (Mw of up to 4.7), have a low b value (Nakamura, 1977), and are sometimes also associated with exceptionally large stress drops (Oberst, 1987). As tectonic quakes on a stagnant-lid planet, they might be comparable to intraplate earthquakes, which exhibit some similar properties (Nakamura et al., 1982).

Knapmeyer-Endrun and Hammer (2015) identified and classified specific seismic event types in the Apollo 16 lunar seismic data by using an HMM based classification system. The original event catalog was composed in 1981 by Nakamura et al. (1981) and contained 13,058 events. More than a half of the events was classified as deep

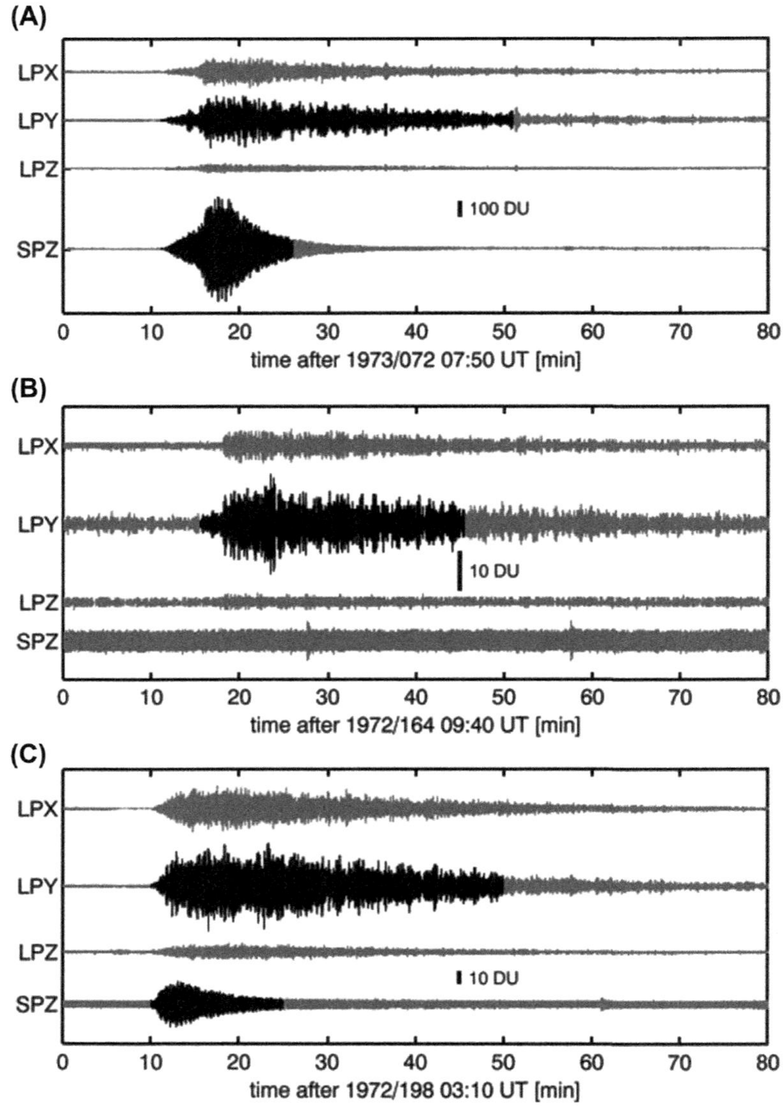

Figure 4.24

Typical lunar seismograms for different event types. The plotted events were used as prototypes in the event classifier. (A) Shallow moonquake. (B) Deep moonquake. (C) Impact. Note the different amplitude scale of the plots as given by the black bars. Black parts of LPY and SPZ components indicate the time intervals used to learn the features of the respective events for the classifier. Traces signed as "LPX (Y, Z)" were recorded with a long-period seismometer having an eigenperiod of 15 s "SPZ" is a vertical short period record (eigenperiod of the sensor 1 s). *From Knapmeyer-Endrun and Hammer (2015).*

moonquakes while shallow moonquakes and impacts represent only a small part of detected events. In contrast to classical terrestrial signals, the lunar events are characterized by longer event durations of up to 100 min. However, not all events, listed in the catalog, are assigned to a specific seismic signal class. Ranging from two third in the initial catalog, this number could be reduced to about 25% through intense multiple analyses throughout the last 25 years (Knapmeyer and Weber, 2015). To test the proposed "learning-while-recording" approach also for nonterrestrial signals, Knapmeyer-Endrun and Hammer (2015) first focus on matching the existing event labels. The automatic system classified more than 80% of the events correctly (i.e., according to the manually given label). As a next step, the previously unclassified signals were fed into the classifier and about 60% could be assigned with a reliable class label. Finally, it was possible to find 210 new events in the continuous raw data which were not recognized before. The additional events can be used for a more detailed study of temporal variations in event occurrence or to improve the cluster stacks for phase picking.

4.8.3.3 Classification of seismic waveforms using dynamic Bayesian networks

An alternative approach providing more flexibility than classical HMMs are dynamic Bayesian networks (DBNs) (e.g., Murphy, 2002; Riggelsen et al., 2007; Riggelsen and Ohrnberger, 2014). As we have seen in Chapter 2, Bayesian networks (BNs) are graphical models that can be used to model static systems. When modeling dynamic systems, such as an observed time series, we also want to include their temporal structure in the model. Here, so-called dynamic DBNs come into play. As well as HMMs, DBNs are temporal generative models, that is, statistical models that are able to model characteristics of patterns that evolve in time while allowing for the interpretation and inclusion of expert knowledge. In contrast to HMMs the hidden nodes are not represented by a single variable but by a set of variables. In simple terms, a DBN represents a BN that is unrolled in time and that probability distributions can change in time. Riggelsen et al. (2007) introduce DBNs for seismic waveform classification. In their study, Riggelsen et al. (2007) model the temporal structure of the spectral evolution of the observed characteristic ground velocity. A continuous wavelet transform (CWT) provides the input for the classification system: features that cover the time-frequency decomposition of the observed seismic signals. Throughout the seismic waveform, CWT coefficients evolve in time but also depend on neighboring frequency bands at the same point in time. Thus, multiple variables are active at a time t and that is what is captured by the DBN.

The performance of the DBN classifier is tested on continuous seismic data recorded by the European broadband network. Riggelsen et al. (2007) are able to achieve an average accuracy of 95%. Later on, the approach was further refined by Riggelsen and Ohrnberger (2014). The main improvement covers the change in the background seismic noise with time. Besides regular daily and seasonal variations, the noise characteristics can also change suddenly due to the installation of a new instrument or other environmental modifications. To account for this, Riggelsen and Ohrnberger (2014) suggest a continuous gradual model adaption. Instead of mixing new and old training samples, the noise DBN is trained on the new samples only but not until convergence. By doing so, previous training samples are implicitly contained in the updated model and the model is adapted gradually to actual environmental conditions.

4.9 Natural hazard analyses—HMMs and BNs

4.9.1 Estimating volcanic unrest

Besides the identification of different seismic event types, potentially reflecting specific seismic source processes, overall seismicity rates are often subject to intense research as they are assumed to correlate to the level of volcanic activity (e.g., McNutt, 1996). The idea of an increasing likelihood of an eruption with an increasing number of events has led to many successful eruption predictions as at Tolbachik volcano, Kamchatka, in 1975 (Tokarev, 1983) or Mt. St. Helens, Washington, in 1980 (Malone et al., 1981). Thus, daily counts of seismic events are a common method of choice for monitoring active volcanoes. In addition, there are various stochastic approaches to describe eruptive volcano activity. This allows to incorporate uncertainties in the prediction of eruptions. Often, repose intervals are suggested to estimate the probability of a volcanic crisis (e.g., Wickman, 1966; Bebbington and Lai, 1996). However, such a static strategy does not allow for changes in the volcanic activity. In other words, their problem resides not just in the lack of detecting change points, but even in the omission of considering changes in the dynamic behavior as we will explain later.

Although the probability of an eruption can be determined by modeling repose intervals, no change point in the system behavior is accounted for. The type of distribution cannot change over time, once determined all periods of activity are assumed to follow this distribution. However, this is not always correct as shown by Varley et al. (2006). Thus, different levels of volcano activity (represented by different states of eruptive behavior) are characterized by different probability distributions (e.g., Wickman, 1966; Mulargia et al., 1987). To combine both aspects Bebbington (2007) introduced HMMs for identifying different volcanic regimes. The idea is based on a volcano switching between a fixed number of different hidden regimes, represented by the hidden states, which are assumed to control the observations such as eruption rates or erupted volume. Besides forecasting the probability of the next onset, this allows to detect points where the system changes into another state. Bebbington (2007) identified different regimes, that is, states with persistent activity, on a timescale of decades to centuries.

Aspinall et al. (2006) applied the HMM approach to the development of volcanic unrest. The method is based on accelerated earthquake occurrence, following ideas of the generic "time-to-failure" or "Material Failure Forecast" concept. Each state in the HMM represents another level of unrest that is characterized by a specific number of daily earthquake occurrences. Thus, identifying transitions between the states enables to detect accelerated earthquake rates. The representation in terms of an HMM offers the possibility to switch backward into an earlier state characterized by a lower earthquake rate. Monitoring the sequence of "visited" states allows to identify specific patterns that can be useful for forecasting. Although not seized by the authors, this approach offers the interesting opportunity to forecast the next day seismicity in a probabilistic fashion. Based on past observations of successive daily earthquake rates transitions between different states can be quantified probabilistically. Thus, being in one state we can give a reliable estimate of residing in this state (i.e., observing today's number of earthquakes tomorrow again) and of changing into each other state, characterized by another specific number of daily earthquakes.

Approaches based on Bayesian analysis exploit similar strategies. Their main advantage is the possibility to include a variety of observations (e.g., number of earthquakes, deformation, elevated temperature) in the forecasting process. In the Bayesian event tree (BET), alternative scenarios are modeled: all starting from a common start point and ending in a final outcome, which depends on the evolution of the volcano in the meantime (Marzocchi et al., 2008). The evolution is described in terms of different observations (e.g., magma ascent, number of earthquakes). Each observation (also called node) has different possible outcomes determining the path through the event tree. Although the BET is characterized by a specific sequence of observations, the approach of BN has no such restrictions (Hinks, 2008). Different nodes do not necessarily follow in sequence allowing more flexibility in the modeling process. However, more flexibility goes side by side with a more complicated construction of the network. Given sufficient data, both structure and parameters of a BN can be estimated purely from observations. However, this is precisely the problem. For volcanological BNs, there are relatively few "samples" of eruption even on a global scale. In this case, expert knowledge must be applied to impose physical and logical constraints enabling the construction of a BNN with sparse training data.

4.9.2 Reasoning under uncertainty—tsunami early warning tasks

The classical way to reveal dependencies between variables are correlation coefficients, which imply that there is at least a monotonic relationship between the variables. However, correlation must not be confused with causality. In climate sciences, so-called causality networks are widely used to infer causality (e.g., Zerenner et al., 2014). In natural systems, often the single and joint effects of the driving forces are not fully understood and this may introduce a large degree of uncertainty into any quantitative analysis. Scarcity of observations is another source of uncertainty. Given these problems and the complexity of the underlying physical process (i.e., the large number of influencing factors), a possible solution may be found in a probabilistic framework such as BNs, which capture most of the described problems. By using a BN, we are able to represent (in) dependencies between involved variables in a graphical network that allows us an improved understanding and direct insights into the relationships and mechanisms of the system. The approach has been successfully implemented for tsunami early warning (Blaser et al., 2011, 2012) and probabilistic seismic hazard assessment (e.g., Kuehn et al., 2011). See Vogel et al. (2014) for an overview. In the following, we demonstrate the application for tsunami early warning tasks (Blaser et al., 2009, 2011, 2012).

Commonly, tsunami early warning systems evaluate seismic source parameters either by a simple rule based model or by using a precalculated database to select the most appropriate tsunami scenario. Those deterministic approaches usually give a warning level for a particular region or the arrival time and the maximum expected wave height for a specific site. Their deterministic nature does not account for the uncertainties arising from incomplete and inaccurate evidences. Blaser et al. (2009) suggest to capture the uncertainties inherent in the tsunami early warning by using a BN. BNs are able to integrate and quantify the uncertainties via the joint probability distribution that represents the dependencies of all variables in a statistically proper way. Blaser et al. (2011) evaluate real-time seismic source parameters in combination with prior information, physical knowledge, and expert judgment to a probabilistic real-time threat assessment. The authors demonstrate a BN-based tsunami early warning system for the region Sumatra, Indonesia. The BN is composed of eight variables: the "tsunami warning level" and seven seismic parameters, where the latter are

hypocenter location (epicenter and depth)
magnitude
rupture direction
centroid location
rupture width and length.

The variable "tsunami warning level" is linked to the expected tsunami height:

no tsunami (wave height at the coast smaller than 10 cm)
minor tsunami (between 10 and 50 cm)
tsunami (50 cm–3 m)
major tsunami (larger than 3 m).

The structure as well as the local probability distributions of a BN can be learned from data. As the historical tsunami database is much too sparse to learn from data alone, a synthetic database is created. Blaser et al. (2009) introduced an approach based on ancestral sampling to generate these data. The available domain knowledge (expert knowledge as well as physical/mathematical laws known to hold in the realm of tsunami generation) together with a priori information about the geological and geographical situation is used to construct a Bayesian network. First, the mathematical formulations of the subprocesses (e.g., about the tsunami excitation and propagation) are ordered hierarchically. The output of one subprocess is used as input in the next one. Together with the assumption of prior probability distributions Blaser et al. (2009) compile in that way a large synthetic database that is in turn used to train a BN.

Given the probabilistic framework of BNs a first estimate on the expected tsunami height can be given as soon as first evidences are available. Generally, they become available at irregular time instances and are updated continuously. New or updated evidences can be integrated instantaneously and the probability of the imminent tsunami risk is estimated straight away. Blaser et al. (2012) tested the implemented tsunami early warning system for 10 large earthquakes offshore Sumatra. The probabilistic tsunami threat assessment shows close agreement to the observations. Blaser et al. (2012) obtain the best performance for threat assessments based on evidences which are derived in postprocessing steps. Nonetheless, the real-time application (evaluation based on parameter estimates available 5 min after origin time) shows satisfying results. Although results are quite promising, Blaser et al. (2012) also note that even though the uncertainties inherent in the process are quantified the results are hardly unambiguous. This leaves the decision maker with some uncertainty with respect to the actions to be taken considering probabilistic warning level.

Appendix 4.1. Normalization issues

In Fig. 4.13, we noticed a rather particular distribution of the ground deformation data, and a similar picture was also obtained for magnetic and gravity data. In geophysics such a distribution is not so uncommon. By their nature, data are often controlled by exponential processes, and perturbations on geophysical fields may rapidly decay as the distance between source and observation point increases.

In Chapter 1, we already mentioned some options for normalization purposes. Here, logarithmic scaling is unfeasible, as part of the values are negative, and accounting only for the absolute data values brings along the loss of precious information on the geophysical field. A way out is given by the logistic function $f(x) = \frac{2}{1+e^{-\alpha x}} - 1$ (see Chapter 2, Eq. (2.11)), which varies between -1 and 1 (see also Fig. 2.11). Similar to this, we may use the hyperbolic tangent function

$$f(x) = tanh(\alpha x) \tag{A4.1}$$

where α is a gain factor which controls the slop of the functions around $x = 0$. For their application of MLP, Nunnari et al. (2001) proposed the following scheme:

$$flag = sign(x) \tag{A4.2a}$$

$$f(x) = (x \cdot flag)^{pow} \cdot flag \tag{A4.2b}$$

where *pow* is a float which is obtained by trial and error (typically somewhere between 0.1 and 0.4). The term $(x \cdot flag)$ is always ≥ 0, whereas through (A4.2b) we warrant that the polarity of the original value is maintained in $f(x)$.

Appendix 4.2. SVM Regression

In regression with SVM, we search the smallest margin width ||w|| which embraces our samples. We allow samples that may fall out of this range to keep the margin width small, even when we have outliers or noise. We restated the SVM optimization problem accounting for the slack variables

$$\underset{\xi_i, \xi_i^*}{\text{minimize}} ||w||^2 + C\sum_{i=1}^{N}(\xi_i + \xi_i^*) \tag{A4.3}$$

$$\text{subject to} \begin{cases} y_i - \mathbf{w}^T\mathbf{x}_i - b \leq \varepsilon + \xi_i \\ \mathbf{w}^T\mathbf{x}_i + b - y_i \leq \varepsilon + \xi_i^* \\ \xi_i, \xi_i^* \geq 0 \end{cases}$$

where $C \geq 0$ determines the importance of errors. Finally, we deal with so-called "insensitive" loss function (Smola and Schölkopf, 2004), that is,

$$|\xi|_\varepsilon = \begin{cases} 0 & \forall \ |\xi| \leq \varepsilon \\ |\xi| - \varepsilon & \forall \ |\xi| > \varepsilon \end{cases} \tag{A4.4}$$

The minimization of the "insensitive" loss function is an optimization problem with constraints.

Similar to the SVM classification outlined in Section 2.5, this optimization problem with constraints is solved using the Lagrange multiplier method, that is,

$$L = 1/2||w||^2 + C\sum_{i=1}^{N}(\xi_i + \xi_i^*) - \sum_{i=1}^{N}(\eta_i\xi_i + \eta_i^*\xi_i^*) - \sum_{i=1}^{N}\alpha_i(\varepsilon + \xi_i - y_i + \mathbf{w}^T\mathbf{x}_i + b)$$
$$- \sum_{i=1}^{N}\alpha_i^*(\varepsilon + \xi_i^* + y_i - \mathbf{w}^T\mathbf{x}_i - b) \tag{A4.5}$$

with the α_i, α_i^*, η_i, η_i^* being the Lagrange multipliers.

Finally, we obtain

$$w = \sum_{i=1}^{N} (\alpha_i - \alpha_i^*) \boldsymbol{x}_i \tag{A4.6a}$$

and

$$f(\boldsymbol{x}) = \sum_{i=1}^{N} (\alpha_i - \alpha_i^*) \boldsymbol{x}_i^T \boldsymbol{x} + b \tag{A4.6b}$$

To determine b we make use of the so-called Karush—Kuhn—Tucker (KKT) conditions (Karush, 1939; Kuhn and Tucker, 1951), after which at the point of the solutions the product between dual variables and constraints vanishes.

$$\alpha_i \left(\varepsilon + \xi_i - y_i + \boldsymbol{w}^T \boldsymbol{x}_i + b \right) = 0$$
$$\alpha_i^* \left(\varepsilon + \xi_i^* + y_i - \boldsymbol{w}^T \boldsymbol{x}_i - b \right) = 0$$

and

$$(C - \alpha_i) \xi_i = 0$$
$$\left(C - \alpha_i^* \right) \xi_i^* = 0$$

From this we conclude that

$$\varepsilon + \xi_i - y_i + \boldsymbol{w}^T \boldsymbol{x}_i + b \geq 0 \textit{ and } \xi_i = 0 \ \forall \alpha_i < C \tag{A4.7a}$$

$$\varepsilon + \xi_i - y_i + \boldsymbol{w}^T \boldsymbol{x}_i + b \leq 0, \quad \forall \alpha_i > 0 \tag{A4.7b}$$

A similar reasoning holds considering α_i^*.

Appendix 4.3. Bias—Variance Trade-off in Curve Fitting

In Chapter 2, we stated that a classifier with supervision is a system that after a learning phase (training) makes predictions using new data, that is, it achieves capabilities of generalization. This implies the assumption that the new data are produced by the same mechanism of the data used for training, with some additional noise component. The crux resides in the fact that we may fit data with an arbitrary accuracy—allowing many degrees of freedom (i.e., number of parameters or coefficients)—but we may run into an increasing risk when we apply the system to new data. This is known as the "Bias-Variance" Trade-off. A model with few degrees of freedom may have a high bias, as it fits the training data poorly but, in turn, it has a small variance. On the other hand, complex systems may have a low bias as they fit the data well, but they come along with a high variance. In the following, we show (see Bishop, 1995; Geman et al., 1992) that we can split the prediction error into two parts: the "bias" and the "variance."

Given:

- the true function, we want to approximate

$$f = f(\boldsymbol{X})$$

- the dataset for training

$$D = \{\dots x_i, t_i \dots\} \text{with target } t = f + \varepsilon, \text{ and expectation } E(\varepsilon) = 0$$

we train an arbitrary neural network to approximate the function f by

$$y = g(\boldsymbol{x}, \boldsymbol{w})$$

The mean-squared error of this network is

$$MSE = \frac{1}{N}\sum_{i=1}^{N}(t_i - y_i)^2$$

Now, suppose that we test a network on an arbitrary number of test points, that is, we pass to the expectation of *MSE*

$$E[MSE] = E\left[\frac{1}{N}\sum_{i=1}^{N}(t_i - y_i)^2\right] = \frac{1}{N}\sum_{i=1}^{N}E\left[(t_i - y_i)^2\right]$$

We scan split the expectation inside the sum with an augmentation trick:

$$E\left[(t_i - y_i)^2\right] = E\left[(t_i - f_i + f_i - y_i)^2\right]$$
$$= E\left[(t_i - f_i)^2\right] + E\left[(f_i - y_i)^2\right] + 2E\left[(f_i - y_i)(t_i - y_i)\right]$$

Recall that $t = f + \varepsilon$, so we can continue

$$E\left[(t_i - y_i)^2\right] = E\left[\varepsilon^2\right] + E\left[(f_i - y_i)^2\right] + 2\cdot\left(E[f_i t_i] - E[f_i]^2 - E[y_i t_i] + E[y_i f_i]\right)$$

Now, we note that

$$E[f_i t_i] = f_i^2$$

because f is deterministic and $E[f_i t_i] = f_i$

$$E\left[f_i^2\right] = f_i^2$$

Finally, we note

$$E[y_i t_i] = E[y_i(f_i + \varepsilon)] = E[y_i f_i + y_i \varepsilon] = E[y_i f_i] + 0$$

which exploits the fact that the expectation of the noise is 0 and it is statistically independent from the prediction of our system. Consequently, we get

$$E\left[(t_i - y_i)^2\right] = E\left[\varepsilon^2\right] + E\left[(f_i - y_i)^2\right] \tag{A4.8}$$

The *MSE* can be decomposed into a variance term induced by the noise, and the variance expressing the difference between the variables predicted by the true model and our estimate.

The term $E\left[(f_i - y_i)^2\right]$ can be further decomposed again using the augmentation trick

$$E\left[(f_i - y_i)^2\right] = E\left[(f_i - E[y_i] + E[y_i] - y_i)^2\right]$$

which becomes

$$E[(f_i - E[y_i])^2 + E\left[(E[y_i] - y_i)^2\right] + 2E[(E[y_i] - y_i)(f_i - E[y_i])\Big]$$

that is

$$\text{bias}^2 + \text{Var}(y_i) + \ 0$$

as the product *2E[..])* vanishes. We therefore can understand the *MSE* as composed of three parts: error introduced by noise, bias introduced by error in the fit, and the variance introduced by the model complexity

$$MSE = E\left[\varepsilon^2\right] + bias^2 + Var(y_i) \tag{A4.9}$$

Note, however, that we do not have any a priori idea of the variance of the noise. The only model-dependent terms are the $bias^2$ and $Var(y_i)$. Let us consider the case in which the true *MSE* corresponds to the variance of the noise. Trying a perfect fit of the data, we have a vanishing bias. Thus

$$MSE = \ E\left[\varepsilon^2\right] = bias^2 + Var(y_i) = Var(y_i)$$

and we may end up to fit noise.

CHAPTER 5

Applications with unsupervised learning

5.1 Introduction

Modern geophysical data analysis entails the accumulation of a large amount of information which often relates processes undergoing changes in time and space. In Chapter 4 we have learned about some applications of learning with supervision, i.e., we assume to have a-priori information, such as the membership of a pattern to a category, or a set of model parameters corresponding to observations made in the field. Supervised learning techniques can be extremely effective as they allow to design functions relating a set of observations ("input") to targets ("output"), whatever may be the complexity of the problem. We have also seen that the undesired phenomenon of overfitting can be fixed applying appropriate test schemes. The success of these methods, however, critically depends on the validity of the target information. This information can be either flawed or—even worse—simply not available, in particular when the observational data set is large and assigning the targets becomes overly tedious. Besides, Earth is an always changing planet, which means that a-priori information found to be valid for some time may become obsolete at some moment. A priori information found to be valid for data collected in a certain region is often found not being applicable in some other place.

Unsupervised learning techniques (such as the ones outlined in Chapter 3) do not depend on a priori targets as they focus on the internal structure of the set of observations. Similarity among patterns is defined on a metrics—the only a-priori information used in unsupervised learning. The choice of the metrics is strongly related to the features, and some data transformation may be necessary in order to obtain an appropriate measure for (dis)similarity among patterns.

The example applications presented in this chapter regards: centroid based clustering using various metrics; fuzzy and density based clustering as well as vector quantification, here in the form of "Self-Organizing Maps" (SOM). Clustering covers a wide field in the framework of unsupervised learning, and is successfully applied when the data set has internal heterogeneities. The possibility to reveal them, however, relies on the chosen method and the underlying structure of the set of observations. Having numerical data—here we widely focus on this kind of features—we may use a metric based on distances like Euclidean, Mahalanobis, Manhattan, etc., which lead to clusters with a regular shape, i.e., with spherical, ellipsoidal or hyperbolic symmetry. Other techniques

Advantages and Pitfalls of Pattern Recognition. https://doi.org/10.1016/B978-0-12-811842-9.00005-4

allow us to treat clusters with irregular shape, for instance density-based approaches. With these techniques we aim at identifying gaps on a local scale which interrupt the connectivity of data ensembles. In the first case, cluster centroids play a key role; in the second case the centroids have no specific meaning. Besides, we discuss data where the metrics regards the direction of feature vectors rather than their magnitude. The orientation of principal stresses in a seismic source is a typical example.

A further question regards the choice of the number of clusters. In partitioning clustering (see Chapter 3) we have to answer this question at the beginning. The definite "right" answer then is obtained in hindsight, i.e., carrying out various experiments with differing numbers of clusters and comparing the results obtained during various trials. In hierarchical clustering our clustering gets finer and finer as we proceed, but we have to decide when to stop. In approaches like density-based clustering we do not set the desired clustering, but we govern the process by setting some control parameters. Again, we have to carry out several experiments and compare the results in hindsight.

Besides detecting heterogeneities, cluster analysis aims at data reduction. This goal is reached at best when there are strong heterogeneities among the clusters. In the techniques with centroid-based measures, we can explain a good deal of the total data variability just by considering centroids. In the ANOVA terminology (see Chapter 3), we would say we can limit ourselves to the dispersion "between"— the dispersion of the cluster centroids only - neglecting the "within" part, i.e., the dispersion measured inside the clusters.

In the wider framework of data reduction, we find vector quantification techniques, in particular the SOM, which we introduced in Chapter 3.2. SOM comprise micro-clusters each of which represents a number of patterns, being prototypes of the patterns assigned. A specific feature of SOM is the projection of the prototypes in a low-dimensional representation space, typically a 2D map. The projection is carried out in a way that prototypes representing similar patterns are found close to each other on the map. The number of micro-clusters can be rather large. On the other hand, the detection of heterogeneities becomes of secondary interest. Vector quantization and SOM are often applied to very large data sets, being them structured or characterized by smooth, transitional regimes. At the same time, due to the projection in a low-dimensional representation space, the visualization of large and high-dimensional data sets becomes extremely simple and effective, providing that the projection fulfills the condition of "topological fidelity" which was discussed in Appendix 3.3.

We present a few applications of unsupervised learning techniques to geophysical data. The first application deals with seismic data recorded at Stromboli volcano, the feature vectors of which are given by spectral density values. We show how cluster analysis can be exploited to identify regimes of volcanic activity. Another application handles density-based clustering for the identification of clusters with irregular shape. These clusters are formed by the distribution of earthquake hypocenters which mark structural elements buried in the Earth crust. Further examples regard SOM for the purpose of data reduction. The possibility of the projection of high-dimensional data in a low-dimensional representation space, such as maps, allows an efficient application to regionalization problems. We show an example using climatic data in Europe, relating the zones defined with SOM to more conventional ones, such as the "Köppen zones". Beside regionalization, the representation of patterns using the SOM allows a straightforward visualization of changing pattern characteristics. Here we discuss the technique in the framework of volcano monitoring and the definition of alert procedures. A final example regards data for which we are interested in the orientation of the feature vectors rather than the absolute values of the components. This is the case when we deal with values normalized between −1 and 1. The greatest difference between two feature vectors is then encountered when they point into two opposite directions. In our example, we present the clustering of seismic moment tensor, trying to identify groups in which stresses acting in the seismic source show similar orientations.

5.2 Cluster analysis of volcanic tremor data

The seismic background radiation of active volcanoes like Mt. Etna or Stromboli provides useful information about the state of activity, even when visual monitoring or field surveys are hindered, for instance, by unfavorable meteorological conditions or during nighttime. At Etna (see Chapter 4), tremor is a persistent signal without clearly identifiable transients, such as explosions or earthquakes. The analysis of this signal is based on the extraction of some key parameters, in particular the RMS (Root Mean Square) amplitude. Besides, the spectral content allows more detailed analyses of the signal characteristics and their relation to volcanic activity.

In the framework of pattern recognition, we may process the spectra as feature vectors whose components are spectral densities measured in various frequency intervals. In other words, spectra are considered as multivariate random variables to which one can apply all the suit of techniques introduced in the Chapter 2 and 3.

To highlight the changes in the spectral characteristics of patterns with time, we may plot spectrograms, such as the one shown in Fig 4.1 (see Chapter 4). The full resolution of spectra would render the graphs difficult to read; therefore, it is convenient to simplify the spectral feature vectors by defining "frequency bins", i.e., averaging over the spectral amplitudes (see Fig. 5.1) measured in predefined frequency intervals. An example of such a simplified representation is shown in Fig. 5.2; the spectra were calculated from the background seismic radiation recorded at Stromboli volcano in 1993.

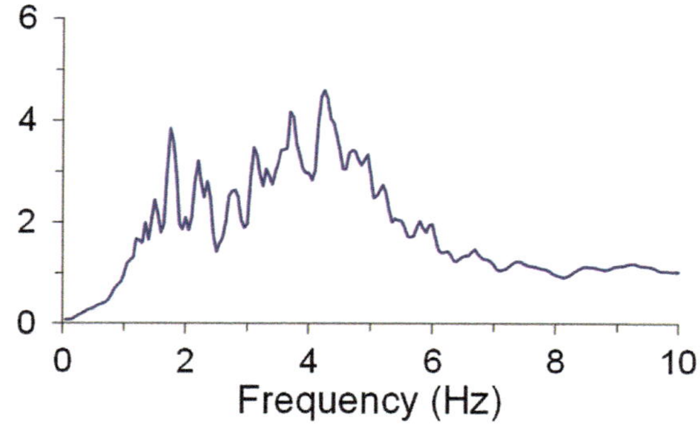

Figure 5.1
Stack of ca. 300 tremor spectra, which were calculated from the signal recorded by the vertical component of a short-period (1 s) velocity sensor at Stromboli in 1993. The original spectral resolution was 0.05 Hz.

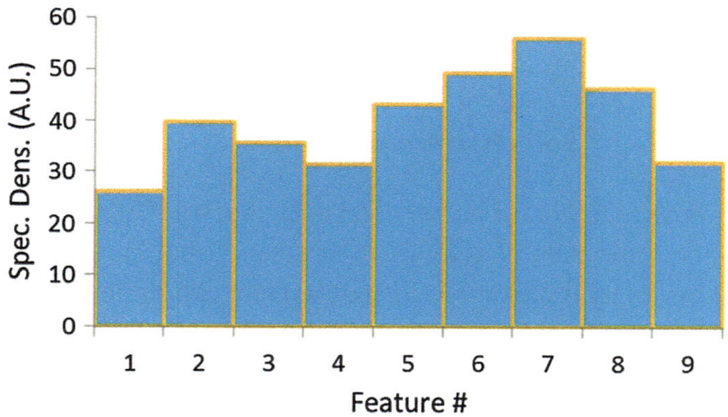

Figure 5.2
Spectral feature vectors as reported by Falsaperla et al. (1998). To reduce the number of components, the Authors considered nine frequency bins averaging over the spectral amplitudes in the intervals: 1–1.5, 1.55–2, 2.05–2.5, 2.55–3, 3.05–3.5, 3.55–4, 4.05–4.5, 4.55–5, and 5.05–5.5 Hz. The chosen overall frequency limits (1–5.5 Hz) covered the most relevant frequency range of the signal.

The use of a limited number of frequency bins also allows the statistical treatment, avoiding problems arising from feature vectors with many dimensions. For example, Falsaperla et al. (1998) considered nine frequency bins and calculated essential statistical parameters of the spectra, i.e., mean values and covariance matrix, for the tremor signal recorded at the vertical component of a short-period seismometer (Table 5.1).

The development of spectral characteristics depicted in Fig. 5.3 reveals changes not only with respect to amplitudes, but also to spectral shapes. For instance, at the beginning of the time window, the components #1 and #9 have the lowest amplitudes, whereas the

Table 5.1: Mean values and covariance matrix of tremor spectra (from Falsaperla et al., 1998).

x	μ	Covariance matrix								
1.00−1.5	26.2	69	86	66	51	49	45	55	42	35
1.55−2.0	39.6		198	122	94	72	48	74	36	30
2.05−2.5	35.6			149	125	132	141	168	127	94
2.55−3.0	31.3				124	132	155	279	239	178
3.05−3.5	43.2					206	353	324	238	21
3.55−4.0	49.2						350	352	324	238
4.05−4.5	56.0							449	347	271
4.55−5.0	46.1								339	248
5.05−5.5	34.2									196

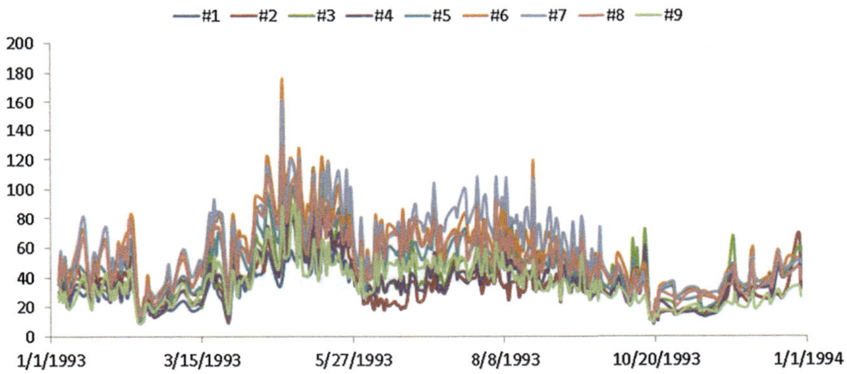

Figure 5.3

Development of the nine spectral features with time considered by Falsaperla et al. (1998). Amplitudes reported on the vertical axis are in digital unit.

component #7 achieves the highest amplitudes. In June 1993, however, the relatively lowest amplitudes correspond to the component #2. We may now ask whether - beside mere overall RMS amplitudes directly available from the raw time series - there was some relation between spectral characteristics (both shape and spectral energy) and the state of the volcano. One can try to answer this question applying cluster analysis to the nine-dimensional feature vector.

It is a good praxis to start with a simple clustering technique. Here we have chosen the conventional K-means. The K-means technique is a partitioning clustering method, for which we have to choose the desired number of clusters a priori. Clusters are formed on the base of the Euclidean metrics (see Chapter 3). Consequently, we suppose that the clusters have (hyper)spherical hulls. The separation of the dispersion follows essential concepts of ANOVA, in this case the "within" dispersion (sum of distances of cluster members with respect to the centroid) in the clusters, which we wish to minimize. This entails that the "between" dispersion, i.e., the distances measured between the cluster centroids, is maximized. The separation of "within" and "between" dispersion has allowed to establish formal criteria for the selection of the number of clusters, which is certainly an important advantage of the K-means clustering. One of the most popular indices is the so-called Davies Bouldin ("DB") index (Davies and Bouldin, 1979), whose definition is given in Appendix 5.1.

First, we carry out a sequence of clustering trials, changing the number of clusters. Eventually, we adopt the number of clusters that achieves the lowest DB index (see Fig. 5.4).

Fig. 5.4 can be obtained using the package "KKAnalysis" coming along with this book. Here we carry out the clustering trials starting with two clusters up to a partition with seven clusters. Based on the results of Fig. 5.4, one adopts a partition with three clusters as a reasonable choice. The sequence of cluster IDs can now be plotted over time (Fig. 5.5).

In Fig. 5.5, cluster #3 prevails in April and May 1993; cluster #1 is widespread almost all time, whereas cluster #2 appears in disrupt time intervals. Even though the occurrence of

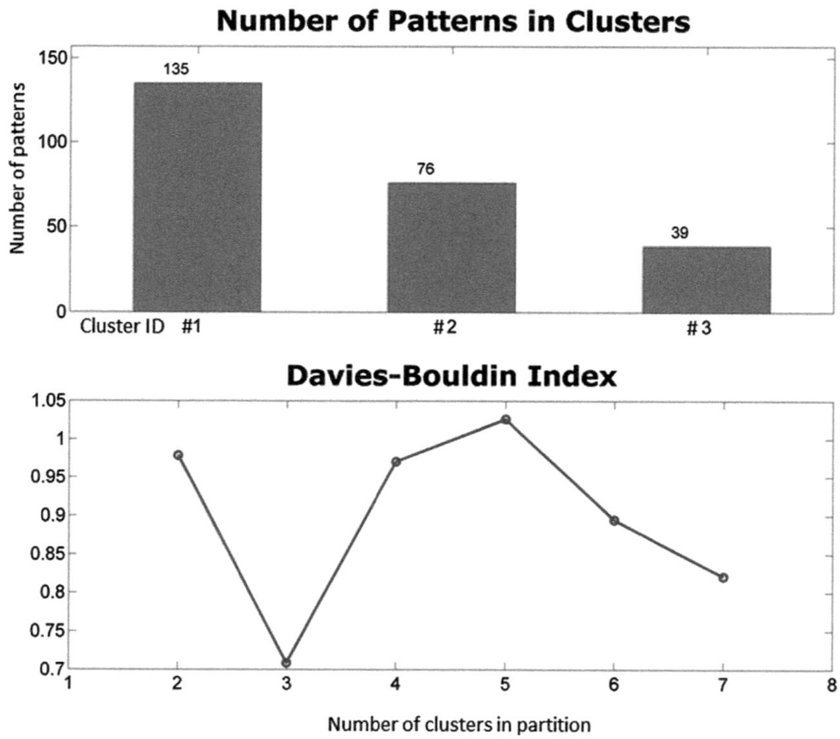

Figure 5.4

Choosing the number of clusters using the Davies-Bouldin index.

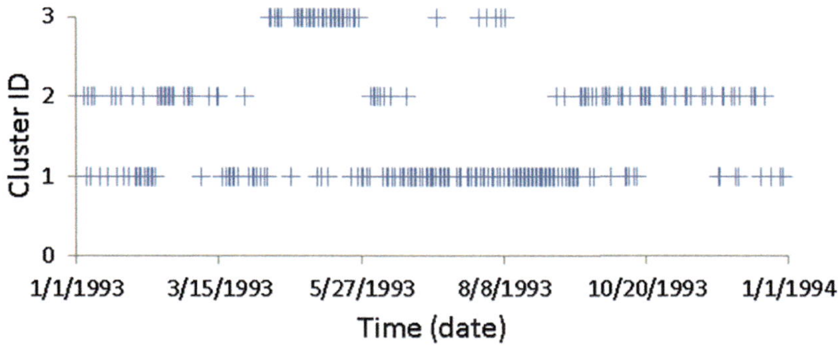

Figure 5.5

Sequence of clusters from K-means. Note that the cluster labels #1, #2, and #3 are categorical. Cluster #1 is the largest one, clusters #2 and #3 are the second and third one with respect to number of cluster members.

cluster #3 coincides with a lava effusion reported between 16 and 18 May 1993 (see Falsaperla et al., 1998), possible prevailing regimes do not turn out clearly. One of the reasons for the somewhat unsatisfactory picture may reside in the fact that data groups do not actually form (hyper)spherical bodies. In other words, there is a change in the spectral shape within groups besides the variations encountered between groups.

In alternative to K-means, Principal Component Analysis (PCA, see Chapter 2) may provide more immediate clues. Table 5.2 reports the three largest eigenvalues and corresponding eigenvectors of the data set (for the full PCA, see Falsaperla et al., 1998).

The largest eigenvalue in Table 5.2 explains ~80% of the total variance in the data set. The largest components of the corresponding eigenvector belong to the features #6, #7, #8, i.e., the spectral values in the intervals 3.55–4 Hz, 4.05–4.5 Hz, and 4.55–5 Hz, which means that the variability in the data set is strongly controlled by these frequency bands. On the other hand, the largest components corresponding to the second largest eigenvalue are found for the features #2 and #3 (1.55–2 Hz and 2.05–2.55 Hz). We therefore surmise that fluctuations in spectral amplitudes do not occur independently, but are mutually related. As K-means forms (hyper)spherical clusters, possible correlations among the

Table 5.2: The three largest eigenvalues and corresponding eigenvectors of the covariance matrix reported in Table 5.1.

λ	Eigenvectors e^T								
1604	0.08	0.12	0.23	0.23	0.35	0.45	0.5	0.44	0.33
304	0.33	0.73	0.41	0.28	0.04	−0.16	−0.11	−0.22	−0.17
58	−0.04	−0.22	0.15	0.42	−0.09	0.41	−0.73	0.20	0.09
..	..	..	..	..	..	..	..	..	..

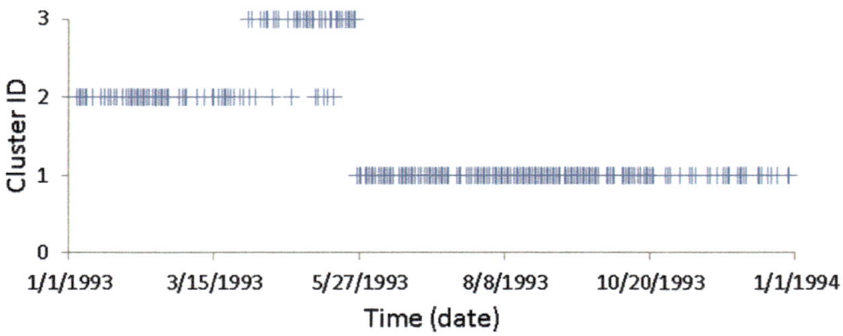

Figure 5.6
Sequence of clusters using the Adaptive Determinant Criterion.

feature components within a cluster are not accounted for. For this reason, the use of Euclidean metric in K-means is not the best choice for the afore-mentioned problem.

In Chapter 3 we introduced clustering metrics that account for correlations among the components of the feature vectors. Let's consider the Adaptive Distance Criterion (Section 3.1.2.1), which exploits such a metrics; in this case, we calculate the individual dispersion matrices $\boldsymbol{S}_j$ obtained for each cluster and minimize the sum

$$D_{tot} = \sum_{j=1}^{k} \sum_{i=1}^{m_j} \left(\boldsymbol{x}_i - \overline{\boldsymbol{x}}_j\right)^T \boldsymbol{G}_j \left(\boldsymbol{x}_i - \overline{\boldsymbol{x}}_j\right) \tag{3.14}$$

where $\boldsymbol{G}_j = (\det \boldsymbol{S}_j)^{1/l} \boldsymbol{S}_j^{\ -1}$ (see Chapter 3).

Recall that by using $\boldsymbol{G}_j$ one gets a metric that resembles the Mahalanobis distance, beside the fact that one works with the inverse of the dispersion matrix instead of the covariance matrix. In this way, we can form clusters whose shape resembles, for instance, elongated ellipsoids.[1] Even clusters which apparently intersect each other can be separated.

In comparison to Fig. 5.5, the sequence of clusters in Fig. 5.6 allows us to identify well developed regimes. Data belonging only to cluster #2 mark the beginning of 1993. Similar to Fig. 5.5, cluster #3 shows up between April and May 1993. This further supports the hypothesis of a correlation with the state of unrest of the volcano, with magma closer to the surface until the start of lava emission on 16 May 1993. From June on, data belonging to cluster #1 are the only ones present (Fig. 5.6). Following Falsaperla et al. (1998), this regime may be related to the return of magma at depth. Explosive sequences at the end of 1993 actually yielded the ejection of rather cold, detritic material, with no juvenile

[1] In general, the elements which distinguish two clusters from each other are given by quadratic functions. See Appendix 2.1 for a more detailed discussion.

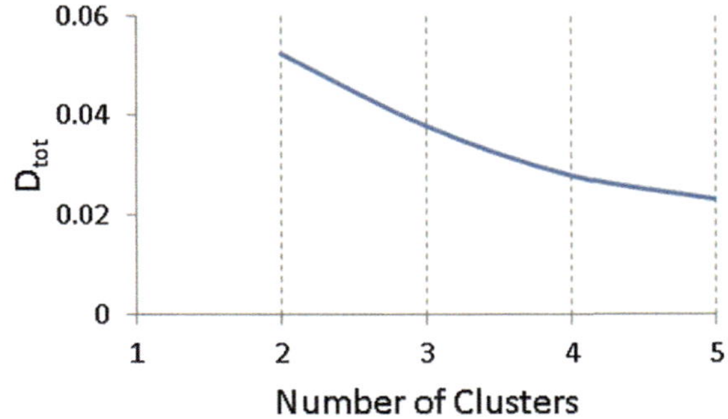

Figure 5.7
Sum of distances (D_{tot}) versus a priori choice of the number of clusters.

magmatic component. In the volcanological context, this is interpreted as an evidence of a relatively low position of the magma column with respect to the previous period.

Though these results look sound, we are still left with the problem of choosing the "right" number of clusters. Formal criteria, such as the Davies-Bouldin Index or alternatives (see Appendix 5.2–5.4), depend on the relation of heterogeneity measured within and between the clusters. In the case of the Adaptive Distance Criterion, there is an individual metric for the distance of a cluster member to its centroid. This entails that for the metrics "between" the cluster centroids we have difficulties to identify a meaningful choice.

In a heuristic approach we may plot the total sum D_{tot} (Eq. 3.14) over the number of clusters (Fig. 5.7). Assuming two clusters as the minimum number to choose, we note that passing to three clusters there is an improvement of D_{tot} of $\sim$0.015 points. With more clusters, the curve in Fig. 5.7 tends to flatten out. We therefore may limit ourselves to a small number of clusters, such as indeed three, but not more than four.

As a further diagnostic, we may look for the detection of crucial, new heterogeneities with substantially modified partitions. For example, comparing the clustering information using partitions with three (Fig. 5.8) and five clusters (Fig. 5.9), we notice that cluster #1 is quite stable and with a low degree of heterogeneity. In both the partitions, this cluster essentially occurs from June on. Conversely, the other clusters are diversely distributed in time (from January to May). The increase in the number of clusters in the partition (from three to four) yields the splitting of clusters in that time span without an effective gain of new information (Fig. 5.9).

5.3 Density based clustering

The clustering with K-means and Adaptive Distance highlights data groups whose hulls have rather simple and mathematically well-described shapes. However, many problems of

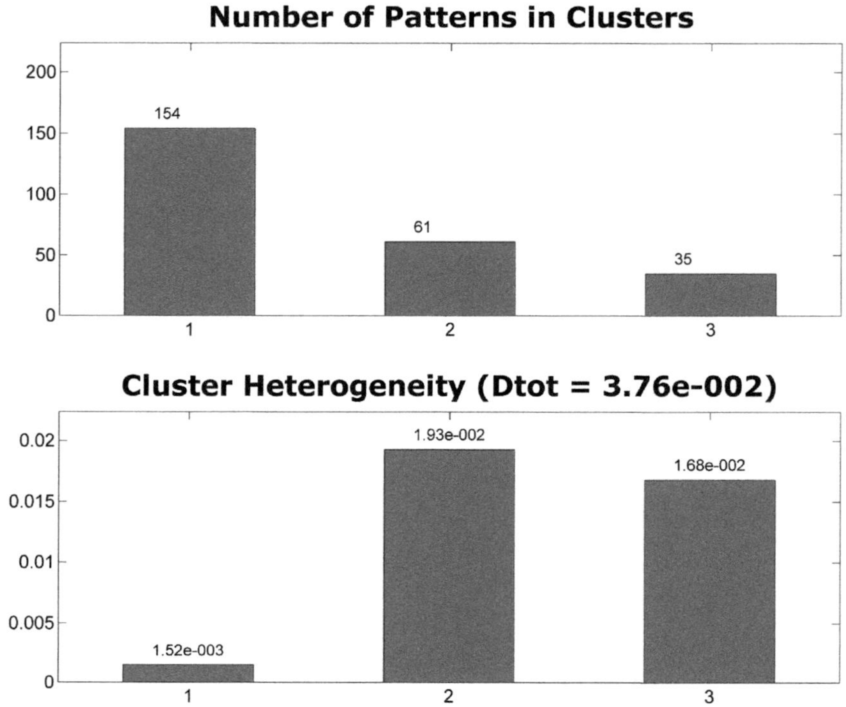

Figure 5.8
Overall clustering information using three clusters. The figure reports the number of patterns and heterogeneity (D_{tot}) in each cluster.

geology and geophysics deal with data samples with irregular-shaped bodies. This is the case of the spatial distribution of seismic sources in the context of seismic hazard, in which earthquake location is a key question.

The occurrence of earthquakes is strongly linked to the geodynamic framework of a region. Strong earthquakes concentrate in places where there are strong deformations of the Earth crust (the so-called solid Earth). The occurrence of strong ground motion outside tectonically active zones is also a matter of concern, as the shaking may have been caused by man-made sources, in particular during nuclear armament testing. The advent of modern seismic monitoring instruments has allowed the record and location of even small magnitude events with high accuracy. Earthquakes with small magnitude occur more frequently than the strongest ones, roughly 10 times for each magnitude unit. On the other hand, even small earthquakes are linked to deformation processes in the Earth crust. This link makes them efficient tracers for active tectonic structures, in particular those buried at depth, invisible to geologists.

Commonly, tectonic faults are depicted as plane elements. Earthquake hypocenters following those elements can be supposed to form groups described in simple

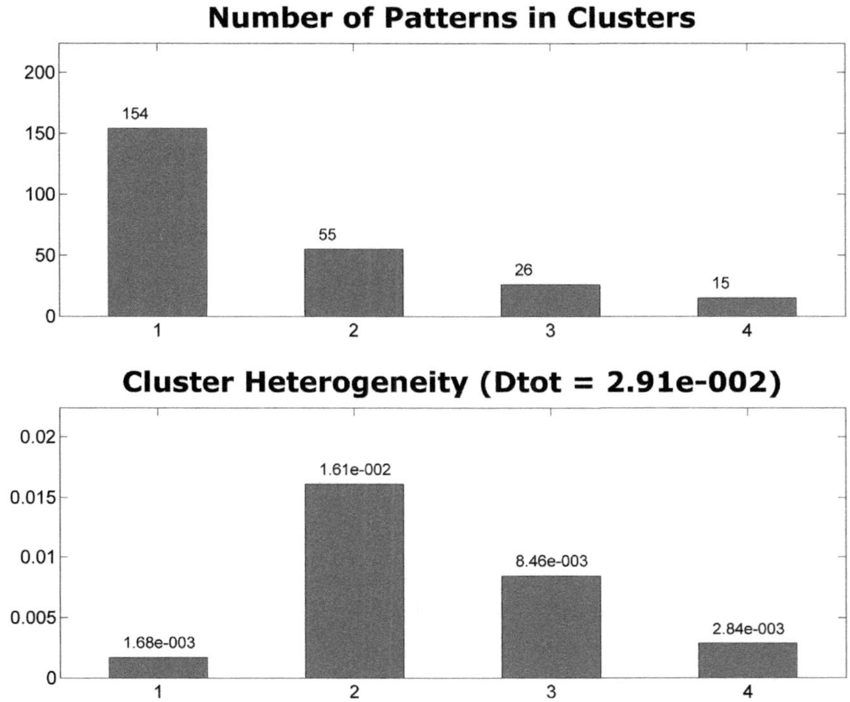

Figure 5.9
Overall clustering information using four clusters. The figure reports the number of patterns and heterogeneity (D_{tot}) in each cluster.

mathematical terms. K-means or Adaptive Distance based clustering may work fine for the identification of groups with such a geometry. Famous examples are the seismicity maps along the San Andreas Fault (USA), where the hypocenters of large and small earthquakes line up on linear elements, following the main fault and secondary tectonic structures. In many cases, however, tectonic elements have quite complex geometries. For example, normal faults are known to exhibit a "listric" geometry, i.e., a shallow dip angle at depth, which steepens up toward the surface, where it may become vertical. Bow-formed structures are typical, for instance, in collisional boundaries with the presence of an indenter, where deformation vectors delineate a fan-shaped arrangement. Around extensional zones, shear fractures may show a geometry similar to a circular rosette (see Fig. 5.10)

Bow-shaped elements on Mt. Etna (Italy) can have origin from the presence of an extensional center in the area of the summit craters. Extension may be caused by the upraise of magma from the deeper parts of the volcano. At the same time, the topography of the mountain itself causes an outward directed load on its flanks, which is particularly intense on the eastern quadrant, adjacent to the Ionian Sea. On the other hand, in a

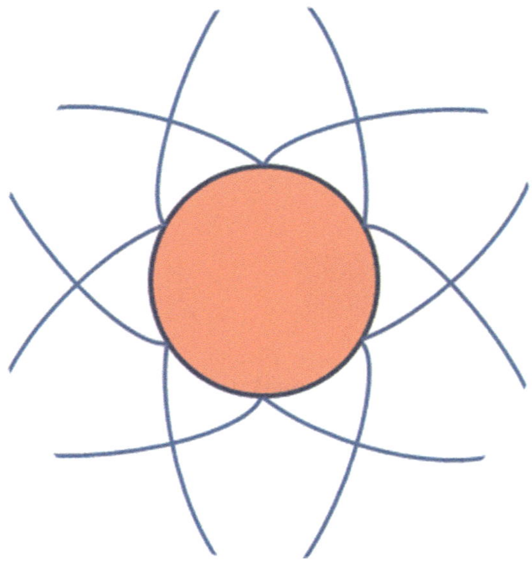

Figure 5.10
Shear trajectories around an extensional circular hole. *Redrawn form Jaeger and Cook, 1979.*

volcanic context like Mt Etna, the deeper structure is characterized by irregular heterogeneities separating various geological bodies from each other. Those bodies can be unveiled by both tomographic investigations as well by the location of hypocenters, highlighting possible pathways for the ascent of fluids and magma, and shedding further light on the feeding system of the volcano. The link between earthquake locations and the 3D structure of the volcano (both revealed from seismic velocities and attenuation parameters) was clearly proven by Chiarabba et al. (2004).

In Fig. 5.11 we apply the DBSCAN-STRATA code to a data set consisting of 328 hypocenter locations of a seismic swarm, which occurred in 2003 in the context of a major eruption of Mt Etna (see Mostaccio et al., 2013). The clustering clearly reveals non-linear elements, such as an arc-shaped group in the northern part of the mountain, aligning along the "Pernicana Fault System". This structure is a shallow fault along which horizontal displacement prevails, and with deformation rate that may reach even several cm per year. The body marked by blue diamonds matches the high-velocity (and low-attenuation) body highlighted by seismic tomographic studies (e.g., Chiarabba et al., 2004).

To obtain the clustering shown in Fig. 5.11, we ran the DBSCAN-STRATA code with $k = 10$ nearest neighbors. This value was proposed as a default by the software after selecting the option "Stratify". The setting of the "epsilon" parameters turns out to be uncritical. The stability of the result with respect to the "epsilon" settings is also claimed in the paper by Cassisi et al. (2013; see also Cassisi, 2013), where one can found more

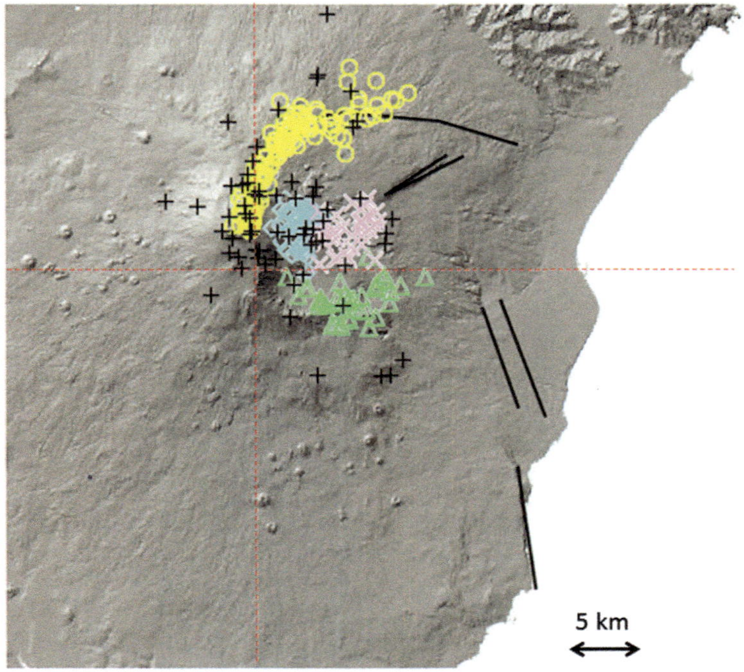

Figure 5.11
Density based clustering of earthquake hypocenters at Mt Etna in October 2003 using the DBSCAN-STRATA code by Cassisi et al. (2013). See https://sites.google.com/a/ingv.it/carmelo-cassisi/home/software/dbstrata. Two dashed lines give the reference x=0 (corresponding to 15° E), and y=80 (corresponding to 37.72° N). We thank Tiziana Tuvè, who provided the hypocenter localizations shown in the figure. The map includes some major tectonic elements, for more details see Azzaro et al. (2012).

details on the software.[2] One of the advantages of this code is that it does not require the a-priori selection of the number of clusters. The formation of new clusters is invoked when a new pattern is encountered which is not density reachable. Using $k = 10$ for the nearest-neighbor parameter we obtain 4 clusters and 70 samples considered as noise. Nonetheless, the picture shown in Fig. 5.11 may change, in particular when the nearest neighbor parameter k differs. In Fig. 5.12 we compare two runs of the clustering, one using $k = 8$ (Fig. 5.12A) and the other one with $k = 12$ (Fig. 5.12B). In the first case, the

[2] The DBSCAN-STRATA program reports an "Approximate" Davies-Bouldin Index, which must be met with some caution here. As clusters have any type of shape, such as bows, snakes etc., the internal dispersion measured with respect to a centroid provides poor information whether the classes are compact, in the sense that there are no important losses of connectivity within a cluster. It is tempting to express the internal heterogeneity by means of some average distances in the unit (hyper) volume (see Section 3.2.3); nonetheless the role of noise remains unclear (such as the black crosses in Fig. 5.11). In fact, we wish to have clusters as compact as possible, but limiting the amount of data assigned to the noise group.

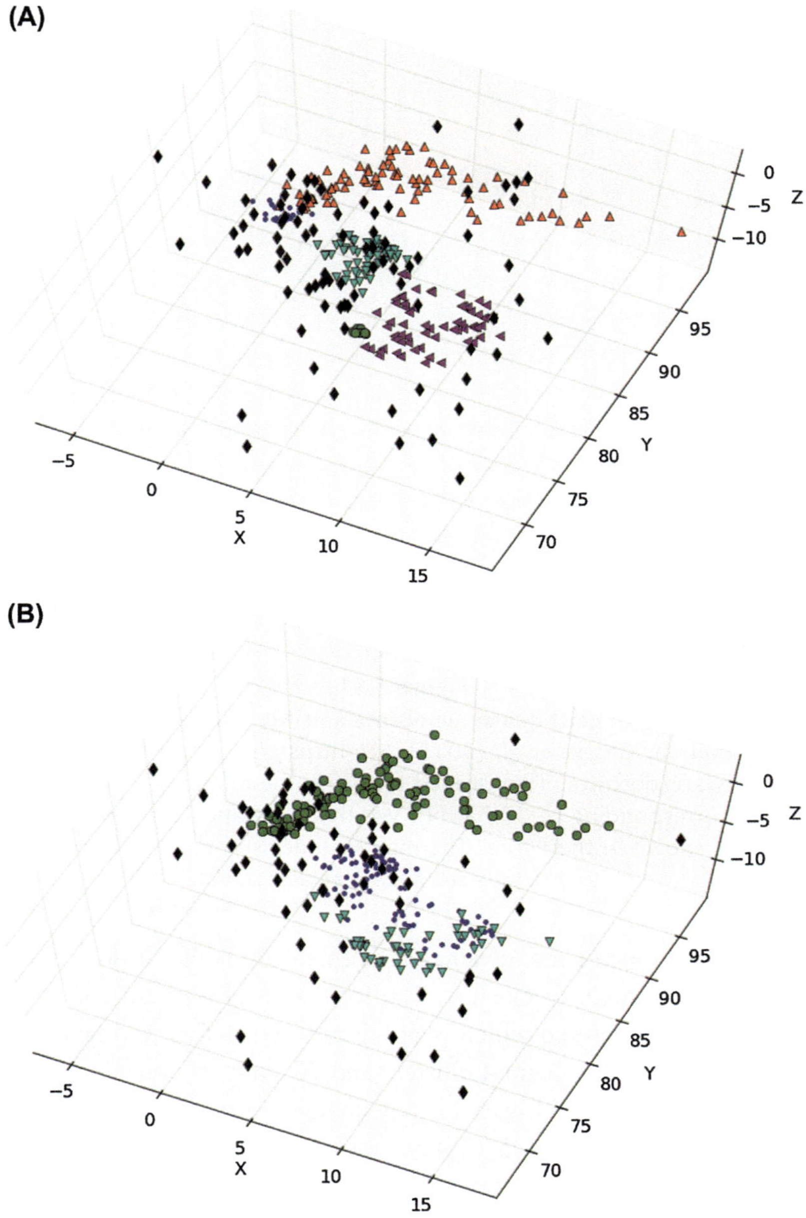

Figure 5.12

DBSCSAN-STRATA clustering with $k = 8$ (A) and $k = 12$ (B). Horizontal coordinates are referenced to 15° E for x = 0 km, and 37° N for y = 0 km.

code finds 5 clusters plus 105 samples assigned to the noise group (black symbols), while it gets 3 clusters and 85 samples in the noise in the second case. Based on these results, our choice of the clustering with $k = 10$ is a good compromise in terms of number of clusters obtained and samples assigned to the noise.

5.4 Climate zones

In Chapter 1 we mentioned weather as an example of all-day multivariate phenomenon characterized by varying data, such as temperature, humidity, presence of clouds or sunshine, and the amount of precipitation. Considering a single site, we define its meteorological conditions as "climate". Culling together all sites where relevant meteorological data are collected allows us to identify climate zones, for instance zones with dry and hot summers and cold winters. Such conditions are commonly referred to as "Continental Climate" as they are often met inside extended lands distant from the sea.

The first attempts of global climate classification date back to the times of Pythagoras, when Greek philosophers developed a first zoning, in particular aiming at gaining some clues about the figure of the Earth (see Sanderson, 1999). In fact, the word "climate" derives from "clima-ata", defined as the "slope of the Earth from equator to north pole" (Sanderson, 1999; Barnhardt, 1957). Moving on from the concept of a spherical Earth, Pythagoras' disciple Parmenides defined five zones, two of which cold, two temperate, and one torrid. The simple Greek system became more and more questionable with the increasing knowledge of the Earth, in particular after the great discoveries made from the 15th century onwards. Indeed, the lands in the New World have climates that cannot be understood as simple extrapolations of European climates, as they exhibit strong irregularities and contrasts with respect to both temperature and moisture supply.

The first comprehensive world-climate classification was proposed by the German-Russian scientist Wladimir Köppen (Köppen, 1900). Köppen, a plant physiologist, surmised that vegetation can serve as a proxy of climatic elements. He chose the five vegetation groups established by the French botanist De Candolle (De Candolle, 1874) to distinguish five climatic zones: "A" with plants typical of the Torrid Zone; "B" with plants of the Dry Zone; "C" with Temperate Zone vegetation, and finally "D" and "E" for the Frigid Zone, where typical plants of the rigid weather grow. With the increasing amount of available numerical weather parameters collected worldwide (mainly temperature and precipitation/moisture), new discrimination rules were established to replace or refine these climatic zones. For instance, the Köppen zone "A" is defined on the base of temperature (high temperature), while a second letter expresses the degree of moisture; for example, "Af" stands for tropical and rainy conditions.

Four out of the five bio-climates defined by Köppen are defined thermally. Monthly temperature maxima of 0°, 10^0, and 18°C form boundaries among: ice, cold winter wet, warm winter wet, and tropical wet climates (Prentice, 1990). The boundary between moist and dry climates lies close to the point where annual precipitation is less than annual potential evapotranspiration. In the most commonly used Köppen-Geiger scheme (see Köppen, 1936), the five primary types are subdivided according to the seasonality of precipitation and the severity of the dry season (when existing). Further distinctions are based on additional thermal criteria, among which the mean temperatures of the coldest and warmest month. For the most recent versions of the Köppen-Geiger scheme see Kottek et al. (2006), and Peel et al. (2007).

The modern versions of Köppen's scheme and its alternatives are based on numerical data, such as temperatures, precipitation, evapotranspiration, etc. However, they come as a classification, based on rules established a priori, fixed in order to reproduce the older vegetation-based scheme using the one defined by numbers. Various efforts have been undertaken exploiting the numerical data in a formal classification scheme rather than by rules and thresholds fixed by the experts. Cannon (2012) proposed a global climate classification obtained by Multivariate Regression Tree (MRT). An MRT is a binary tree with nodes of simple "below/above" decision rules applied to predictors (our climate parameters). They have a hierarchical structure similar to the hierarchical clustering described in Chapter 3. Cannon's study aims at formulating a rule- and computer-based global climate classification, comparing it to the homogeneity of the zones defined with the Köppen-Geiger classification. Cannon applied rules that describe the Köppen-Geiger scheme in terms of variables derived from long-term climate normals of monthly mean temperatures and precipitations.

At the base level, Cannon (2012) divided the data into two subgroups using the criterion of the "mean annual temperature" (MAT threshold), for instance set up at 12°C. In the next level, the search climbs up the tree, setting splits for new branches. At each split node, a decision is made on which branch is followed for a specific pattern. The decision process continues until patterns are assigned to their corresponding class. Throughout an exhaustive search for the best splits, each value of the predictors is considered as a possible threshold. The gain in homogeneity after each hypothetical split is monitored considering

$$\Delta WSS = WSS_A - (WSS_B + WSS_C) \tag{5.1}$$

where WSS_A is the dispersion measured in the parent class and WSS_B, WSS_C are the dispersions measured in the two child classes created during the splitting.

From the formal point of view, binary decision trees come with the advantage of simplicity and ease of understanding. They work fine when classes can be separated linearly, and decisions are based only on one component of the feature vector. In other words, the discriminating element is parallel to the axis of this component. As class boundaries are rarely that simple, many nodes of decision trees may be needed in order to mimic more complex class boundaries.[3]

Unsupervised learning and clustering may offer a more efficient answer to the identification of climate zones, as these techniques are data driven. Decision boundaries are ideally detected along heterogeneities between groups of patterns; at the same time we can define centroids around which many patterns are concentrated. The characteristics of those centers, together with some measure of dispersion around them, lead to a considerable data reduction facilitating the task of zoning. Zscheischler et al. (2012) used multivariate statistics and clustering to a sequence of parameters measured at a monthly rate worldwide. Beside the classical parameters already considered by Köppen - temperature and precipitation - the authors considered the "Downward Short-Wave Radiation", "Enhanced Vegetation Index" (known to be responsive to structural variations in the canopy) and the "Fraction of Absorbed Photo-synthetically Active Radiation" (related to the primary productivity). The authors measured the quality of clustering by the amount of dispersion explained in terms of the sum of "within" dispersions encountered inside the clusters $\mathbb{C}_i$, i.e,

$$WCSS_k = \sum_{i=1}^{k} \sum_{\boldsymbol{x}_j \in \mathbb{C}_i} ||\boldsymbol{x}_j - \boldsymbol{\mu}_i||^2 \tag{5.2}$$

with $\boldsymbol{\mu}_i$ being the centroid of $\mathbb{C}_i$. We get the degree of explained variance from the ratio

$$EV = 1 - \frac{WCSS_k}{WCSS_1} \tag{5.3}$$

which is 0 for $k = 1$ (all data in one cluster) and is one for $WCSS_k = 0$. The choice of the suitable number of clusters remains open. Dealing with a K-means clustering, one could opt for the Davies-Bouldin Index or the Dunn Index. The authors used the "Variation of Information" (*VI*) as a distance measure of clustering (see Appendix 5.5). This allows them to evaluate the stability of clustering, for instance, restarting it varying the initial conditions (such as the often randomly defined initial clustering).

Zscheischler et al. (2012) performed resampling experiments in order to evaluate the stability of clustering. K-means clustering was applied to a subset of data selected

[3] Imagine a straight line dividing two areas in a sheet. We can use the equation of the line for the description of the discriminating element. In a binary strategy we compose the line as a sequence of steps in vertical and horizontal direction. Many small steps may be necessary to mimic the line. However, the binary strategy has its appeal when highly non-linear discrimination functions are needed. It adapts in a mathematically simple way to those non-linear elements.

randomly. In a subsequent step, the left-out samples were assigned to the clusters considering the closest centroid. In this way, a full clustering was obtained from a part of the whole subset. We can compare this to the original clustering using the *VI* index. In a boot-strap like approach, it is possible to repeat this procedure, every time carrying out a new random selection of events used for clustering. We therefore get an idea on the uncertainty of the *VI* distance estimation. From a comparison between the *VI* indices and the instability, Zscheischler et al. (2012) inferred a cluster number of 12 or 13 as a reasonable choice, even though difficult to handle as described in the following.

Using the PCA the authors investigated how well the data matched the original Köppen classification (see Fig. 5.13). For the sake of simplicity, they referred to the five major Köppen zones (A to E), to each of which they assigned a color code. Only the two eigenvectors with the largest eigenvalues—the first two Principal Components—were considered from the results of the PCA, as the further principal components had small eigenvalues and could be neglected. Then the authors plotted the colored symbols

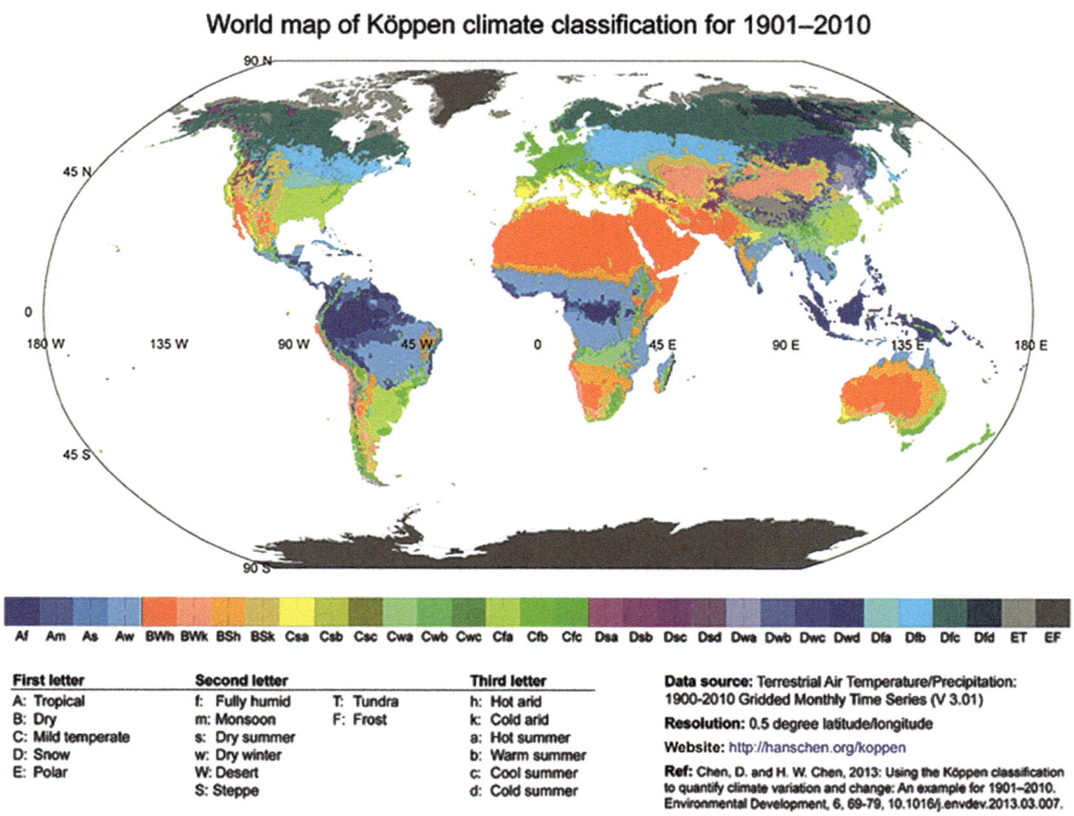

Figure 5.13
The Köppen classification (see Chen and Chen, 2013).

representing the Köppen zones in a system of axes with the two most relevant principal components. They obtained an immediate graphical information on how the data used in clustering and PCA were related to the Köppen zones. It turned out that colored pixels formed well identifiable areas but no strong heterogeneity could be distinguished. The authors concluded that the Köppen zones—in terms of the variables considered in their study—form rather transitional regimes, without erratic changes in the parameters when passing from one zone to another (see Fig. 5.14).

The clustering methods shown so far are based on identifying groups along heterogeneities, or gaps between objects. Patterns having a small distance from each other form clusters, and it may be sufficient to deal with parameters describing those clusters rather than considering the patterns one by one. In this way, we considerably reduce the amount of data, neglecting only a minor part of the information. Beside the identification of heterogeneities, data reduction is indeed a key issue in unsupervised learning. Actually, in contexts with transitional regimes, the aspect of data reduction becomes the main target, whereas revealing heterogeneities turns into a minor issue. In our example of the climate data, we probably face with such a situation. Having more than four or five clusters produces a picture which is difficult for the analysist to understand. Therefore, in their comparison of their PCA to the original Köppen zones, Zscheischler et al. (2012) limited

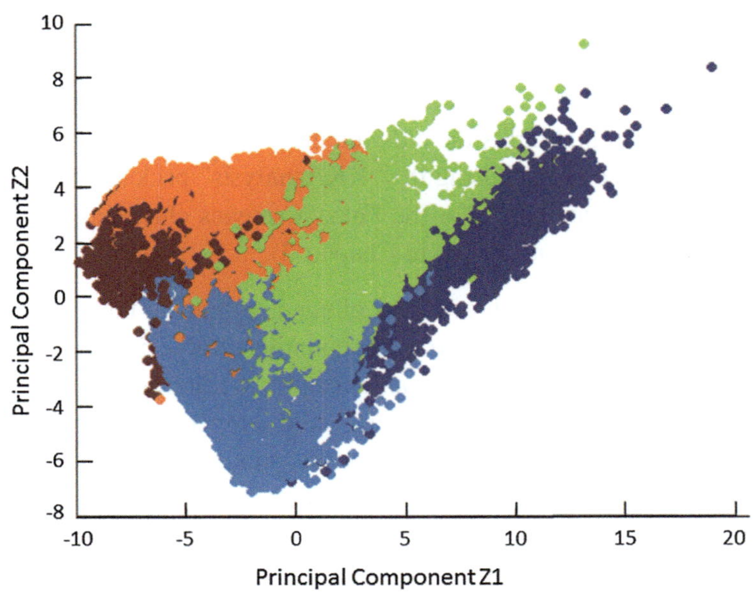

Figure 5.14

Projection of the data analyzed in Zscheischler et al. (2012) onto the first two principal components. Data points of the northern hemisphere are shown. Their colors were selected according to the five main climates of the Köppen-Geiger classification to which they belong. *Modified from Zscheischler et al. (2012).*

themselves to a simplified representation, considering only the main five climate regimes from A to E (instead of 12 or 13 previously suggested as a reasonable choice).

In case of transitional regimes, we may adopt fuzzy clustering as an alternative to the crisp schemes applied in our previous examples. In Chapter 3 we mentioned Fuzzy C-means, which—similar to K-means—is based on the Euclidean metric. However, instead of assigning a pattern exclusively to a single category, it allocates a class membership vector that expresses the varying degrees of similarity of a pattern to each cluster. Even though fuzzy clustering may be a convenient choice where no strong heterogeneities exist, we are still left with the problem to represent the results of clustering - here the geographical distribution of the class-membership values. Colored symbols are a possibility, especially when the number of clusters is small. For example, a classification with three clusters can be envisaged for the European Continent. Here we may identify three climate zones considering the Mediterranean region as one macro-unit, Western and Northern Europe as a second one, and finally a third region encompassing the Northern and Eastern part of the continent. We provide a graphical example of this separation based on the application of Fuzzy C-means to a data set freely downloaded from Iten (2016; http://www.iten-online.ch/klima/klimatabellen.htm). It consists of data collected at 348 stations deployed in Europe. Table 5.3 reports the data of a few stations, located in Denmark, Germany, and Italy.

Three is a "magic number" for our classification purposes as we can assign an RGB color code to each of the classes, so that cluster 1, cluster 2, and cluster three control the saturation in red, green, and blue, respectively. Mixing the three contributions, we design symbols the color of which highlights the class membership degrees. Fig. 5.15 depicts the results of the classification obtained using the "KKAanalysis" package by Messina and Langer (2011), which offers an option for the Fuzzy C-means clustering with a fuzzy exponent $q = 2$ (see Eq. (3.15) and (3.16) in Chapter 3).

In Fig. 5.15 we clearly identify the three major climate regimes in the European continent. The red triangles dominate in the Mediterranean countries; Table 5.3 reports a few data from the station Otranto (Southern Italy) as an example. Mediterranean sites are characterized by hot temperatures in summer, and mild temperatures in winter; precipitations mainly occur in fall and winter. For the station Otranto the class membership value for red is ~ 0.94. Blue colors dominate in Central and Eastern Europe (see the values in Table 5.3). The blue cluster covers regions with moderately warm summers and cold winters; precipitations (rain and snow fall) occur all the year round. The class membership corresponding to the blue cluster for the station Berlin is ~ 0.87. The area around the Northern Sea has dominant green cluster. The Station Roskilde (see Table 5.3; class membership for the green cluster 0.86) is a typical example for such patterns. Compared to the station Berlin, summer temperatures are lower, winters are less

Table 5.3: Climate data set collected at sites in Europe. The parameters reported are the maximum temperature (during daytime), minimum temperatures (during nighttime), duration of daily insulation ("Sun(h)"), and days with Precipitation ("Pr".(d)") encountered within a month.

Roskilde (Denmark), Latitude 55.6419 N, Longitude 12.0878 E												
Mon	Jan	Feb	Mar	Apr	May	Jun	Jul	Aug	Sep	Oct	Nov	Dec
Day	2	2.1	5	10.4	16.1	19.4	21.8	21.2	17.5	12.1	7.3	4.2
Night	−2	−2.5	−0.8	3.1	7.5	11.2	13.6	13.5	10.5	6.7	3.3	0.7
Sun(h)	1.2	2	3.8	5.4	7.9	8.2	7.7	6.7	5.2	2.8	1.1	0.6
Pr. (d)	17	13	12	13	11	13	14	14	15	16	16	17
Berlin (Germany), Latitude 52.520 N, Longitude 13.404 E												
Mon	Jan	Feb	Mar	Apr	May	Jun	Jul	Aug	Sep	Oct	Nov	Dec
Day	1.8	3.5	7.9	13.2	18.6	21.8	23.1	22.8	18.7	13.3	7	3.2
Night	−2.9	−2.2	0.5	3.9	8.2	11.4	12.9	12.4	9.4	5–9	2.1	−1.1
Sun(h)	1.5	2.6	3.9	5.2	7.1	7.4	7	6.8	5.2	3.6	1.7	1.2
Pr. (d)	10	9	8	9	10	10	9	10	9	8	10	11
Otranto (Italy), Latitude 40.4643 N, Longitude 17.2470 E												
Mon	Jan	Feb	Mar	Apr	May	Jun	Jul	Aug	Sep	Oct	Nov	Dec
Day	12.1	12.8	14.6	18.1	22.3	27	30	29.9	26.7	22	17.3	13.7
Night	6.3	6.6	8.1	10.9	14.8	19.1	21.8	21.8	19.4	15.4	11.6	8.2
Sun(h)	4.1	4.6	5	7.4	9.2	10.5	11	10.5	8.2	6.4	4.7	3.5
Pr. (d)	7	6	6	5	4	3	1	2	4	6	6	8

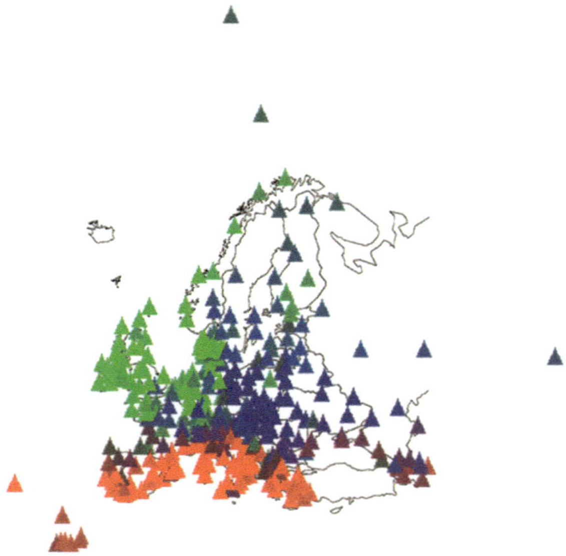

Figure 5.15
Fuzzy clustering of climate data in Europe. Mediterranean sites are characterized by a prevailing 'red' cluster, blue colors are found at sites with continental conditions, whereas the green cluster membership prevails in NW Europe.

cold, and rain (or snow) fall happens all over the year, but more frequently. Beside these prototype zones (with class membership values predominantly belonging to one cluster), there are other zones, for instance around the Black Sea, with climate conditions in between the Mediterranean and Eastern Europe continental areas. They are marked with colors of the symbols varying from dark red to purple, a mixture of blue tones typically assigned to the continental climates, and red tones of the Mediterranean zones.

Even though the representation of pattern characteristics using the fuzzy cluster membership values in our example looks good at a first glance, some caveats must be kept in mind. First of all, such a method may work only when the number of clusters is limited, say three or four. In addition, Fuzzy C-means may give surprising results when applied to outliers. In Table 5.4 we report data from the station in Murmansk (Northern Russia), which is the northernmost harbor in that country, and known for a rather harsh climate. The fuzzy cluster membership vector is given by the tuple

$$[0.1267, \quad 0.4184, \quad 0.4548]$$

which could be (mis)understood as a climate half-way between continental and northwestern, as the class membership corresponding to the blue and green clusters are similar. Actually, the values just report that the pattern has the same distance to the centroid of the blue and green class, but does not tell anything with respect to the absolute

Table 5.4: Example data set for an extreme climate.

Murmansk (Russia), Latitude 68.9585 N, Longitude 33.0827 E												
Mon	Jan	Feb	Mar	Apr	May	Jun	Jul	Aug	Sep	Oct	Nov	Dec
Day	−8.3	−8	−3.2	1.4	7.5	14	17.5	15.2	9.6	3.2	−2.6	−6.2
Night	−14.8	−14.3	−10.3	−5	0.8	5.8	8.9	7.8	4.4	−1.1	−7.7	−12.3
Sun(h)	0.1	1.1	3.8	6.4	6.3	7.7	7.5	4.9	2.8	1.4	0.1	0
Pr. (d)	9	7	6	7	7	10	9	11	11	11	10	10

distance, which is large indeed. Langer et al. (2011) show an example of fuzzy clustering behavior for patterns far away from the centroids. We can leave it as an exercise to examine how fuzzy clustering behaves in the presence of strongly outlying patterns, where distances are large to all cluster centroids.

A further issue is the limitation to three or four cluster centroids, when we wish to create colored symbols. Common visualization systems allow a color coding according to the RGB scheme; modern systems may allow four color components.

An alternative responding to the needs of data reduction and efficient representation of pattern characteristics, is given by vector quantization methods, in particular the Self-Organizing Maps (SOM) that we already presented in Chapter 3. SOM are more robust for a priori choices than clustering; at the same time, the color coding in SOM mirrors the numerical characteristics of a pattern better than its membership to a cluster. Finally, SOM do not lead to misinterpretations, such as the Murmansk case afore-mentioned.

Recall the two decisive steps in SOM applications:

(i) we identify prototypes of feature vectors, each of which represents a number of patterns. Similar to clustering, the prototypes are defined as the centroid vectors of the patterns they represent. The metrics to decide which pattern is assigned to a prototype is given by Euclidean distances, similar to K-means clustering. Note that the number of prototypes created in SOM can be high (several hundreds or even more).

(ii) Having so many prototypes, we must find a code allowing an effective representation. As we have seen in the application to the World Poverty Map (see Chapter 3.3, Fig. 3.17), this is achieved by the projection of the prototype feature vectors in a low-dimensional representation space, ideally a 2D map. At the same time, the projection of the prototype vectors, the "Best Matching Units" (BMU in the SOM jargon), is carried out so that BMUs being close to each other in the original data feature space, i.e., representing similar patterns, are placed close to each other on the map. Finally, applying a color code to the BMUs allows us to represent the prototypes (or BMUs) by colored symbols. The color of a BMU depends on its position on the map. Contrary to the fuzzy class membership, BMUs with a highly saturated color represent border

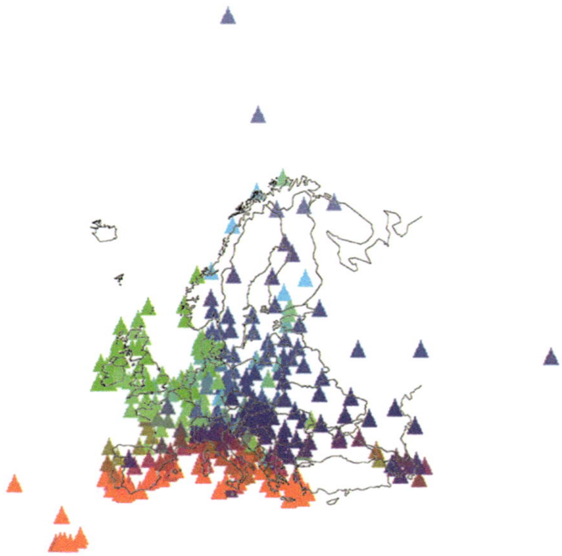

Figure 5.16
SOM representation of climate data in Europe. For the sake of visualization, we interchanged red and green components in the output files produced by the KKAnalysis software. Thus the Mediterranean sites with hot and dry climates are represented by red symbols. Green tones prevail in the NW part of the continent, whereas sites in the Central and Eastern part are characterized by blue tones. At single sites, such as those situated in the Alpine Mountains, we may notice specific conditions with strong differences with respect to the surrounding areas.

or extreme positions rather than positions close to the centroid (i., e., the average vector) of a cluster.

On the whole Fig. 5.16 shows tendencies similar to those observed in Fig. 5.15: a clear regime in NW Europe, where essentially green tones prevail; red tones in the Mediterranean part; blue tones in Central and North-Eastern Europe. Compared to the representation with Fuzzy C-means, saturated colors represent extreme climate conditions more faithfully. For instance, the color of the symbol for our Murmansk example is given by the RGB tuple

$$[0.4395, \quad 0.5605, \quad 1]$$

i.e., a full saturation in blue. That means the associated pattern takes a position at a border of the SOM. On the other hand, the first two values (given by the contribution of red and green in the RGB tuple) highlight other climate influences. In fact, even though being so far in the Northern part of Russia, Murmansk is the only harbor of that country in Europe being ice-free all the year round.

Transitional zones are recognizable for instance around the Black Sea. The symbols for these sites show dark red to purple tones, which stand for the presence of some elements of the Mediterranean climate, such as high temperatures in summer and little rain. On the other hand, the influence of the large surrounding land mass is evident. Winter temperatures are generally low; besides, there is a strong difference between day and night temperatures. Finally, mountain areas, such as the Alpine mountain belt, are marked by symbols with colors clearly differing from those for the sites in the nearby areas.

5.5 Monitoring spectral characteristics of seismic signals and volcano alert

In Section 5.2 we discussed the application of cluster analysis to the seismic signal radiated by volcanoes, known as volcanic tremor. In the example concerning Stromboli, it turned out that spectral characteristics of this signal undergo well identifiable changes according to the activity of the volcano. Similar observations were also made at Mt Etna (see, e.g., Alparone et al., 2003; Falsaperla et al., 2005). Fig. 5.17 summarizes amplitude and spectral content at Etna in 2001, when the activity of the volcano showed numerous paroxysms, with violent episodes of lava fountaining from eruptive centers situated in the summit area, but also flank eruptions with explosive phases and lava effusions. In particular, Falsaperla et al. (2005) observed that with increasing volcanic activity also the signal amplitude augmented, but the increase in energy mainly affected the frequency band between 1 and 3 Hz (see spectrograms of 4 and 5 July in Fig. 5.17).

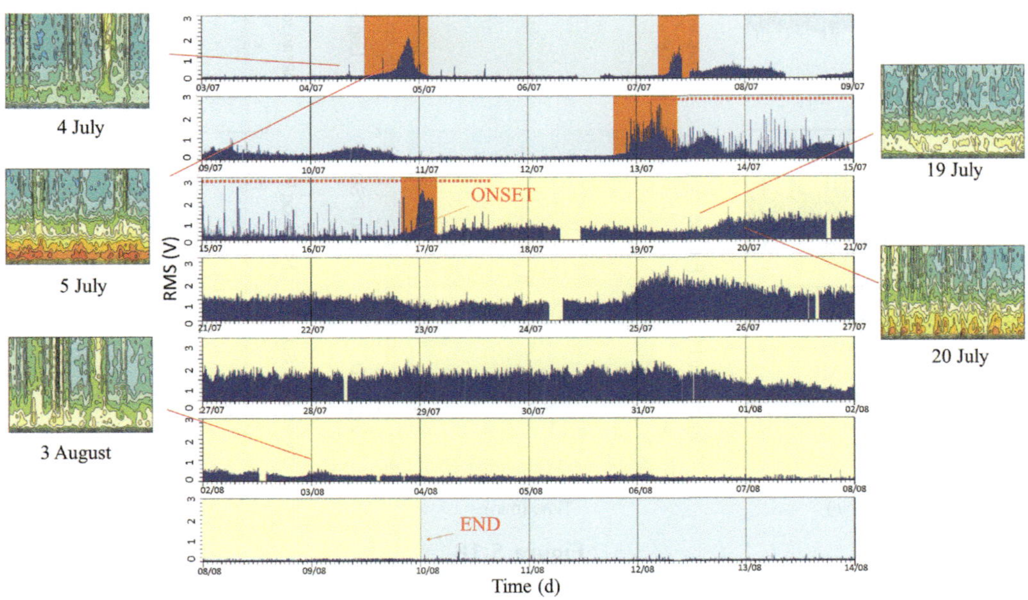

Figure 5.17

RMS Amplitude of the tremor signal at Etna recorded from July 3 until August 14, 2001. Spectrograms provide information on the spectral content in selected time spans.

In other words, changes in the state of volcanic activity came along with changes in the spectral shape that became increasingly narrow. Moving on from these findings, Langer et al. (2009) explored the possibility to reveal a formally reproducible link between the style of volcanic activity and the spectral characteristics. They applied supervised learning techniques, namely MLP and SVM, as well as unsupervised classification to data collected before and during the flank eruption of Mt Etna which started on July 17, 2001 and ended 21 days later, on August 8. From the application of supervised classification, four neatly distinguished regimes could be identified: (i) a Pre-eruptive ("PRE") period (from July 1 to July 16); (ii) episodes of lava fountaining ("FON"), five of which occurred shortly before the flank eruption and one at its beginning (July, 17); (iii) the (flank)eruptive ("ERU") phase (from July 17 to August 8); and (iv) the post-eruptive ("POS") period with data collected in the time span between August 9 and 15, 2001. Fig. 5.18 outlines the signal characteristics encountered during these four regimes. We recognize slight differences both in amplitude and spectral shape between the pre-eruptive and

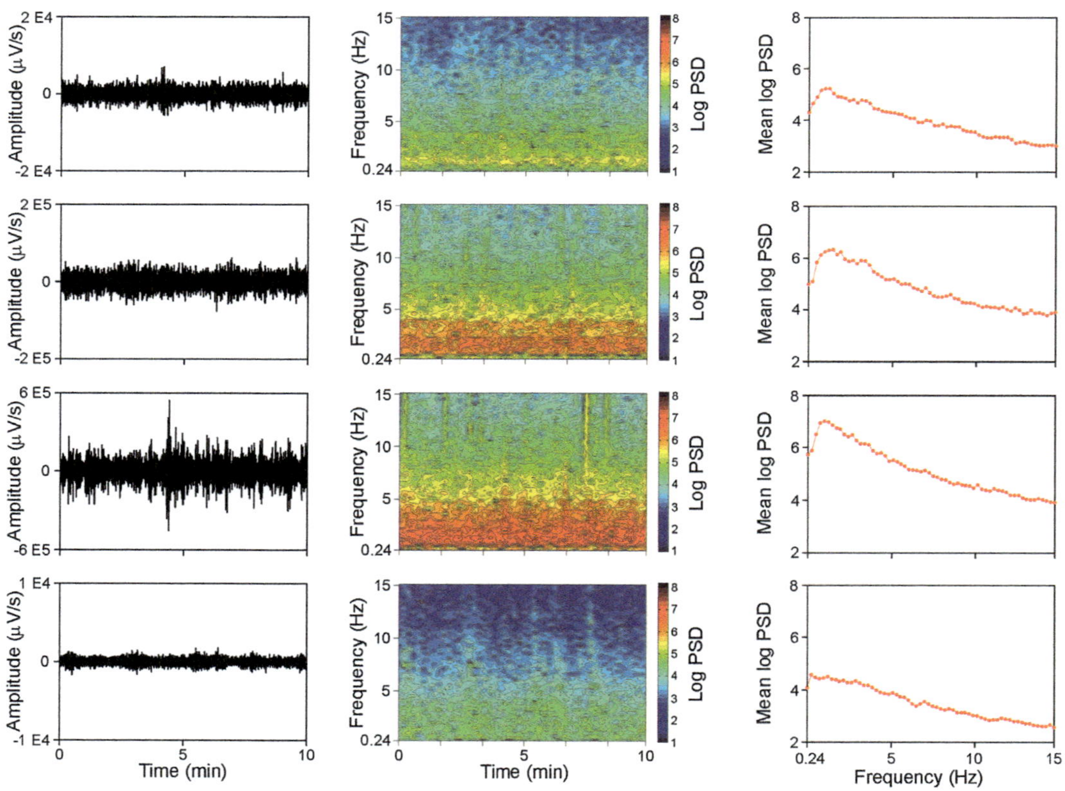

Figure 5.18

From top to bottom, examples of pre-eruptive, lava fountain, effusive and post-eruptive patterns: (left-hand column) time-series, (middle column) spectrograms, and (right-hand column) 62-D feature vector. The examples are referred to the *Z* (vertical) component of the seismometer. PSD stands for power spectral density. *From Langer et al., 2009.*

post-eruptive patterns, the latter having the flattest spectral shape among all groups. Spectra of signals recorded during the flank eruption exhibit a well-developed peak around 1 Hz, whereas in the spectra related to lava fountains a second peak around 3 Hz can be recognized.

For the definition of the feature vectors, Langer et al. (2009) calculated the Fast Fourier Transform of the tremor samples and obtained spectrograms from successive time windows of 1024 points, with overlap of 50%. The spectrograms had a range of frequencies between 0.24 and 15 Hz, and a frequency spacing resolution of approximately 0.24 Hz. Undesired transients, in particular frequent earthquakes during the seismic swarm from July 12 to 17, had to be excluded manually, which led to a reduction of the length of the time-series in a few cases down to ca. 2 min. Typically spectrograms were calculated over 10 min with frequency–time dimensions of 62 × 145 points. For the sake of having a homogeneous number of features—as requested by the classification techniques adopted—the rows of each spectrogram were averaged, ending up with a feature vector of 62 components. All the three components (Z, NS, and EW) of a seismic station were used, even though this might imply a certain degree of redundancy of the patterns. However, the use of this information accustomed the classifiers to recognize the same embedded information within different patterns, contributing to improve its performance. Such a strategy resembles to those procedures where new patterns are generated from existing ones just adding random perturbations (see, e.g., Freeman and Skapura, 1992). Overall, the authors considered 425 data vectors, 284 of which were obtained from Z and NS (142 for each component), and 141 data vectors were from the EW component.

Supervised classifications were tested using a "leave-one-out" - technique for testing, i.e., setting aside one sample for testing and using the remaining ones for training. As all the 425 patterns had to be tested, the classification had to be run 425 times. Finally, the authors found ~95% and 82% of match using SVM and MLP, respectively. Misclassifications were essentially encountered during the transitions from one stage of volcanic activity to another.

Using a partition with five clusters, the results resemble to those obtained with four classes. However, the partition using five clusters is more intriguing as it identified groups, which could be easily related to the a priori labeling (Fig. 5.19A–C). The partition revealed that the majority of the patterns with a priori classification POS formed a cluster on their own. Clusters of the period PRE correspond to '1' and '2'. Changes among the two do not occur randomly but are concentrated before and after eruptive episodes (lava fountains–FON - and before the flank eruption ERU). Besides, the clustering suggests a distinction within the patterns encountered during the flank eruption, with the presence of a cluster '3a' at the beginning and toward the end of the ERU, and a cluster '3b' throughout its most energetic phase.

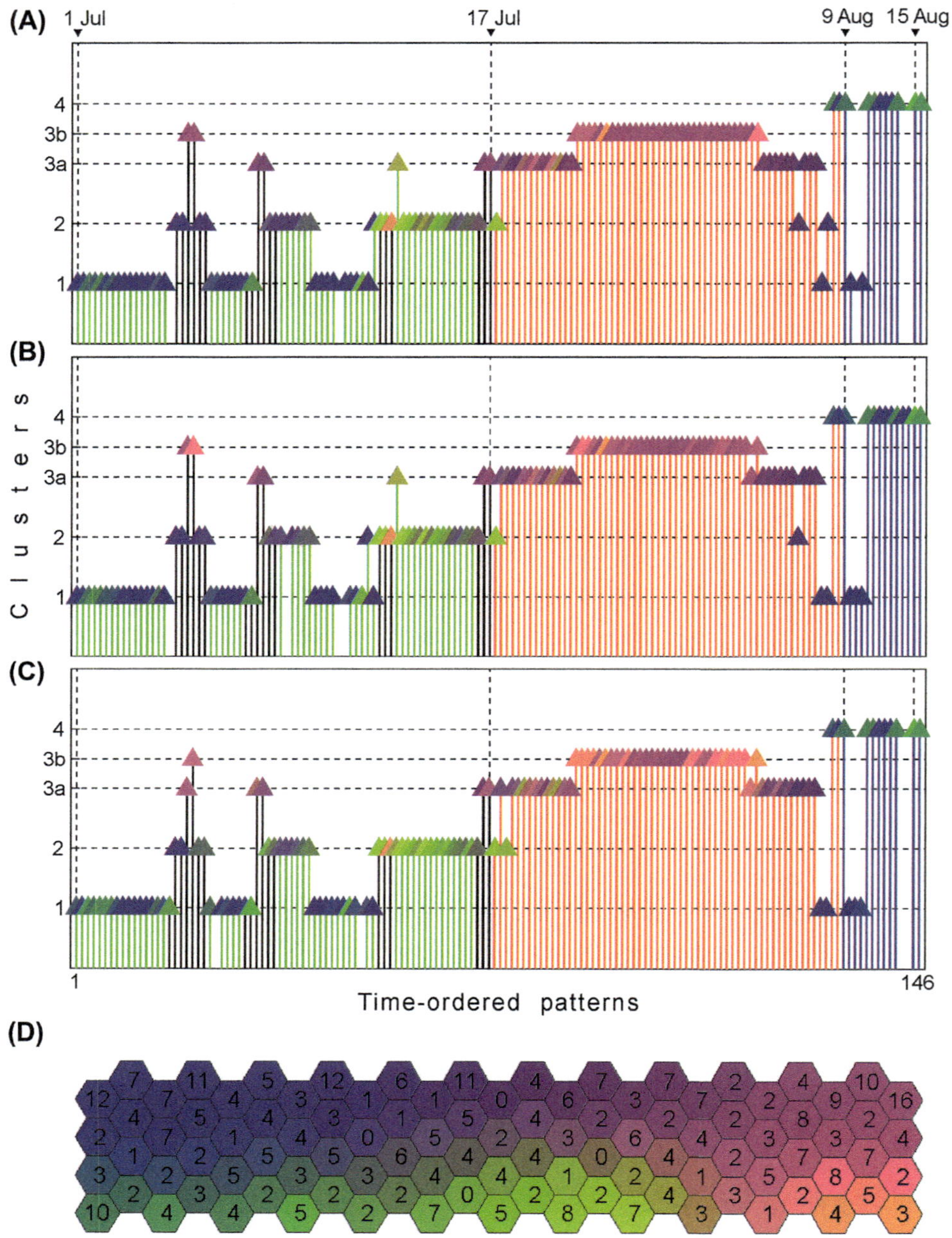

Figure 5.19

Synopsis of clustering (using the Adaptive Distance Criterion) and SOM for (A) EW, (B) NS and (C) Z-component (*from Langer et al., 2009*). Each triangle represents a pattern. The position of the triangles on the vertical axis corresponds to the labels assigned to the patterns during the cluster analysis. Green, black, red, and blue colors of the vertical bars mark the a priori classification, i.e., PRE, FON, ERU and POS, respectively. Patterns are ordered with respect to time; gaps indicate times where a component is missing. (D) Hexagonal grid of nodes making up the SOM. The number inside the hexagons reports the number of patterns for which the node was identified as best matching unit (BMU). The colors of the single triangles in (A), (B) and (C) correspond to the ones of the BMU in (D) to which the patterns belong.

The time distribution of the patterns, with changes between clusters concentrated before or shortly after an eruptive event (FON or ERU), suggests the presence of transitional regimes, that is, signal characteristics that change gradually rather than abruptly. As noticed before, crisp clustering does not catch these transitions as the decision whether a pattern belongs to a class is defined by some threshold. Transitional regimes can be effectively visualized using SOM colors, as the slow changes in pattern characteristics mirror in the sequence of colored symbols. Such a result can be also achieved applying fuzzy clustering to the tremor features, similar to what we discussed for the climatic data example. Both techniques allow to highlight even small changes related with an impending criticality at an early stage of a volcanic unrest. Such a strategy was proposed by Langer et al. (2011; see also D'Agostino et al., 2013). On the base of seven episodes of volcanic unrest at Mt Etna between 2007 and 2008, the authors learnt how to identify criticalities well before the onset of an eruptive phenomenon. In particular, they noticed that SOM had blue colors before an eruptive event, which turned to an increasingly red component (in the RGB color code) when the onset of the eruption approached. Similar changes also occurred in the fuzzy class membership values, which clearly mirrored the gradual transition from a pre-eruptive to an eruptive condition. A typical example for such an episode — a lava fountain occurred on March 29, 2007 - is shown in Fig. 5.20. Lava

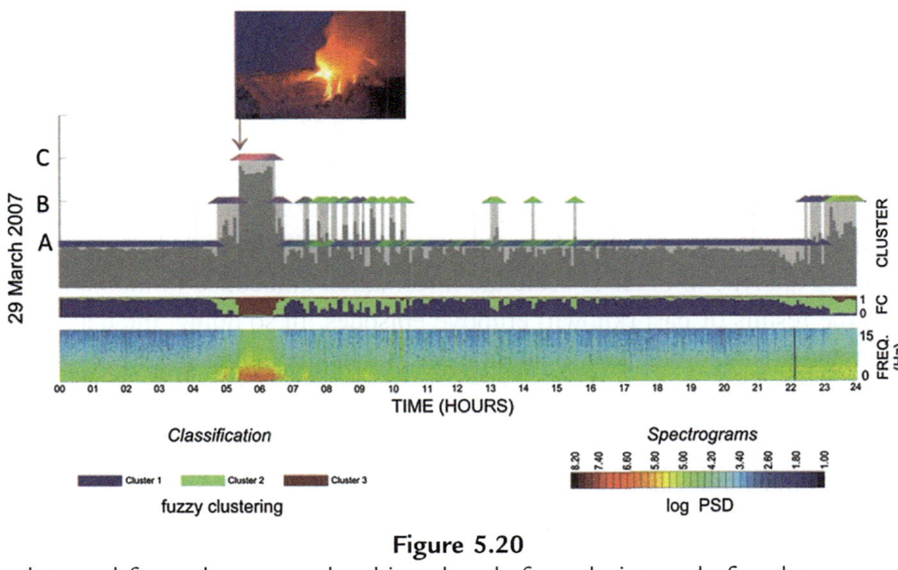

Figure 5.20

SOM colors and fuzzy cluster membership values before, during and after the onset of a lava fountain event. SOM colors switch from blueish tones to purple and red when the unrest develops. Prevailing cluster membership values switch from 'A' to 'B' and 'C'. Beside the increase of spectral amplitudes, we also notice changes in the shape of the spectra which tend to become narrower during a paroxysm. Noise is often characterized by green SOM colors. *Modified from Falsaperla et al., 2014.*

emission started at 05:20 UT (see arrow in Fig. 5.20); it was preceded by changes in SOM colors as well as fuzzy cluster membership from ~04:15 UT, that is about 1 hour before the eruptive event.

As repeated episodes of lava fountaining may occur at any time, the link between changes of pattern characteristics and status of volcanic activity was exploited for setting up an early warning system for Mt Etna. At this aim, Langer et al. (2011) first developed a scheme for automatic feature extraction, which is summarized in Fig. 5.21. The first step of the scheme requires the calculation of the Short Time Fourier Transform (STFT) with a gliding window of 1024 points applied to the whole time series. Each short-time spectrum forms an element in a spectrogram. To reduce the number of features, frequency bins are averaged over the spectral amplitudes in a selected frequency band. Instead of considering directly the short-time spectra as feature vectors, a further simplification is obtained by considering an ensemble of 60 short-time spectra. Such an ensemble corresponds to a time span of 5 min. The authors also preferred to consider the 10% (bottom) percentile, focusing on the lower amplitudes encountered in the 5-minute time span. In this manner one widely eliminates the effect of transients like wind gusts or local earthquakes that are considered as noise in the context of volcanic tremor analysis (see also Di Grazia et al., 2006).

After completing the scheme for automatic feature extraction, the further issue was the link to the applications of unsupervised classification for the identification of criticalities. Now suppose an online system, in which patterns of the last 24 h are processed by SOM and fuzzy clustering. These methods "adjust" their results according to the data set considered. Consequently, for example, fuzzy clustering will use the full number of clusters (three in the case discussed here) for each of the 24-hour time frames, even in the absence of a paroxysm. Similarly, in the SOM some of the patterns will be assigned fully saturated red or blue colors regardless the presence of a paroxysm or a perfect quiescence of the volcano. To overcome the problem, Langer et al. (2011) included a reference data set in the data to analyze in order to scale the information embedded in the new patterns according to the results of past (known) eruptive episodes. In so doing, the data set also encompassed patterns recorded during previous eruptions, including short pre- and post-eruptive periods.

The reference data set ideally represents the parent population of the patterns under study. The new data (in this case, the samples corresponding to the last 24 h) are stored in a ring buffer, in which the less recent pattern is removed any time a new pattern arrives. For the classification purposes, the merged data (reference and new patterns) are pooled in such a way that a red symbol produced by SOM will correspond to a paroxysm, and a blue symbol will signal a "quiet" condition.

In Langer et al. (2011), the fuzzy cluster membership values of clusters 'A' and 'B' marked relatively "quiescent" periods. An example is shown in Fig. 5.22; the eruptive

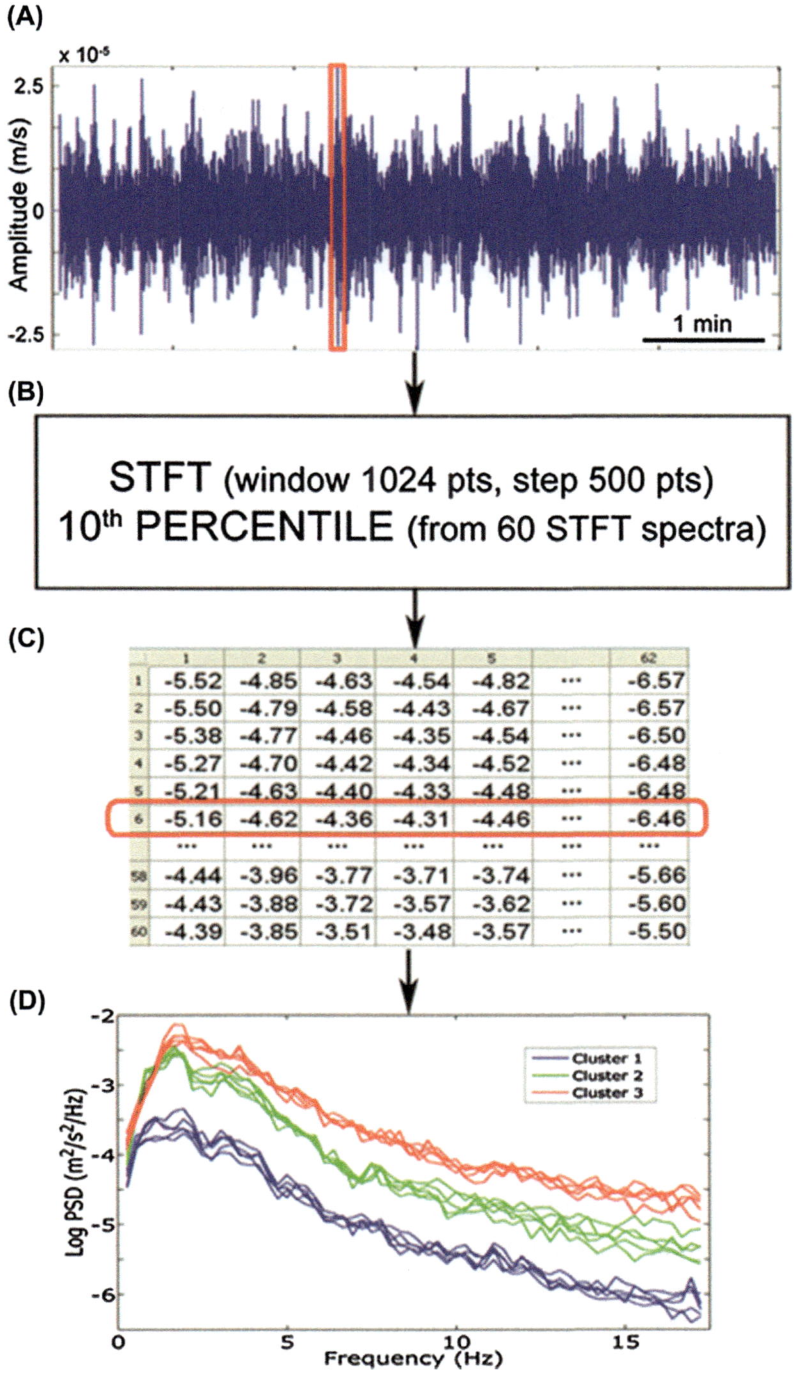

	1	2	3	4	5		62
1	-5.52	-4.85	-4.63	-4.54	-4.82	...	-6.57
2	-5.50	-4.79	-4.58	-4.43	-4.67	...	-6.57
3	-5.38	-4.77	-4.46	-4.35	-4.54	...	-6.50
4	-5.27	-4.70	-4.42	-4.34	-4.52	...	-6.48
5	-5.21	-4.63	-4.40	-4.33	-4.48	...	-6.48
6	-5.16	-4.62	-4.36	-4.31	-4.46	...	-6.46
	...	...	...	...	...	...	...
58	-4.44	-3.96	-3.77	-3.71	-3.74	...	-5.66
59	-4.43	-3.88	-3.72	-3.57	-3.62	...	-5.60
60	-4.39	-3.85	-3.51	-3.48	-3.57	...	-5.50

Figure 5.21

Preprocessing scheme for the creation of patterns. (A) Volcanic tremor signal recorded at a seismic station; the red rectangle highlights a time window of 1024 points. (B) Short Time Fourier Transform (STFT) carried out in a gliding window scheme. During the analysis the window is shifted in time steps of 5 s. (C) Sorted ensemble of 60 individual spectra from STFT. Each column in the table corresponds to a frequency bin with a width of ~0.25 Hz. The sixth row (highlighted in red) corresponds to the 10th percentile spectrum, which is the feature vector describing the pattern. (D) Examples of patterns used in the classification. Colors represent the prevailing cluster membership encountered in fuzzy cluster analysis.

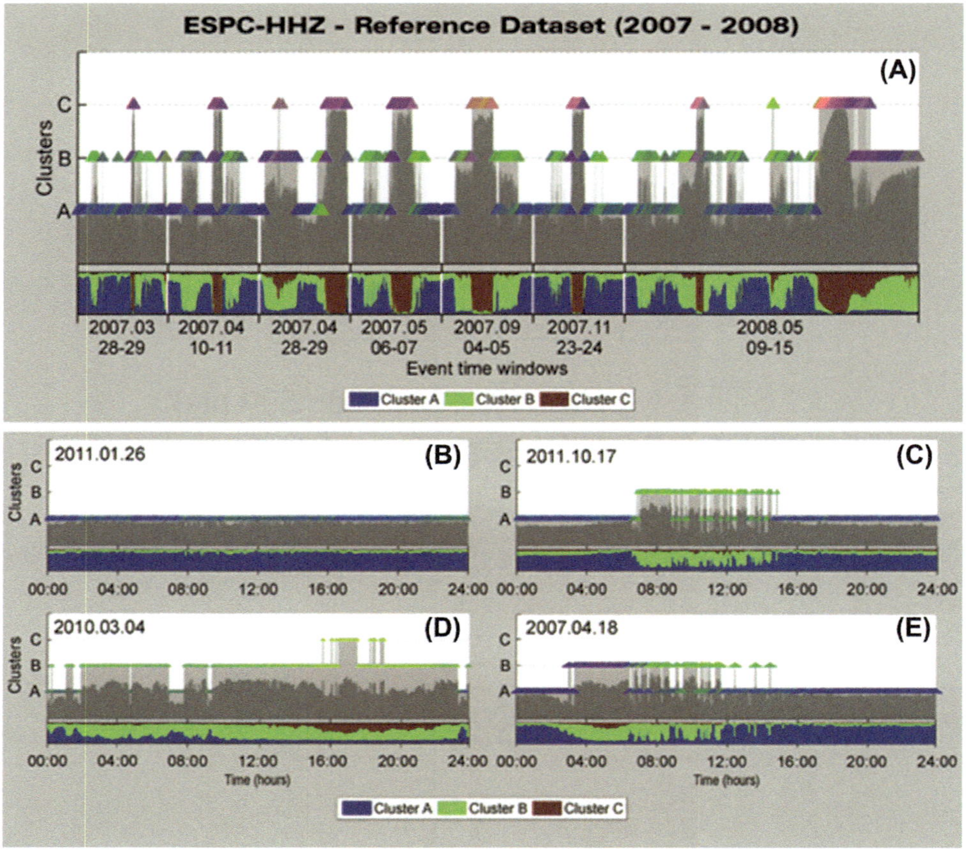

Figure 5.22

Examples of unsupervised classification during periods in which volcanic activity is low or absent (see Langer et al., 2011). SOM colors and cluster membership vectors of the reference data set are shown in (A). B–D are examples corresponding to: a day with quiet conditions (B); anthropogenic noise during daytime (C); a stormy day (D). The case shown in (E) belongs to episodes of presumably failed (aborted) eruptions occurred in the time span between February and April 2007 (Falsaperla et al., 2014).

episodes occurred in 2007 and 2008 made up the reference data set (Fig. 5.22A). The other panels in the figure depict SOM colors and fuzzy cluster membership during: a "quiet" day (5.22B); a day with noisy conditions (5.22C); a stormy day (5.22D); and finally, a mild unrest episode that did not lead to an eruption (Fig. 5.22E).

Cluster membership values belonging to cluster 'C' showed up during the passage of a strong storm (Fig. 5.21D). In this case the SOM colors remained green and indeed no volcanic activity occurred.

To explain the functioning of the alert system, let us consider the episode that corresponds to the lava fountain occurred on August 29, 2011 (Figs. 5.23A, B and C, see also Fig. 5.20).

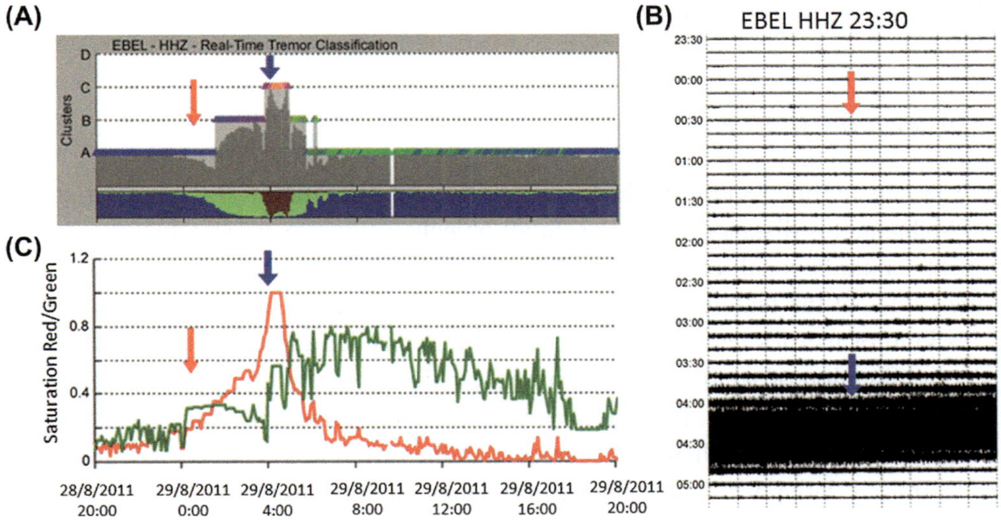

Figure 5.23

Application of the classifier during the unrest on August 29, 2011. The upper left panel displays the image seen in the monitoring room at INGV-OE. Each colored triangle represents a pattern of 5 min of tremor. As in Fig. 5.20 and 5.22 fuzzy cluster membership are also shown. Panel (B) depicts the original time series. Panel (C) depicts the degree of red-green saturation as a function of time. The red arrows indicate the time when the automatic alert was issued, the blue ones the onset of the lava fountain.

Soon after midnight of August 28, there was a mild increase of red tones in the SOM, accompanied by fuzzy cluster membership changes (Fig. 5.23A). Accordingly, the development of the saturation degrees with respect to 'green' and 'red' changed (Fig. 5.23B). The alert system for the recording site 'EBEL' (a seismic station located closed to the active craters) was tuned to flag a criticality with the 'red' saturation' beyond 0.2. Note that the tuning of the alert criteria is based on empirical considerations. The analyst has to formulate criteria which should accomplish a high-sensitivity system minimizing 'false alarms' due to noise, meteorological conditions, or even regional earthquakes/teleseisms whose wave trains may last many minutes. Therefore a second parameter relevant for the alert issue is the difference 'green-minus-red' saturation, which should not go beyond 0.25 in that specific site (see D'Agostino et al., 2013, for more details). A further issue is the declaration of the end of a criticality. Fuzzy cluster membership information has proven to be important for this purpose. In particular, for the 'EBEL' site it is required that the cluster membership values switch from 'B' to 'A'.

Fig. 5.23 documents the efficient performance achieved by such an alert system. Following the established criteria, the alert was flagged shortly after midnight of August 28 (see the red arrows in Fig. 5.23A and C). The signal amplitude was still rather low in that moment (Fig. 5.23B). The alert flag provided precious time to INGV staff to inform competent authorities, as the lava fountain started at ~04:00 UT, that is 3 hours later (see blue arrows in Fig. 5.23).

5.6 Directional features

In the examples outlined so far we have dealt with Euclidean metrics or some variants, like the Mahalanobis distance or similar. In all objects we had feature vectors in which the components could be small or large; consequently, the distance of two objects was large when the values of the feature vectors were small for one pattern and large for the other one. This kind of metrics leads us off-road when we consider angular data, such as for instance the direction of wind, the position of the pointer on a watch, or strike and dip of a fault in geology. For example, the angle between the wind directions 350 degrees and 10 degrees is 20 degrees rather than 340 degrees, and the difference between 360 degrees and 0° is null. Indeed, large differences in angular values do not automatically mean that two objects are very different from each other. The consequences for the application in unsupervised classification can be severe. Consider the case of a SOM with sheet geometry: Data in which the angle is given by low values would be projected along—let's say - the lower margin of the sheet, whereas higher angles end up at the opposite margin. Their distance on the map is large (corresponding to 340 degrees) even though it is small (20 degrees) in the original data space. In other words, this case violates the request of topological fidelity. The problem of losing topological fidelity may also occur with normalized data. For instance, when we compare the spectral shapes neglecting the absolute amplitudes of the components. The similarity among those data is then given by the correlation coefficient among the components of the feature vectors. In some way this is a situation similar to the angular data, as the correlation coefficient corresponds to the angle between the direction cosines between two feature vectors.

Directional data with feature vectors given by angles are quite frequent in geology and geophysics. For instance, the orientation of a tectonic element—such as a planar fault—is given by its strike and dip (see Fig. 5.24). Earthquakes are commonly understood as being caused by a sudden shear fracture occurring along a tectonic element. In earthquake seismology such a rupturing element is addressed to as the seismic source. During rupturing, elastic energy is radiated in the form of transversal shear waves and longitudinal compressional waves. In seismotectonic analysis, the distribution of seismic energy recorded at receivers deployed around the fault can be exploited to gather useful information. In fact, the amount of energy measured at a specific receiver depends on its position with respect to the seismic source, the orientation of the fault, and the slip direction. Having enough receivers deployed around the source, one can observe characteristic "radiation patterns", which allow the identification of the orientation of the fault and the slip vector.

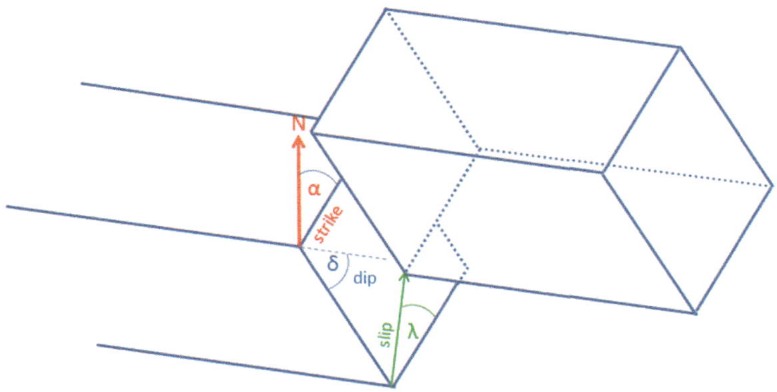

Figure 5.24

Geometry of faulting. The orientation is described by the strike angle α, i.e., the angle between the geographical North and the line formed by the intersection of a tectonic element with a horizontal plane. The dip angle δ represents the slope of the element, and is measured perpendicular to the strike. The slip angle λ describes the orientation of the dislocation, and it is measured to a line on the fault parallel to the strike direction.

As we show schematically in Fig. 5.25, the distribution of radiated seismic energy around a source can be exploited for the identification of the type of dislocation occurred during an earthquake. A simple and robust method is based on the distribution of the polarities of the first arrivals (P-wave onsets, see the wavelets depicted in Fig. 5.25). For a source plane with arbitrary orientation and oblique slip vector the interpretation is not as simple as shown in Fig. 5.25, as the polarities depend not only on the azimuth, but also on the angle against the vertical under which the signal of the first arrival is emitted from the source. For this reason, the distribution of the polarities is represented on a stereographic net, accounting both for the azimuth and angle of emission (see Fig. 5.26). To identify the nodal planes on the stereographic net, the fields of positive and negative polarities are separated by delineating two perpendicular great circles, one of which corresponds to the orientation of the fault.

An alternative form for representing the geometry of dislocation in the seismic source is given by the identification of the principal stress axes. In the case shown in Fig. 5.26A (a horizontal dislocation along a vertical plane) the axis of maximum compression ‘P’ is found at an azimuth of 45 degrees measured clockwise against North. Correspondingly, the axis of maximum distension ‘T’ is perpendicular to the former. Both ‘P’ and ‘T’ are parallel to the Earth surface. The third intermediate axis ‘B’ is orientated perpendicular to ‘’P’ and ‘T’, i.e., perpendicular to the surface. In general, its position corresponds to the point where the nodal planes intersect. The three principal axes ‘P’, ‘T’ and ‘B’ can be understood as the normalized eigenvectors of the moment tensor

$$\boldsymbol{M} = \begin{bmatrix} m_{11} & m_{12} & m_{13} \\ m_{21} & m_{22} & m_{23} \\ m_{31} & m_{32} & m_{33} \end{bmatrix} \tag{5.4}$$

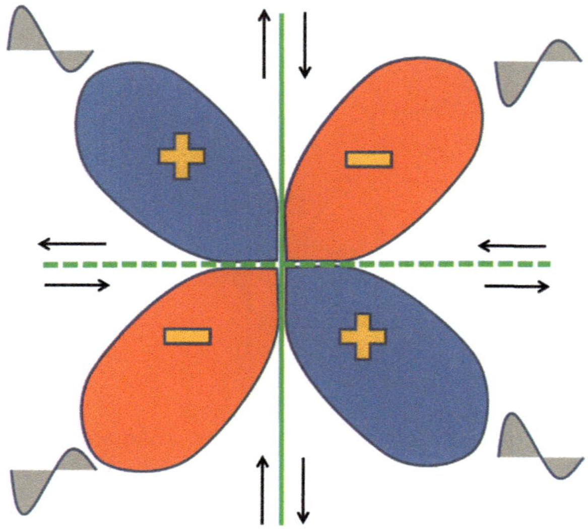

Figure 5.25
Schematic representation of seismic P-wave radiation around a vertical fault (green solid line) with horizontal dislocation (black arrows parallel to the green line). The fault forms a nodal plane, i.e., P-wave amplitudes measured in that direction vanish. Moving clockwise around the source we notice P-wave amplitudes forming lobes shaped similar to Four-leave clover. Amplitudes reach their maxima at angles of 45, 135, 225 and 315°. Besides, one identifies four distinct fields characterized by the polarity of first onset of the wave. Two of them (depicted as red fields) have negative polarities, the two remaining (blue fields) show positive polarities. The seismic radiation forms the elastic response of the shear stress release during rupturing; therefore, there are two nodal planes, one given by the fault plane itself, while the second complementary nodal plane (dashed green line) is perpendicular to the first one.

(see Kagan, 2007). Here the m_{ij} indicate the strength of the stress measured along the vertical and horizontal axis. For instance, a moment tensor for the case shown in Fig. 5.26A has the form

$$\boldsymbol{M} = \begin{bmatrix} 0 & m_{12} & 0 \\ m_{21} & 0 & 0 \\ 0 & 0 & 0 \end{bmatrix} \tag{5.5}$$

The moment tensor is symmetric, so we have $m_{12} = m_{21}$.[4]

[4] In reality the moment tensor is given in terms of normalized vector components, with values varying from -1 to 1, and a scalar part, expressed in physical units, dyn·cm or N·m. The scalar part is related to the earthquake strength; consequently, a so-called moment magnitude Mw is calculated from the scalar values of the moment tensor. Here we are interested in the orientation of the forces; therefore, we consider only the normalized vector components of the tensor as features.

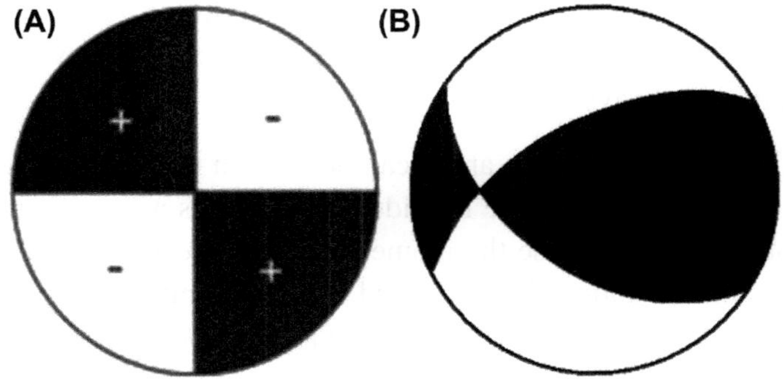

Figure 5.26

Fields of polarities in a stereographic net. Panel (A) corresponds to a horizontal dislocation along a vertical fault, in the jargon a "horizontal strike slip". Panel (B) corresponds to a dipping fault plane with an oblique slip vector. "Beach balls" like the ones shown here are addressed to as "focal mechanisms" or "fault plane solutions". For more details, we address the interested reader to classical seismological text books, e.g., Aki and Richards (1980). Negative polarities ("down") mean displacement toward the source, i.e., compression; positive polarities indicate displacement away from the source, i.e., dilatation.

There are various options to express the difference between two focal mechanisms. Kagan (2007) proposes a metric which considers the principal axes of two fault plane solutions P, T, B and P′, T′, B′. The metric is based on the angle by which a system P, T, B must be rotated so that the axes fall in the same position as P′, T′, B′. The orientation of axes is given by the orientation matrix

$$\boldsymbol{D} = \begin{vmatrix} t_1 & p_1 & b_1 \\ t_2 & p_2 & b_2 \\ t_3 & p_3 & b_3 \end{vmatrix} \tag{5.6}$$

where each vector $\boldsymbol{t}, \boldsymbol{p}, \boldsymbol{b}$ is given by plunge and azimuth angles. For example

$$t_1 = \cos(\alpha_t)\cos(\beta_t) \tag{5.7a}$$

$$t_2 = \cos(\alpha_t)\sin(\beta_t) \tag{5.7b}$$

$$t_3 = \sin(\alpha_t) \tag{5.7c}$$

(see Kagan, 2007). The rotation angle is obtained from

$$\Phi = \arccos\left[\frac{1}{2(|\boldsymbol{t}\cdot\boldsymbol{t}'| + |\boldsymbol{p}\cdot\boldsymbol{p}'| + |\boldsymbol{b}\cdot\boldsymbol{b}'| - 1)}\right] \tag{5.8a}$$

for $\Phi \leq 90°$ and

$$\Phi = \arccos\left[\frac{1}{2(\boldsymbol{t}\cdot\boldsymbol{t}' + \boldsymbol{p}\cdot\boldsymbol{p}' + \boldsymbol{b}\cdot\boldsymbol{b}' - 1)}\right] \tag{5.8b}$$

for $\Phi > 90°$. The angle Φ is known as the "Kagan" angle. In theory, up to four solutions may be obtained, in practice the minimum angle is used. It can be shown that the minimum Kagan angle is always $\leq 120°$.

In the context of clustering, the Kagan angle can be used in the framework of a clustering strategy, in which the angle replaces the Euclidean distance as a metric. In alternative to the Kagan angle, one can directly use the moment tensor as feature vector as proposed by Cesca et al. (2014). Various options are proposed for the definition of the metrics. The independent components of the tensor can be treated as vectors, and the difference is then obtained by applying L^1 or L^2 norms. Besides, a metric can be defined on the base of the correlation, among the components, i.e.,

$$d = 1 - \sum m_i n_i \Big/ \left(\sqrt{m_i^2} \sqrt{n_i^2} \right) \tag{5.9}$$

where the m_i and n_i are the components of the two moment tensors to be compared. In comparison to the Kagan angle, the use of moment tensor components in classification and clustering has a few advantages. As Cesca et al. (2014) pointed out, uncertainties can be accounted for by simply applying a weight, which has an inverse - for instance, reciprocal - relation to those uncertainties. A further issue is that the Kagan angle is measured using the principal axis obtained for a pure double couple mechanism, which we introduced earlier. Though being of outstanding importance in earthquake seismology, the double couple mechanism is only a specific case, as it describes sources with a shear dislocation (see Figs. 5.25 and 5.26). Other source types (tensile or compressive cracks, with isotropic expansion or compression) do not follow a scheme which can be coped with the Kagan angle (see Box 5.1). Besides, centroid based clustering using the Kagan angle may become cumbersome as efficient upgrade rules for the centroids and dispersion inside a cluster (see Box 5.1) are not available.

For the moment-based clustering, Cesca et al. (2014) applied the DBSCAN method (see Chapter 3.1.2.3). Considering the micro-seismicity in a coal mine in the Ruhr area (Germany), the authors identified six families of signals, each of which was distinguished by the orientation of the stress axis, the degree of the contribution of the tensional/compressional component as well as the isotropic deformation. The major part of the events was classified as "noise".

Moment tensor components can be dealt with centroid-based clustering. As an example, we present the application of SOM to moment tensors calculated for earthquakes in the Central Mediterranean area (see http://rcmt2.bo.ingv.it/Italydataset.html; Pondrelli et al., 2006). The choice of SOM is appropriate as we can deal not only with compact clusters (where members are similar to each other), but also with clusters the characteristics of

Box 5.1 Double Couple and Other Mechanisms

Most part of seismic energy is released by earthquakes caused by shear fracturing along a tectonic fault. The forces acting in the source are described by a so-called "double couple" as shown in Fig. 5.25, where shear forces are present parallel and perpendicular to the fault plane. Beside this classical model, seismic sources such as explosions or implosions can release isotropic energy, i.e., energy that is equal in all directions. Such an isotropic source has no nodal lines separating fields of compression and extension. In monitoring nuclear tests, the differences in the radiation pattern between tectonic earthquakes and explosions are an important criterion for the distinction of the two types of events.

Actually, even tectonic earthquakes often do not come as pure shear fracture events due to the complex distribution of stress around the crack tip (see left hand side of the figure below).

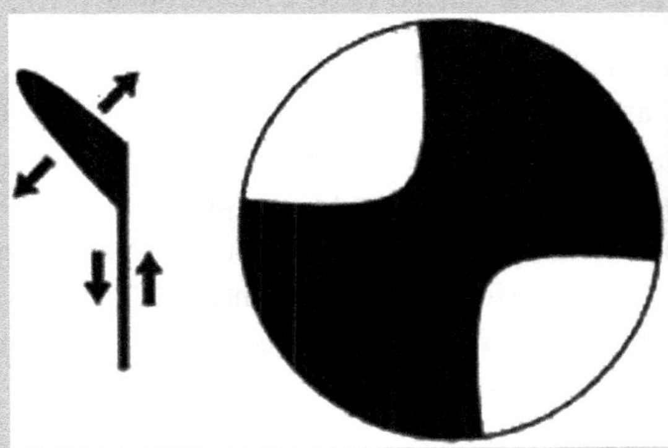

In those mixed cases, we get a beach ball like the one shown on the right hand side of the figure (courtesy of G. Foulger; see also Miller et al., 1998; Foulger, 2018). We still recognize fields of compression and extension; we can also define a P and T axis. However, the B axis (given by the intersection of the nodal planes) is not properly defined. Modern techniques aim at identifying the forces acting during the source process by a so-called "moment tensor inversion" that includes the modeling of waveforms. In the results of such an inversion, the analysts distinguish: the "double couple" part (reported as "DC"), the isotropic part ("ISO"), and the tensile/compressive part ("CLVD").

which scatter to a large degree. The latter will be found in parts of the map where little populated BMUs appear.

In general, a careful inspection of the SOM allows us the identification of highly populated BMUs and fields of BMUs which represent compact groups with similar characteristics. On the other hand, parts on the map with poorly populated BMUs - found in distant places - mirror patterns having little similarities to other members of the data set. Those pattern can be addressed to as outliers.

Using moment tensor components for classification, we deal with normalized data. Large differences between normalized feature vectors do not mean that we compare small to large values, but deal with feature vectors whose orientation points in opposite directions. Consequently, using a SOM with a classical 2D representation space - the 'sheet' geometry - is inappropriate, as such a geometry is in contrast with topological fidelity. A way out can be a Torus geometry for the representation space rather than a sheet. In Fig. 5.27 we show the SOM applied to moment tensor components obtained for 628 Mediterranean earthquakes, using a Torus geometry. Note that with such a geometry the full saturation of the RGB code of BMUs is not necessarily found along the borders of the map. At the same time, units situated at the lower left hand corner of the map are similar to those on the upper left. As we will see in the following, patterns whose BMUs lie in opposite margins of the map can be indeed close to each other.

Fig. 5.28 depicts the beach balls obtained for patterns of BMUs #12, #13, #14, #27, #28, # 131, #132, #133, #134 and #135. These BMUs are placed close to each other on the Torus map shown in Fig. 5.27. We also notice the similarity of their colors, essentially characterized by a strong saturation in red, and some presence of green. From a seismotectonic viewpoint, the mechanisms of this group belong to the type of "reverse faulting", with maximum compression (P axis) and intermediate axis (B) in horizontal direction, and more or less vertical dilatation (T axis). The mechanisms reflect a compression of the crust in NE–SW direction.

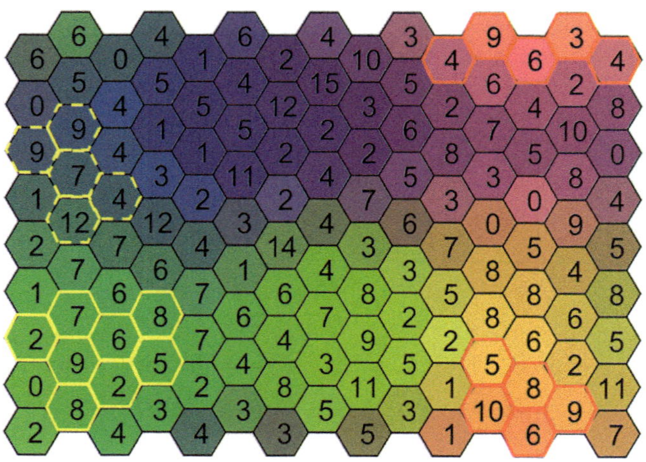

Figure 5.27

SOM of moment tensors obtained for 628 Mediterranean earthquakes. To identify the index of BMUs, we start from the lower left corner, i.e., the green node carrying a '2' which corresponds to the BMU #1; the dark yellow node on the lower right corner is BMU #15; the purple one in the upper right is BMU #135. Beach balls corresponding to BMUs highlighted with yellow and red margins are shown in Figs. 5.28 and 5.30.

Figure 5.28
Moment tensors (in the form of beach balls) corresponding to patterns at the upper and lower right margin of the SOM in Fig. 5.27. The correspondent BMUs in Fig. 5.27 are highlighted as hexagons with red margins.

On the other hand, we may gather patterns for which the BMUs have color code dominated by blue (over 90%). The corresponding beach balls are depicted in Fig. 5.29. Most of them represent horizontal strike slip motions, again with a P axis in NE direction, but dilatation axis more or less horizontal.

Beach balls with a strong concentration of green BMU colors are shown in Figs. 5.30A and B. Both groups are – in terms of seismotectonic interpretation – so-called normal faulting mechanisms, with a principal axis P essentially vertical. As the two groups share similar mechanisms, their chromatic differences are negligible. Nonetheless once can still notice an evident distinction: Nodal planes tend in NNW direction in Fig. 5.30A, whereas they are generally NW or WNW oriented in Fig. 5.30B.

The results of classification so far described focus on the most compact groups, encompassing ~200 out of the 628 earthquakes considered. The distribution of their seismic sources follows the main direction of the Apennine chain, which forms the backbone of the Italian Peninsula. The reverse faulting mechanism shown in Fig. 5.28 is Northeast the Apennines, i.e., in the foreland of the Alpine mountains, in the Adriatic Sea, and adjacent the Dinarides

Figure 5.29
Patterns where saturation of blue is larger than 90%.

mountains. The pattern of crustal deformation inferred from the fault mechanisms discussed here is in good agreement with findings from geological and geodetic studies (see, e.g., Serpelloni et al., 2007; Anzidei et al., 2014; Heidbach et al., 2016).

Appendix 5

Appendix 5.1 Davies-Bouldin index

The Davies-Bouldin Index (DBI) (Davies and Bouldin, 1979; see also Stein et al., 2003) provides a simple guideline for the choice of a suitable number of clusters in partitioning clustering algorithms. The DBI for the partition is obtained by comparing the average similarity encountered among all clusters to the largest one.

The similarity between clusters i and j is given by

$$R_{ij} = (s_\mathrm{i}+s_\mathrm{j})/\|\mathbf{c}_i - \mathbf{c}_j\| \tag{A5.1}$$

where s_i and s_j are the variance measures in each cluster; $\mathbf{c}_i$ and $\mathbf{c}_j$ are the corresponding cluster centroid vectors. With $R_i = \max\ (R_{ij})$, the DBI is obtained from the average of the R_i taken over all k clusters, i.e.,

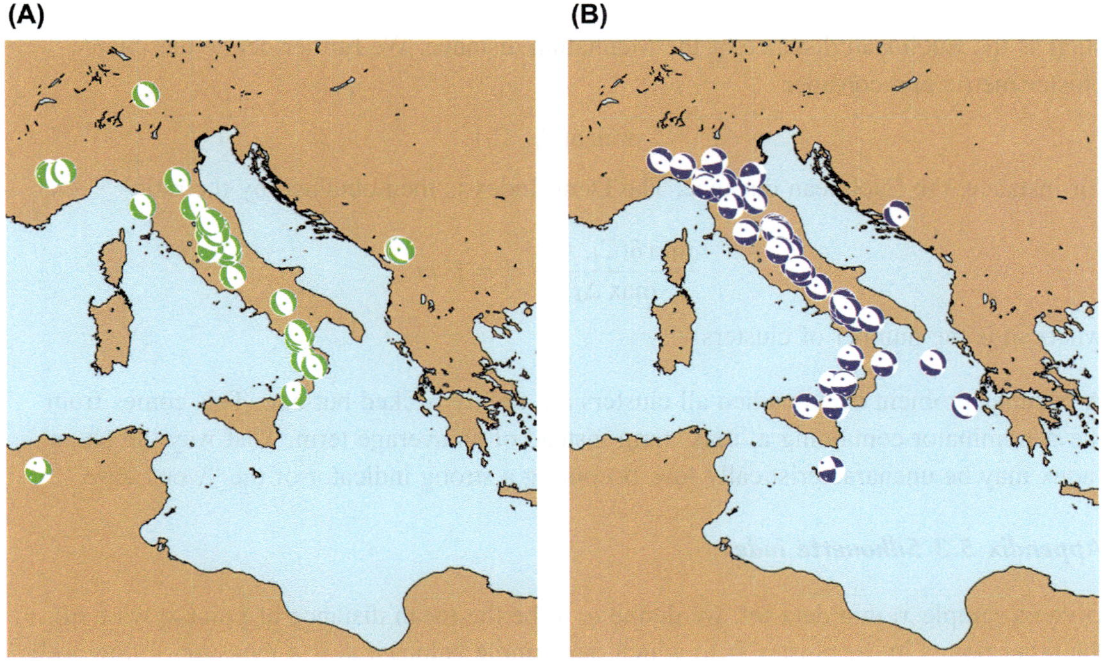

Figure 5.30
Patterns with a strong saturation in green and low blue (A), and stronger blue (B). Patterns in (A) correspond to BMUs #2, #17–19, #31–34, i.e., close to the lower left corner of the map in Fig. 5.27 (hexagons in Fig. 5.27 evidenced with solid yellow margins). Patterns in (B) correspond to BMUs #62, #77–78, and #91–92 (hexagons in Fig. 5.27 evidenced with yellow dashed margins).

$$\mathrm{DBI} = 1/k\sum\nolimits_{\mathrm{i}} R_i \tag{A5.2}$$

The DBI is used as a standard in the KKAnalysis package by Messina and Langer (2011), which is provided along with this book.

Appendix 5.2 Dunn index

Here we compare the degree of heterogeneity encountered in the clusters ("within") to the minimum of the distances between the cluster centroids (Dunn, 1973). Formally we write

$$\Delta_i = \max d(\boldsymbol{x}, \boldsymbol{y}),\ \boldsymbol{x}, \boldsymbol{y} \in \mathbb{C}_i$$

An alternative is

$$\Delta_i = \frac{1}{N_i} d(\boldsymbol{x}, \boldsymbol{\mu}),\ \boldsymbol{x} \in \mathbb{C}_i$$

with $\boldsymbol{\mu}$ being the centroid vector of the i-th cluster $\mathbb{C}_i$. Note that d stands for some metric, such as the Euclidean distance or the Manhattan distance. We further define the inter-cluster metric and consider

$$\min \delta(\mathbb{C}_i, \ \mathbb{C}_j)$$

for instance, the Euclidean distance. The Dunn Index is then obtained by the ratio

$$\frac{\min \delta(\mathbb{C}_i, \ \mathbb{C}_j)}{\max \Delta_k}, \ k = 1..m$$

where m is the number of clusters.

A peculiar problem occurs when all clusters are tightly packed but one. This comes from the denominator containing a 'max' term instead of an average term. That way the Dunn Index may be uncharacteristically low, becoming a strong indicator of the "worst case".

Appendix 5.3 Silhouette index

Given a sample x_i in a data set, we define a_i to be the mean distance of point x_i w.r.t. all the other points in the cluster $\mathbb{A}$ to which our sample belongs. a_i is a measure of how well the point is assigned to the cluster. The smaller the value, the better the assignment.

We define b_i as the mean distance of our sample w.r.t. other points of its closest neighboring cluster $\mathbb{B}$. The cluster $\mathbb{B}$ is a cluster to which our sample is not assigned. Among all clusters to which the sample does not belong, cluster $\mathbb{B}$ has the closest distance to our sample x_i .

For each x_i we obtain a silhouette value s_i

$$s_i = (b_i - a_i)/\max(b_i, a_i)$$

s_i lies in the range of [-1,1]. An overall measure of clustering quality is obtained from the average over all silhouette values. High b_i represent a good separation of our sample, while small a_i indicate that the sample is well assigned to the cluster to which it belongs. The better the overall clustering quality, the higher the average of silhouette values.

Besides, one can calculate average silhouette values for each of the clusters to distinguish well defined clusters from the others. For more details see, e., g., Kaufmanm and Rousseeuw (1990).

Appendix 5.4 Gap index

In the "Gap Statistics" we start with the sum of the distances $d(x_i, x_j)$ between the patterns in a cluster

$$D_{ij} = \sum_{x_i \in \mathbb{C}_q} \sum_{x_j \in \mathbb{C}_q} d(\boldsymbol{x}_i, \boldsymbol{x}_j)$$

and use $W_M = \sum_{q=1}^{M} D_q/(2n_q)$, where M is the number of clusters. Clearly, a low value of W_M corresponds to a clustering with compact clusters. This is compared to W_M^r, which is obtained from n reference data set with uniformly distributed points located in the same data space as our samples $\boldsymbol{x}_i$. The Gap index is given by

$$Gap_n(M) = E_n\left(\log\left(W_M^r\right) - \log(W_M)\right)$$

(see Tibshirani et al., 2001).

Appendix 5.5 Variation of information

We have a total of n patterns, grouped in K clusters, and n_k patterns in each cluster $\mathbb{C}_k$, i.e.,

$$n = \sum_{k=1}^{K} n_k$$

Consider a second clustering $\mathbb{C}'_k$, with cluster sizes n'_k. We now create a confusion matrix, reporting the class memberships to both clusterings:

$$n_{kk'} = |\mathbb{C}_k \cap \mathbb{C}'_{k'}|$$

Picking a sample at random with equal probability, the probability that our sample belongs to cluster $\mathbb{C}_k$ is

$$P(k) = \frac{n_k}{n}$$

The uncertainty of the picking is expressed by the entropy

$$H(C_1) = -\sum_{k=1}^{K} P(k) log P(k).$$

denoting the first clustering experiment by C_1. This experiment leads to the aforementioned cluster $\mathbb{C}_k$. Similarly, we denote as C_2 the clustering leading to cluster $\mathbb{C}'_{k'}$.

We can define the joint probability $P(k, k')$ that a sample is found in the clusters $\mathbb{C}_k$, $\mathbb{C}'_{k'}$

$$P(k, k') = \frac{\mathbb{C}_k \cap \mathbb{C}'_{k'}}{n} = n_{kk'}/n$$

and the information as

$$I(C_1, C_2) = \sum_{k=1}^{K} \sum_{k'=1}^{K'} P(k, k') log \frac{P(k, k')}{P(k)P(k')}$$

Finally, we obtain the *VI* from

$$VI(C_1, C_2 = H(C_1) + H(C_2) - 2I(C_1, C_2) = H(C_1, C_2) - I(C_1, C_2)$$

(see Zscheischler et al., 2012; Meilă, 2007).

PART III

A posteriori analysis

CHAPTER 6

A posteriori analyses—advantages and pitfalls of pattern recognition techniques

6.1 Introduction

In the previous chapters, we have seen that researchers may apply a variety of methods in Geophysics and related fields to handle large datasets. Those data can be used to support or decline a theoretical model developed to explain observations. For instance, the motion of planets follows well-established physical laws, which can be exploited to predict the future behavior of astronomical objects with great precision. On the other hand, theoretical models may not be available or are insufficient to match a complex reality. In Geology and Geophysics, we often face such a condition. As a consequence, researchers try to draw conclusions from the observation as is, establishing empirical relations in regression schemes or setting up rules formulated in "if … then … else" terms to provide guide lines whether to take action, for instance. Furthermore, they may be interested in the identification of structures within datasets to find out where observations tend either to cluster themselves or lack in a feature space. Indeed, researchers can exploit data even in the absence of clear-cut physical models following a data-driven approach, similar to concepts in statistics. In case of a data-driven approach, the meaning and reliability of the drawn conclusions are not based on the logic or plausibility of a model, but must be verified on the data themselves. For example, in supervised learning we identify the parameters of the model simply by comparing the calculated output to the target, validating the model when the error is acceptable.

In unsupervised learning, we face the question whether we should consider meaningful the structures revealed in the data. Cluster analysis offers rules that give a formal answer to this question. These rules, however, are not generally applicable to all clustering methods. Some heuristic approach may be necessary to explain why a specific clustering result is accepted or preferred to an alternative solution. Finally, one may be interested just in reducing the data, for instance by identifying a limited number of prototypes without taking care of heterogeneities separating one group from another. Here, the choice of features and metrics becomes a key issue.

In the following, we present methods and considerations regarding the a posterior analysis of supervised and unsupervised learning techniques. Among these, we compare various

Advantages and Pitfalls of Pattern Recognition. https://doi.org/10.1016/B978-0-12-811842-9.00006-6

methods of cross-validation. The uncertainty of tests can be assessed by specific techniques, such as the "Receiver Operation Curve" and the "Kappa-Statistics." We highlight the importance of appropriate target information, as in the methods we use we cannot remedy flaws in the chosen target; nonetheless, we may bring out deficiencies in the a priori knowledge. Furthermore, features may not be properly chosen, for instance, when they do not allow us to describe objects in a unique way, or when they introduce effects bringing us off road as not being related to the phenomena of interest.

As we have seen in the previous chapters, pattern recognition offers efficient methods to handle large and complex datasets. Nonetheless, in this chapter, we invite the reader to assume a critical attitude when validating the success of an application. A high score of correctly identified patterns does not automatically mean that the method is truly effective, for instance, when a random guess leads to a high score as well. At the same time, users should not despair when an approach does not lead to the desired success. A sound a posteriori analysis on the reasons for an apparent failure may provide interesting insights into the problem, such as an inappropriate definition of the targets, inadequate features, etc. Often the problems can be fixed just by adjusting some choices regarding targets, features, metrics, etc. Sometimes, a change of strategy may be necessary to achieve a satisfying result.

6.2 Testing issues

Supervised learning strategies furnish methods for the formulation of models, which allow us to establish a mathematical relation between input (our observations) and output, that is, target values that are a priori defined on the base of expert knowledge. Different from models based on some theory, data-driven models do not offer intrinsic clues on their validity. In this context, the successful application of such models to new data becomes an important criterion. As previously described in the Chapters 2 and 4, we face the risk of "overfitting," which means that a model is exactly valid for the data considered during the learning phase, but is unsuitable when applied to new data. Chapter 4 provided examples about how to test the validity of a model. Testing is carried out on a dataset not used during the learning phase. The model is considered valid when the test achieves good results. One may also request that the results for the test dataset have a similar achievement as the results obtained for the learning dataset. From a statistical point of view, the test error is a random variable itself (see Section 4.2). Recall Eqs. (4.1) and (4.2), which allow us to quantify the uncertainty of the estimated mismatch. In our example for the Stromboli explosion quakes, we obtained an estimated mismatch of 17% of the patterns. This value has its own uncertainties. In fact, the 95% confidence interval for the number of misclassifications is defined by 8% (lower boundary) and 38% (upper boundary).

To limit the degree of uncertainty in the assessment of the results, it is commonly proposed to conduct a cross-validation by resampling data used during learning and testing. This strategy, which is known as N-fold cross-validation, yields N models or classifiers instead of a single one. Eventually, we have to decide which of the various models should be used. Each single application may have a considerable scatter, and each model is tested on a different test dataset. A possibility to reduce the scatter is to consider all models and identify, for each pattern, the result for which most models vote. A procedure like this, however, requires a further independent test set, not used for any of the models. This implies the partition of data into three datasets: one for learning, a second one for the cross-validation, and a third unified test set, to which all learned models are applied. The triple partition of data may work well when sufficient samples are available, but may be difficult to apply for small datasets. An alternative, which comes with the cost of a considerable computational effort, is found in jack-knifing or "leave one out" schemes. Langer et al. (2009) applied such a scheme to feature vectors derived from volcanic tremor spectra (see also Section 5.5). In their application, they considered 425 spectral patterns and repeated MLP (*M*ulti *L*ayer *P*erceptrion) or SVM (*S*upport *V*ector *M*achine) training 425 times, always setting aside one pattern for test, and using the remaining 424 patterns as a training set. Each cycle gave either one success or one mismatch. The final, overall score was then obtained summing over all successful classifications during the 425 cycles. The choice of the best one among the n (425) classifiers was not critical in this case, as they were obtained for very similar training datasets and were supposed to differ only to a minor degree from each other. A further advantage of this procedure is that you can monitor each pattern, identifying the ones where an error occurs. This allows a posteriori analysis of all patterns one by one, highlighting patterns that may exhibit peculiarities, for instance, with respect to the data quality.

6.3 Measuring error

In classification, we are interested in the number of patterns for which the output of our model matches the a priori given target categories. Typically, the results are reported in so-called confusion matrices, such as the ones in Table 4.1. Relating the numbers in the diagonal elements to the total of patterns, we measure the overall score of success (e.g., 83% in Table 4.1b). Recall that, rather than counting the number of patterns falling on the desired side of a distinguished element as in SVM, in MLP learning is based on minimizing the RMS error between calculated output and target. As discussed in Chapter 4, we train MLP on the base of the RMS error, but we report its success considering the sum of the diagonal elements in the confusion matrix, defining a success when the maximum of the calculated scores is found for the category of the target. As an alternative, one could consider a success when the maximum is at least 0.5, for instance, and label as uncertain category or noise all patterns where none of the calculated output values reaches such a score.

The score governs the degree of certainty to which a pattern is assigned to a class. For binary classification problems—like the distinction between earthquakes and nuclear tests (see Section 2.1)—there are well-established procedures for determining a proper choice. In a two-class problem, they reduce it to a yes/no issue, that is, give a "1" or "yes," when the pattern is assigned to, let us say, class $\mathbb{A}$, otherwise "0" or "no." In this case, the 2 × 2 confusion matrix is

$$\begin{bmatrix} True\ Positives & False\ Positives \\ False\ Negatives & True\ Negatives \end{bmatrix}$$

Correctly classified patterns are represented in the diagonal elements of the matrix, mismatching classifications in the off-diagonal elements. In this context, a "False Positive" is a pattern for which the classifier gives an output "yes" (class $\mathbb{A}$); whereas, its a priori membership is "no." A "False Negative" is a pattern classified as "no," even though it belongs to class $\mathbb{A}$. Based on these rules, one defines the following parameters (see, e.g., Fawcett, 2006; Hossin and Sulaiman, 2015)

- True positive rate: $Tp = \sum True \frac{Positives}{\sum Positives_{a-priori}}$
- False positive rate: $Fp = \sum False \frac{Positives}{\sum Negatives_{a-priori}}$
- Accuracy: $\left(\sum True\ Positives + \sum \frac{(True\ Nagatives)}{(\sum Positives_{a-priori} + \sum Negatives_{a-priori})} \right)$
- Precision: $\sum \frac{True\ Positives}{(\sum True\ Positives + \sum False\ Postives)}$

In practice, it is desired to maximize the Tp maintaining Fp at the lowest possible value. In the context of the earthquake/nuclear test problem, for example, it is of paramount importance to catch the nuclear events at best; however, a high Fp comes with the cost of diplomatic trouble when a country is repeatedly blamed for nuclear testing without reason.

Playing with the threshold score, we can draw a diagram called "Receiver Operating Curve" (ROC) (see, e.g., Metz, 1978)), which can give us a general measure for the quality of a classification scheme. The plot of an ROC is depicted in Fig. 6.1. It is obtained from a sequence of Tp versus Fp pairs, each of which corresponds to a chosen threshold.

Here, we illustrate the principles of ROC analysis for the example of earthquake—nuclear test discrimination described in Section 2.2.3. Recall the discrimination criterion in that example, which was obtained by rotating the dataset using the PCA. The critical threshold to obtain the best separation of the two datasets was given by the value that fell in the middle of the centroids of the two datasets. As in the use of ROCs we are interested in the

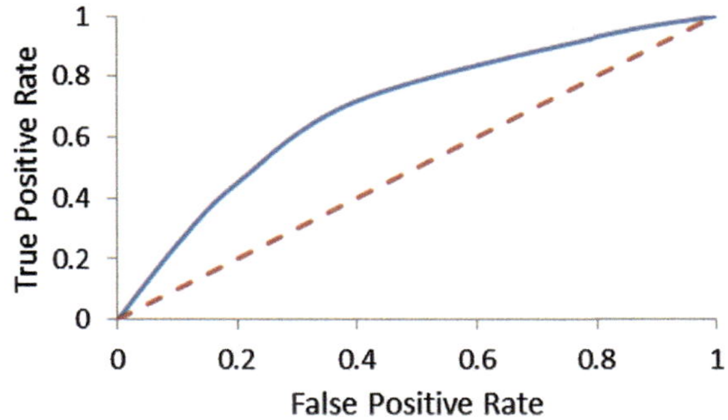

Figure 6.1
Receiver operation curve (ROC). *Modified from Spampinato et al. (2019).*

trade-off between *Tp* and *Fp*, we can now introduce some bias motivated by the need to add some extra certainty in declaring a sample as a nuclear test. In fact, a false detection entails considerable troubles in the test ban treaty context, as a country accused of nonauthorized testing can be subject to sanctions and expensive procedures, such as "On-Site-Inspections" aiming at verifying the ground truth.[1] In case of a perfect separation taking the dataset as is (Fig. 2.1), *Tp* would be equal to 100% without even a single False positive (Fig. 6.2A), that is, $Fp = 0$. Adding more certainty would just increase *Fp* without any gain in *Tp*, as this is already at its maximum. The problem posed by Fig. 2.4 (see Chapter 2) is more interesting. In addition, a set of $M_b - M_S$ pairs from earthquakes and nuclear tests is considered, assuming that the original values are affected by additional noise.

As we notice in Figs. 2.5 and 6.3, the two datasets cannot be separated perfectly with a linear discrimination function. However, here we decide to conduct a linear discrimination for its simplicity and for the paucity of the available data. What is the price to pay for this choice? In addition, we can play with a threshold, adding a positive or negative bias. Increasing the number of earthquakes erroneously declared as nuclear test (in other words, accepting a higher *Fp*), most or all tests will be detected. On the other hand, if we wish to avoid troubles coming along with false detections (low *Fp*), then some tests will escape our attention. The diagram shown in Fig. 6.2B summarizes the $Fp - Tp$ trade-offs applying biases to the threshold from -0.4 to 0.4.

The overall quality of discrimination can be expressed by a parameter called *AUC* (area under the curve), which is obtained by integration of the ROC. With the ROC shown in

[1] In reality, the decision whether to declare an event as a nuclear test—a violation of the Comprehensive Test Ban Treaty (CTBT)—is based on a multidisciplinary analysis, including investigations on the emission of radioactive elements. Once the CTBT will be in force (not yet), a country may present an "On-Site-Inspection- OSI" request at the CBTO, the organization watching the treaty. An OSI request cannot be refused by the accused state. As the phenomena related to tests are volatile, an OSI has to be carried out swiftly and must be well prepared.

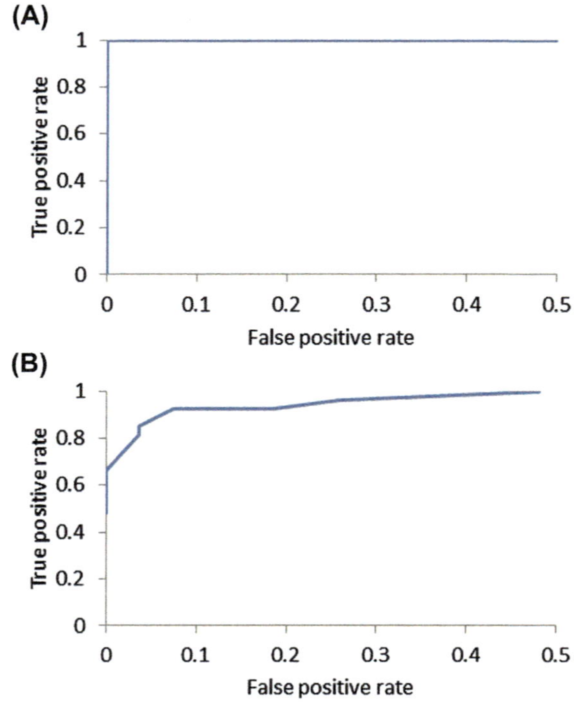

Figure 6.2

ROCs for the discrimination problem of nuclear tests from earthquakes on the base of the M_b-M_S criterion (see Section 2.2.3). The ROC in (A) is obtained for the original data in case of a perfect separation, whereas (B) represents the case in which some additional noise is present.

Fig. 6.2A, for which the quality of the separation is maximum, we get an *AUC* equal to 1. The quality is still very good in Fig. 6.2B, with the *AUC* over 0.9. One recognizes some robustness also in Fig. 6.3, as *Tp* remains high and *Fp* is rather low for a wide range of biases. Conversely, a poor classification quality is encountered when the *Tp* versus *Fp* pairs in the ROC follow a diagonal line ($AUC = 0.5$); such a situation is schematically represented by the red dashed line in Fig. 6.1. In a diagram such as the one in Fig. 6.3, the samples of the two categories (here earthquakes and nuclear tests) would be grouped around centroids situated close to each other and scattered in similar ranges. In other words, their distinction would fail.

The extension of ROC to multiclass problems, like the ones discussed in Sections 4.2–4.4, is not straightforward. Instead of managing trade-offs between *Tp* and *Fp*, we deal with *m* rates of true and $m^2 - m$ rates for errors, namely the off-diagonal elements in the confusion matrix (Fawcett, 2006). Choosing a high threshold score, that is, a high degree of certainty in assigning a category to a pattern, there will be patterns for which no score would be above the threshold for either class. As a way out, you may split the multiclass problem with *m* categories into *m* binary One-against-All classifications. Each of these

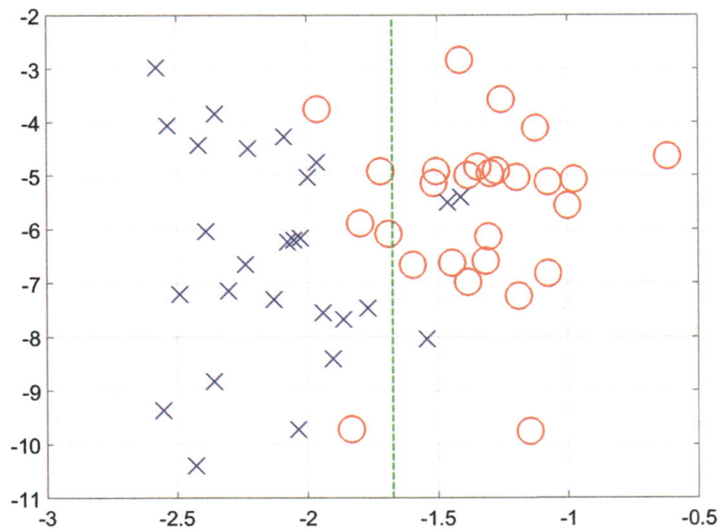

Figure 6.3

Rotation of data shown in Fig. 2.5. The green line marks the discrimination threshold obtained by a linear discriminant analysis, analog to Fisher's *L*inear *D*iscriminant *A*nalysis (LDA). Shifting the discrimination threshold to the right (positive bias) increases the detection rate of the blue samples (nuclear tests). Following a more conservative choice, the shift of the threshold to the left (negative bias) causes the missing detection of more tests.

classifications is quoted with respect to their quality parameters, such as accuracy and precision. Finally, some global measure can be defined, for instance by forming the averages of the parameters themselves or calculating the sum of all true and false positives encountered during the m classifications.

Some caveats must be kept in mind. As we classify one class against a global complementary rest, the composition of the complementary class changes to some degree. Implicitly, some patterns may be always classified as "false," that is assigned to the complementary class. The number of those patterns is expected greater for higher thresholds—a higher certainty is requested in assigning a pattern to a category. The patterns with an uncertain class membership could be labeled as "noise," but there is no straightforward concept about how to treat such a group in the context of quality assessment of classification.

An elegant way to evaluate multiclass confusion matrices is given by the κ parameter (Cohen, 1960), which provides an overall measure for the success. It accounts for the possibility that a correct classification of a pattern is achieved by a mere random guess (see, e.g., McHugh, 2012). Let us consider the confusion matrix $\boldsymbol{C}$ of Table 4.1b.

	A priori K	A priori L	A priori M	A priori N	Total
Calculated K	10	1	1	0	12
Calculated L	0	10	0	0	10
Calculated M	1	0	6	2	9
Calculated N	0	0	1	4	5
Total	11	11	8	6	*/*

where the elements c_{ij} reported the results of the 36 patterns in the test set. Counting the diagonal elements, we have in total 30 successes. Class K has in total 11 patterns, class L 11, M 8 patterns, and N 6. At first glance, one could start with the a priori hypothesis that all classes are equally probable. Running a random predictor 100 times on all 36 patterns, we expect 275 "K," 275 "L," 200 "M," and 150 "N," that is, a total of 900, which means nine random guesses during each application to the 36 patterns in the dataset. We obtain κ by subtracting the 9 random guesses from the 30 correct classifications, which is 21, and form the ratio with respect to possible extra successes, $36 - 9 = 27$, which is 77% instead of the raw rate of 83% obtained just summing up the diagonal elements in the confusion matrix.

As the random guess confusion matrix is created for equal a priori class probabilities, a validation based on the κ parameter is inappropriate in case of strongly inhomogeneous classes. Indeed, in datasets where one or a few classes outnumber the other ones, a strong random success will be achieved just by voting always for the category (or categories) with the largest number of samples. Consequently, the random guess confusion matrix will create guesses such that both marginal sums (total of calculated classes and total of a priori classes) are reproduced. The elements $\widetilde{c}_{ij}$ of the random confusion matrix are obtained by

$$\widetilde{c}_{ij} = \sum_J c_{ij} n \Big/ \sum_I c_{ij}$$

where i indicates the column, and j the row. Our random expected confusion matrix is then

	A priori K	A priori L	A priori M	A priori N	Total
Calculated K	3.66..	3.66..	2.66..	2	12
Calculated L	3.055..	3.055..	2.22..	1.66..	10
Calculated M	2.75	2.75	2	1.5	9
Calculated N	1.5277..	1.5277..	1.11..	0.833..	5
Total	11	11	8	6	*/*

which yields 9.55 matches obtained randomly. For instance, the element $\widetilde{c}_{11}$is $11 \cdot 36/12 = 3.66$.

The general relation for obtaining κ is

$$\kappa = [P(a) - P(expected)]/[1 - P(expected)] \tag{6.1}$$

with $P(a)$ being the rate of raw success (here 0.833), and $P(expected)$ the expected random success rate (here 0.265). Using a simplified relation, the 95% confidence interval of κ is obtained from its standard deviation

$$SD(\kappa) = \sqrt{[P(a)1 - P(a)] \Big/ [1 - P(expected)]^2} \tag{6.2a}$$

$$SE(\kappa) = SD(\kappa)/\sqrt{n} \tag{6.2b}$$

and

$$\kappa - 1.96\, SE(\kappa) < \kappa < \kappa + 1.96\, SE(\kappa) \tag{6.2c}$$

(see McHugh, 2012; Cohen, 1960; Fleiss et al., 1969; Flack et al., 1988).

The κ parameter (or Kappa-statistics) may prevent an overly optimistic interpretation of the classification results, especially when strong differences in the size of classes exist. We met such a case in Chapter 4, Table 4.2. Although the raw success rate was over 88%, we get a κ of only 0.64, with confidence intervals given by the limits 0.56 and 0.716 (one can easily do these calculations with a little routine available on http://vassarstats.net/kappa.html). This κ value corresponds to a moderate or substantial agreement (see Viera and Garrett, 2005), but is certainly far from being outstanding as the 88% raw success rate may suggest. In a similar way, one might argue the 78% success rate of event classification achieved by Langer et al. (2006), as for the confusion matrix reported in their Table 4, κ is 0.6 with confidence intervals 0.55 and 0.65, respectively. From the viewpoint of the Kappa-statistics, the success is moderate. An interesting case is offered by the classification of rock samples reported in Table 4.5 (Chapter 4). Here the raw success is ca. 72%, whereas κ is 0.68 (upper and lower confidence limits at 0.647 and 0.711), that is, the discrepancy between raw success and κ is small. From the viewpoint of the κ parameter, the quality of the rock classification is better than that obtained for the infrasound events mentioned in Chapter 4, even though the raw success of the latter was almost 90% (see Table 4.2). A reason for the differences in classification qualities is certainly given by the distribution of classes in the dataset. In the rock samples, the classes were quite equally represented. Having eight classes, the expected rate of random success is limited, some 13.5%, which brings κ close to the raw success rate. Besides, working with a high number of classes (eight in the rock dataset, only three in the infrasound example) renders the classification problem more difficult for MLP, SVM as well as classifiers working randomly. To conclude, κ can be understood as a measure of the additional value a classification method provides with respect to random guess, similar to a statistical test that complements the estimation of a statistical parameter. In this light, our suggestion is to consider both the raw success together with κ.

6.4 Targets

As discussed in Chapters 2 and 4, modern supervised learning schemes are able to identify a formalism such that any relation between observations (input data) and a priori defined targets (output) can be matched. On the other hand, having achieved an excellent agreement between input and output does not mean that this relation is true or generally applicable. In Chapter 2, we mentioned the phenomenon of overfitting, which means that — in an extreme case — you can even fit noise, establishing a formal relationship that does not exist in reality. Testing is therefore necessary, keeping in mind all the caveats we mentioned earlier.

In some cases, one may fail to achieve appreciable results from the classifier, even changing methods of learning, for instance passing from MLP to SVM or HMM. Inappropriate features may be one of the reasons of such a failure. Another aspect regards the definition of the targets. The role of the target definition was discussed by Langer et al. (2006) in an application of supervised learning to seismic data recorded at Montserrat, an island belonging to the Lesser Antilles.[2]

The island is known for the Soufrière Hills volcano, which focused the scientific interest when it resumed its eruptive activity in 1995 after 400 years of relative quiescence (Druitt and Kookelaar, 2002). The volcano provided—and provides—a large volume of data, boosting the study of volcanic processes in the framework of various research projects.

The volcano complex has been regularly monitored since 1992, when enhanced seismic activity was locally recorded. The monitoring at Soufrière Hills volcano encompasses visual observations, ground deformations, seismic radiation, and geochemical analyses (e.g., Aspinall et al., 2002). Continuous seismic monitoring has proven to be a fundamental key for surveillance purposes (e.g., Young et al., 1997). Beside the social impact of its activity, the volcano is an interesting object of study for the variety of seismic signals recorded. On volcanoes like Soufrière Hills with andesitic and SiO_2-rich magma, seismic radiation is dominated by transient signals. The general categories of transient seismic signals proposed by McNutt (2000; see also Fig. 4.1) apply well to this volcano. In particular, one distinguishes "High Frequency Events" (also called "Volcano Tectonic Events"), "Long-Period Events," "Hybrid Events," "Regional Tectonic Earthquakes" (relatively distant tectonic earthquakes, not directly linked to volcano dynamics), and "Rockfalls," which are seen as a consequence of a growing lava dome. For their application of pattern classification to the aforementioned events, Langer et al. (2006) started using a dataset with the original a priori classification defined by the staff of the

[2] Hammer et al. (2012) applied hidden Markov models to data recorded on this volcano (see Chapter 4). As they used a simplified classification scheme and considered only a seismic station, their results are not comparable to the ones published by Langer et al. (2006).

Montserrat Volcano Observatory during their routine analysis. Applying an MLP, the authors got stuck with an overall raw success of ∽70% (and κ amounting to 0.57) both during training and test, even though devoting considerable efforts with respect to a proper selection of features—choosing autocorrelation functions and high-order statistical moments of amplitudes—and quality assessment of the signals. Not being able to improve the overall performance of the MLP classifier, the authors decided to give a closer look to the original classification of the signals. They observed that a part of the a priori defined target categories had flaws. Those errors arose as the staff in some cases was overwhelmed by the amount of events to handle daily. After a revision of the targets, the MLP classifier achieved a success of 78% for the test set and over 80% for the training samples.

Although the revision of the target classification led to a considerable improvement of the success rate, some questions regarding the 20% of mismatches remained. Beside the fact that even their revised targets may be not free of flaws, the authors pointed out intrinsic difficulties that hindered a better performance of the classifier. Uneven representation of the signal categories leads to relatively high errors for underrepresented classes. The confusion of some classes, especially "Hybrid Events" and "Long Period Events" being confused with "Rockfalls," was explained by the heterogeneity of the category "Rockfalls" that was in reality composed of three subclasses, in part indeed resembling the "Hybrid" or "Long Period" class members. The fact that these mismatches showed up in a very similar way both in the training and test sets was a strong hint of some intrinsic contradiction that no classifier would resolve.

Supervised learning schemes can be successfully applied in the classification of rocks, such as in the example reported in Chapter 4. Lacassie et al. (2006) used MLP to distinguish basalt, andesite, dacite, and rhyolite of different volcanic arcs based on their geochemical features. They obtained better results than those with the classical total alkali silica scheme, that is, diagrams where the concentration of Na_2O and K_2O in a rock sample is plotted against the concentration of SiO_2. Corsaro et al. (1996) applied an MLP to geochemical data of Mt. Etna, considering exclusively major elements of the bulk rocks belonging to the entire stratigraphic sequence of the volcano (about 600,000 years). The target classes were defined considering the paleoeruptive centers that formed Mt. Etna. However, the variability of those compositional data led to just a fair score (64% of match) of the MLP classifier. In supervised learning, it is necessary that the targets are free from intrinsic contradictions; in turn, the lacking success may shed light on those contradictions. In the case study proposed by Corsaro et al. (1996), the definition of targets identifying the eruptive center was misleading. An alternative to the previous approach came from Corsaro et al. (2013), who applied unsupervised learning (in particular, *S*elf *O*rganizing *M*aps, SOM hereafter) to a set of patterns defined on the base of the geochemical composition of volcanic products (see Table 3.1, partly repeated below).

Mg#	SiO_2	K_2O	Ca/Al	Th	La	Nb	Nd	Sr	Tb	Cr	Ni	Rb/Nb
0.48	47.84	2.14	0.556	7.97	57.0	42.18	46.87	1153	0.99	22.58	21.04	1.16
0.47	48.12	2.19	0.564	8.07	57.2	41.49	46.49	1133	0.97	22.15	20.54	1.16

The authors were able to show that volcanic products emitted by the same eruptive center undergo considerable compositional variations (see Fig. 6.4). It is known in petrology that the magma "evolves," changing its composition with time. For example, the formation of crystals during the cooling process carries away a few elements from magma, leading to a depletion of these components in the remaining liquid. The classification results obtained by Corsaro et al. (2013) document that an eruptive center is a questionable target as it is not uniquely related to a given geochemical composition (or range of composition) even over a relatively short period of time.

Consider for instance the volcanic products erupted at the South East Crater throughout 10 years (SEC; Fig. 6.4). Bluish and green SOM colors are associated with their petrochemical characteristics in the years preceding 2000. In terms of the original

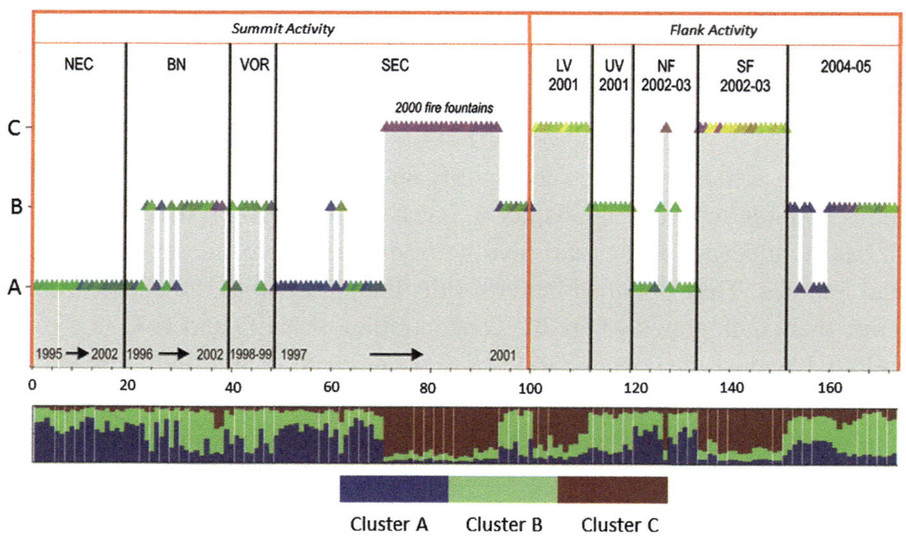

Figure 6.4

Variability of petrochemical characteristics of volcanic products erupted from various centers at Mt Etna, as seen from SOM and fuzzy clustering. Labels on the vertical axis report the cluster to which a pattern prevalently belongs. Numbers on the abscissa indicate the pattern ID. Summit activity is linked to the summit craters "NEC" (North East Crater), "BN" (Bocca Nuova), "VOR" (Voragine), and "SEC" (South East Crater). The flank activity regards eruptions in 2001, 2002–03, and 2004–05, the acronyms "LV," "UV," "NF," and "SF" indicate local eruptive centers that were active during the eruptions 2001, and 2002–03. For more details, see Corsaro et al. (2013). *Modified from Corsaro et al. (2013).*

components, this corresponds to a composition rather rich in SiO_2 along with a minor concentration of Mg#, CaO/Al_2O_3, a low ratio Rb/Nb, etc., indicating a so-called "evolved" magma, which has undergone changes before being erupted. The lava fountains occurred at the SEC in 2000, however, erupted rather "primitive" products, being rich in Mg# and Al_2O_3, having a high ratio Rb/Nb, but relatively low concentration of SiO_2. Those primitive products are indicative of magma rapidly ascending from depth without undergoing changes on the way to the surface. In the context of our discussion, the observation that compositional variations also occurred to nearby craters (Fig. 6.4) gives support to the assessment that SEC and the other summit craters were not a suitable target in this supervised learning scheme. Conversely, the approach with unsupervised learning brought out elements that helped the interpretation of the processes the magma underwent over the study period (see Corsaro et al., 2013). Regarding for instance the SEC, the authors point out that products corresponding to this crater showed the strongest compositional variations, with rather evolved products in 1997–99, followed by the eruption of the most primitive products of the whole summit dataset in 2000. This evolution is efficiently represented applying SOM to the petrochemical patterns.

The aforementioned example highlights the importance of targets as a key aspect for the success of classification methods. In our next case study, we consider applications of supervised learning schemes in inversion problems and how they can guide in the choice of appropriate target parameters. In Chapter 4, we presented two examples for such applications. Both cases dealt with synthetic simulations of geophysical field data, obtained assuming a physical model with given parameters. In the example studied by Nunnari et al. (2001), the parameters were the size, orientation, and location of a magma-filled dyke causing deformations at the Earth's surface as well as changes to the magnetic and gravity field. As the model parameters used for the generation of synthetic data are known, targets are in this case clearly defined. The simulated data are considered as input during the inversion, and the MLP is trained with the aim to reproduce the model parameters at best. It turns out, however, that for some parameters the output calculated by the MLP shows a considerable error. Nonlinear optimization methods, such as simulated annealing, reveal that the lack of accuracy obtained for given model parameters is partly an intrinsic problem, in the sense that the inversion cannot be resolved uniquely for those parameters. Those criticalities lead to peculiarities in the shape of the mapping function created by the MLP. A paper published by Barron (1993) may help to understand this phenomenon.

Recall Chapter 2 where we formulate the output of a nonlinear MLP with one hidden layer, consisting of N_H units. This forms a regression function that maps the input vector $\boldsymbol{X}$ to the output vector $\boldsymbol{Y}$

$$y_k(\boldsymbol{x}) = \sum_{j=1}^{N_H} c_j \left[\sigma\left(\boldsymbol{w}_j^T \boldsymbol{x} \right) + t_j \right] + c_k \tag{2.13}$$

We request that

$$\mathcal{F} = \int_R |\omega| \cdot |y(\omega)| d\omega \tag{6.3}$$

with $y(\omega)$ being the Fourier transform of the theoretical function $y(x)$, and R (denoting the space that is explored) is finite. Then, the maximum error of an optimally trained network is

$$||\hat{y} - y|| \leq 2\sqrt{N_I} \cdot \mathcal{F} \cdot \frac{1}{\sqrt{N_H}} \tag{6.4}$$

where N_I gives the number of units in the input layer. From this equation, which is known as Barron's result (Barron, 1993), one learns that the approximation should generally improve as the number of neurons in the hidden layer is augmented. This is similar to conventional regression approaches, where the explained variance increases with the complexity of the regression function, however, at the cost of a decreasing significance. At the same time, we obtain a small error for a smooth function y as the contribution of large frequencies ω to the integral $\mathcal{F}$ is minor.

A nonuniqueness of the inversion implies that there is more than one model that could generate an observation vector $\boldsymbol{X}$. A weaker assumption is that certain parameters may have a stronger influence on the observed data than others. We may express the sensitivity by the derivative

$$s_k = \frac{\partial \tilde{f}(\boldsymbol{a})}{\partial a_k}$$

with $\tilde{f}$ being the function that relates our model parameters a_k to our data vector $\boldsymbol{X}$. Our data will strongly depend on the parameter a_k if the sensitivity of our model with respect to this parameter is high. $\tilde{f}(\boldsymbol{a})$ can be understood as the inverse of $y(\boldsymbol{x})$. Accounting for Eq. (6.4), we understand that $y_k(\boldsymbol{x})$ will have a flat shape in case of high sensitivities. In the opposite case, $y_k(\boldsymbol{x})$ is characterized by steep slopes or high values of the derivatives $dy_k/d\boldsymbol{x}$, as the factor ω in the integral renders the factor $\mathcal{F}$ sensitive to the presence of high frequencies. In the case of a nonuniqueness of the inversion in a strict sense, the derivative $dy_k/d\boldsymbol{x}$ becomes singular and the integral $\mathcal{F}$ does not converge. It is often not feasible to figure out rigorously whether an inversion problem is nonunique in a strict sense. In practice, nonuniqueness arises for less significant parameters in the presence of disturbances, such as noise, deficiencies of the model, etc. The inversion of parameters with little influence on the observations thus becomes truly ambiguous. Scarcely identified parameters from properly trained MLP are therefore always an indication for an ill-conditioned problem, and should give rise to a sound analysis of the reasons rather than frustration for the unsatisfying result.

The evaluation of the supervised learning schemes critically depends on the reliability of our a priori information, as we evaluate the system performance by comparing its prediction to the a priori definition of a target, being it a category in classification or some set of values in regression and inversion. The question to be asked is: "Is the a priori classification correct?" This question touches two different aspects. The first one can be summarized by "Is the distinction into different classes based on the underlying physics?" In our example of rock classification with SVM (see Chapter 4), we started from the mineralogical composition of a sample, represented in the "Streckeisen diagram," and asked whether we can achieve a similar classification on the base of geochemical features. In this case, the a priori definition of the target is uniquely defined, given the position of a sample in the "Streckeisen diagram." Similarly, in inversion problems, our targets are given on a set of physical model parameters, and the quality of our learning regards the question how uniquely the targets can be related to the observations. Considering seismic waveforms, the a priori class separation could be done by using the epicenter distance of an event to the station (Beyreuther and Wassermann, 2008). In that case, we can justify the distinction on the base of the underlying physical process. Therefore, it is possible to evaluate the overall system performance correctly.

Often, we do not have clear physical criteria to establish unique and reproducible rules for the a priori classification. Recall the explosion quakes on Stromboli in Section 4.2.1. Here, the targets were defined considering the waveforms; this is not a criterion based on physics, but needs the exercised eye of an expert. Can the human analyst be assumed as an error-free reference? Often, corresponding ground-truth data are not available. In seismic signal analysis, the problem is exacerbated by the fact that there is no clear separation among the waveforms of different signal classes. Even though the separation in different event types is justified by true underlying processes, the recorded waveforms are often blurred. The signal characteristics vary with propagation, as paths with greater attenuation produce a depletion of high frequencies, and the duration of the signals tends to lengthen.

In case of a missing classification scheme, we have to define event classes manually. By reasoning about causality, an objective systematization of events would become available at hand. Reasoning about causality, it could be done by using so-called relevance or influence networks (i.e., Bayesian networks). Thus, using such a data-driven approach may allow the discrimination of different signal classes based on their inherent nature. Combined with the possibility to incorporate domain knowledge at any step of the procedure, a well-founded classification scheme may become available.

6.5 Objects

In supervised learning, we collect information of different type, defining objects on the base of a set of observations. In the seismological examples, we considered the observed ground motion and created the feature vectors on the base of these geophysical data. Besides, in the example regarding the magma-filled dyke, we created the feature vectors combing information stemming from different sources, that is, ground deformation, gravity, and magnetic field. As shown in the paper by Nunnari et al. (2001), this can be a winning strategy. The highest accuracy of the output calculated by the MLP was achieved when data from all the three geophysical fields were included. The aforementioned deficiencies of poorly conditioned inversions were limited to a few parameters. In general, multidisciplinary analysis is highly appreciated and is often a decisive add-on in fundraising for research projects.

Falsaperla et al. (2014) presented an application of unsupervised learning to a set of multidisciplinary data recorded during peculiar episodes of volcano unrest at Mt Etna. The research was motivated by the fact that the automatic alert systems, which we outlined in Chapter 5, occasionally signaled criticalities in volcanic tremor even though no manifest activity could be observed. Similar "false alerts" are disturbing as they jeopardize the credit of any institution involved in monitoring and warning issues. In this context, it is of primary interest to verify whether those warnings are linked to noise, instrumental disturbances, bad weather conditions, or the system does unveil some hidden activity that cannot be observed at the surface. For example, during bad weather conditions, fog and clouds may hinder the visual observation even in the presence of significant phenomena of volcanic activity. Evidence of unrest may also arise when magma moves inside the volcano edifice but does not make it to the surface, occasionally intruding laterally the surrounding rock. Falsaperla et al. (2014) used the term "failed" or "aborted eruptions" for such a phenomenon. To shed light on the source of the critical status signaled by the alert system, the authors looked for additional evidence of a change in the internal dynamics of the volcano. Gas emission, routinely monitored at Etna, was considered in this context also in terms of emission of in-soil radon and ambient parameters.

Radon is a radioactive, noble gas emitted from soil and soluble in water. Especially in volcanic/geothermal areas with vigorous circulation of magmatic or hydrothermal fluids at shallow depth, in-soil radon emissions are the surface expression of convective flows of gases along fractures. Convection facilitates the transport of radon from depth to the surface, and the abundance of the radon transported also depends on the abundance of uranium-bearing rocks. Radon emission is sensitive to barometric pressure. Besides, under steady flow conditions, water condensation may cause both an increase in soil temperature due to heat loss and an apparent increase in radon activity in the condensing zone, due to the lower proportion of residual water vapor and the higher proportion of noncondensable gases (particularly CO_2) after water condensation. For this reason, pressure and temperature are simultaneously monitored together with the emission of radon, allowing to account for anomalies being caused by these phenomena rather than the volcano's activity.

During one of the unrest episodes that Falsaperla et al. (2014) analyzed, there was an explosion at a summit crater of Mt Etna that was preceded by high tremor amplitude and a sudden depletion of radon emission together with an increase in temperature. The idea behind their likely connection was that the phenomena were linked to a gas-pulse reach in water vapor, which diluted the concentration of radon and temporary increased the temperature. The authors applied unsupervised learning, in particular SOM and fuzzy clustering, to feature vectors obtained by a combination of tremor spectra with concurrent radon, atmospheric pressure, and temperature measurements at the same site. The raw application of SOM and fuzzy clustering to the combined feature vectors turned out somewhat disappointing. The overall results were indeed very similar to those obtained from the tremor spectral data alone (see Fig. 6.5A and C), whereas radon and ambient parameters seemed to affect the overall picture to a very minor degree (Fig. 6.5B). The similarity of the panels (A) and (C) in Fig. 6.5 led the authors to suspect that the feature vectors were not well balanced. Indeed, the metrics used in SOM and fuzzy clustering encompassed 62 summands of the spectral data, but only three corresponding to radon and ambient parameters. The authors fixed the problem by reducing the dimension of the spectral features. In a preprocessing step, they applied a PCA to these data and kept the three largest eigenvalues. These remaining three components were combined with the radon-related data, obtaining a six-dimensional feature vector with equal weights. The new result of classification with SOM and fuzz clustering is presented in Fig. 6.5D. We notice that the combined features remained stationary from March 14 to March 18, 2007. In the afternoon of the same day, slight but evident variations occurred until the sudden changes shortly before the explosive event.

Although multidisciplinary analysis is intriguing, this example highlights some caveats and pitfalls when data from various fields are mixed together. A further issue is that the underlying source processes are not always the same. For example, Falsaperla et al. (2017, 2018) proposed a conceptual model in which the emission of radon at a probe close to the Etna's summit craters may be linked to both the magmatic source and the occurrence of tectonic earthquakes, even at distance from the site of measure. The mix of multidisciplinary data entails therefore the risk of putting together objects that do not represent the same type of source process.

6.6 Features and metrics

In the previous sections, we frequently addressed the problem of a suitable selection of features and how to measure the degree of similarity of objects. In learning with supervision, we get an immediate feedback whether our choices are effective. If not, the success of the chosen method remains unsatisfying. Inappropriate features and metrics are the most probable candidates for being responsible of failure. In unsupervised learning,

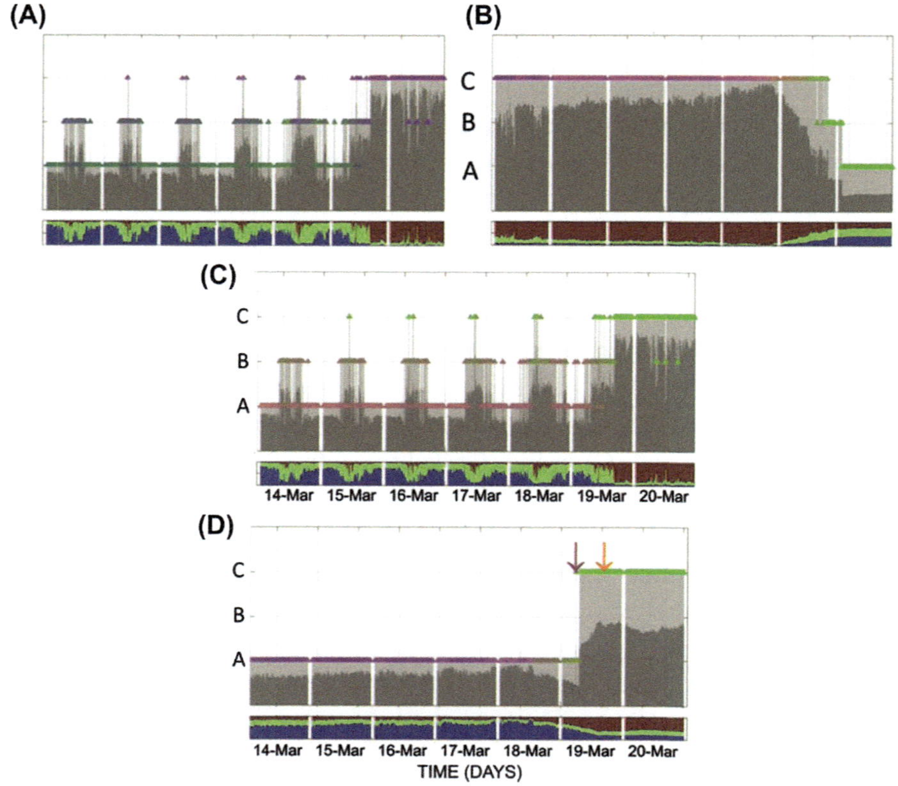

Figure 6.5

SOM and fuzzy clustering (considering the three clusters A, B, and C) applied to volcanic tremor spectra and data related to radon emission. The time span starts on March 14, 2007 at 00.00 UT and ends on March 20, 2007 at 24.00. (A) Results of SOM and fuzzy clustering for 62 spectral components of tremor; (B) results for the three radon-related components (radon emission itself, temperature, and pressure); (C) combination of the 62 spectral components of volcanic tremor with the three radon-related components; (D) combination of three components related to tremor (obtained from the original 62 after a PCA) again with the radon-related components. The purple and orange arrows mark the change from cluster "A" to "C" and the occurrence of an explosive event, respectively. *The figure was redrawn from Falsaperla et al. (2014).*

there is no direct information whether our choice is meaningful. This has to be verified a posteriori on the base of our own needs, for instance, whether we obtained a clear distinction among clusters and achieved a good reduction of data without losing too much information.

In Chapter 1, we have discussed the problem of feature vectors with many components. One of the drawbacks resides in the "curse of dimensionality," with the major part of the patterns concentrated on the margins of the feature space. Besides, Barron's result (Eq. 6.4) states that the maximum error of an MLP increases both with the number of nodes in the hidden layer and the input layers, that is, the dimension of the feature vectors. In "nearest neighborhood" approaches, the curse of dimensionality becomes a serious drawback as the samples have an increasing number of near neighbors, which entails a considerable computational burden in algorithms based on neighborhood considerations (see Chapter 1). At the very end, we conclude that any effort to limit the dimensionality of the problem is justified. As a minimum condition, we may require that the number of samples is greater than the number of dimensions. The choice of a suitable number of features is often carried out by trial and error, A more formal approach was proposed by Köhler et al. (2009), who used SOM to select the most relevant features for discriminating objects. They outlined a two-step procedure. First, the number of features is reduced by testing their relevance. This is done by estimating the randomness of individual features by the Wald-Wolfowitz runs test. Features that do not provide significant discrimination among objects, that is, are random, are rejected. The second step aims at reducing redundant information. Strong dependencies among features decrease the discriminative power of nonredundant features. Evaluating so-called component planes (each component corresponds to a particular component of the feature vector) allowed Köhler et al. (2009) to visualize dependencies between individual feature vector components. In that way, the complete feature set can be reduced to the most relevant subset.

In the examples regarding volcanic tremor (see Chapter 5), spectral features turned out as a reasonable choice, allowing us to link the identified structures to some physical concept. The dimensionality of the feature vector was limited by forming frequency bins instead of using the original components obtained by the Fourier Transform. In the Stromboli example (Section 5.2), two different metrics were considered. The use of the Euclidean metrics in K-means clustering has the advantage that formal criteria, such as the Davies-Bouldin Index, are available for choosing the appropriate number of clusters. However, the relation between the resulting clusters and the activity of the volcano was somewhat unclear in that case. A metric defined on the base of the adaptive determinant criterion offered a better relationship, as it complied with the observation that both amplitudes and spectral shapes change with the state of activity at Stromboli. Accordingly, the clustering with adaptive determinant criterion mirrored the dynamics of that volcanic system in a more appropriate way than K-means. In Section 5.2, the choice of the metric was clearly based on the a posteriori analysis of the results rather than formal criteria of clustering quality.

In a similar way, a posteriori considerations should follow density-based clustering. The technique focuses on local heterogeneities measuring the distance between single patterns; therefore, the choice of the metrics is typically Euclidean. Methods like "OPTICS" (see Section 3.1.2.3) or "DBSCAN-Strata" (Sections 3.1.2.3 and 5.3) reduce the problem of the a priori setting of the parameters (the latter requires the choice of the nearest neighbors) without the necessity of the a priori choice of the number of clusters. Nonetheless, some

Figure 6.6
C-shaped clusters and their centroids (marked by small circles). It turns out that centroids of such clusters may be closer to members of other clusters rather the ones for which they were calculated. They are inappropriate as prototypes.

experiments must be carried out for the comparison of the results to decide the kind of clustering to adopt. Classical concepts applied for the K-means, such as the Davies-Bouldin index or other options mentioned in Appendix 5, are questionable. In general, they require the definition of cluster centroids. Given the irregular shape of clusters identifiable with density-based clustering, such a centroid is often meaningless.[3] The lack of meaningful centroids brings along a further problem: density-based clustering is of questionable value when used to identify a prototype, that is, a feature vector that is close to all members of a cluster. Indeed, forming centroids in the classical way—averaging over the feature vectors of the cluster members—may lead to paradoxical situations, such as the one shown in Fig. 6.6. Here, the centroids of the two clusters fall inside the area of the other cluster. In other words, they are similar to patterns that do not belong to the cluster for which the centroids were defined. From this example, we can see that density-based clustering is powerful for the identification of separations, that is, realms in the data space where no or few patterns are encountered. On the other hand, its application for the purpose of data reduction can be argued.

The methods of pattern recognition are essentially data-driven approaches; they "see" feature vectors just as pieces of information (frequently numbers) useful to describe

[3] Imagine a C-shaped cluster. Its centroid falls outside its body, that is, it is not density reachable.

whether patterns are similar to each other. This brings along some risk, as a pair of items in a feature vector may be inappropriate to establish a metric, even though it describes the patterns well in a physical sense. This was the case in Section 5.6, where we discussed the classification of directional data, in particular focal mechanisms of earthquakes. In Figs. 5.24–5.26, we demonstrated how one can describe the orientation of the forces acting in a seismic source by indicating three angles: strike, dip angle of the fault plane, and direction of motion along the fault plane. From a physical point of view, these three angles are sufficient to uniquely characterize the source mechanism—at least as long as we are interested only in the contribution of shear, that is, the "DC" part in the moment tensor. Such a characterization of the source mechanism is likely to fail in pattern recognition. This can be immediately understood from Figs. 5.25 and 5.26 where we identify two nodal planes orientated perpendicular to each other. Each of the two planes is sufficient to fully describe the mechanism with a physical difference NULL, even though the two planes span up an angle of 90 degrees. Such a discrepancy between the physical difference (NULL) and the one measured in the angles (here 90 degrees) renders them inappropriate as feature vector components in the context of pattern recognition, in particular in unsupervised learning. On the other hand, in Section 5.6, we learnt about the Kagan angle, which provides a reasonable metric for the distinction between two source mechanisms. It can be used, for instance, in density-based clustering where the focus is on local differences between two single patterns. However, its application in centroid-based clustering, such as K-means, fuzzy clustering, and adaptive determinant criterion, is more complicated as it requires additional computation to define the centroids of the clusters. In the example mentioned in Section 5.6, we bypassed the problem by using the moment tensor components. In that case, strongly different mechanisms showed up without ambiguity in the metrics. At the same time, the moment tensor components allow us to distinguish also other types of source mechanism apart from shear (see Box 5.1). Nonetheless, one has to be aware of the peculiarities that appear working with normalized features. Differences among normalized feature vectors correspond to the angle included by the two vectors. A conventional sheet geometry of the SOM is questionable for the reasons stated in Section 5.6, but the problem can be fixed by choosing a toroidal topology of the SOM.

6.7 Concluding remarks

6.7.1 Multilayer perceptrons

One obvious advantage of MLP over SVM is that artificial neural networks may have any number of outputs, while SVM originally have only one. The most direct way to solve the multiclass problem with SVM is to create multiple support vector machines and train each of them one by one. On the other hand, the multiclass problem with MLP can be tackled in one go. Effects of overfitting can be an issue, but can be controlled designing the optimization problem correctly. It also depends on the training examples if they scan correctly and uniformly the search space.

6.7.2 Support Vector Machines

Unlike the development of MLPs which followed a heuristic approach, with applications and extensive experimentation preceding the theory, SVMs involve sound theory first and then implementation and experiments. A significant advantage of SVMs is that while MLPs can suffer from multiple local minima, the solution with SVMs is global and unique. Two more advantages of SVMs are that they allow us a simple geometric interpretation of results and give a sparse solution. Different from MLPs, the computational complexity of SVMs does not depend on the dimensionality of the input space. MLPs use empirical risk minimization, whereas SVMs use structural risk minimization.

6.7.3 MLP and SVM in regression analysis

In nonlinear regression, the calculation of the error in MLP can be directly considered as a measure of the fit, as the targets in this case are not given by categories but by floating point vectors. On the other hand, in the case of regression with SVM, first we must apply the trained SVM to our data, and compare the prediction with the target. Recall that optimization in SVM regression is based on the number of patterns one finds correctly inside a tube (see Chapter 4). In short, MLP fitting in classification problems is measured on RMS, but a posteriori validation is found using Boolean information; in regression fitting and a posteriori validation is based on RMS. SVMs in classification are based on fitting Boolean information (pattern on the right site—yes or no), and so is the a posteriori validation. In regression with SVM, fitting is again based on Boolean information (inside the tube or not), but the a posteriori validation carried out by the user is based on classical definition of the errors like RMS.

6.7.4 Hidden Markov models and Bayesian networks

MLP and SVM can be understood as "error-based" techniques, where we minimize the distance of the patterns from separating elements. On the other hand hidden Markov models (HMMs) and Bayesian networks (BNs) represent probability-based schemes. HMM and BN do not necessarily apply numerical metrics, such as an Euclidean or Mahalanobis distance. HMM and BN are based on identifying models that are determined by some transition probability from one state or observation to another.

The objects we are interested in cannot be characterized by single patterns, but require to consider their interrelations. Speech analysis is a frequently mentioned problem. Here, we first characterize sounds or vowels, and then we form words and expressions as a sequence of such vowels. Sentences are again a sequence, a sequence of words in this case. In Chapter 2, we mentioned numbers that can be understood as a sequence of digits, such as 293 being composed of "2," "9," and "3." No doubt that for a correct interpretation we must read them in the right sequence. In this simple example, the numerical meaning of the digits is irrelevant. At the end, we ask ourselves which HMM is most likely to produce a sequence of observations. HMMs and BNs deal with problems where a distribution cannot be simply described by a parameter vector, but also requires probabilistic or causal dependencies. This is a crucial difference with respect to the other techniques such as SVM, MLP, or even classical discrimination analysis. In the latter technique, each pattern is considered on its own. Although HMMs become of interest to handle a process, the more advanced BNs come into play when we would like to model a large number of variables and their relationships, especially to evaluate inconsistent and incomplete data. Examples are a volcano with an increasing or decreasing level of unrest, or the passage of a cyclone that is characterized by a sequence of meteorological phenomena. HMM and BN allow a straightforward handling of categorical data. On the other hand, their application in regression and inversion is by far not so obvious.

6.7.5 Supervised and unsupervised learning

Even though unsupervised learning follows distinct strategies from the learning techniques with supervision, they can sometimes be used in a complementary manner. For instance, the application of HMM to data on Soufrière Hills volcano, Montserrat, by Hammer et al. (2012; see Chapter 4) include an a priori clustering following the expectation maximization (see also Box 3.2 in Chapter 3), which allow to fix the number of states of the HMMs. In the classification of infrasound signals, Cannata et al. (2011) used the results of clustering as targets. Even though such a mix of supervised and unsupervised may be argued, it can be justified as the clusters are found in different locations around the summit area of Mt Etna. That way, locations could replace the cluster membership as targets that avoid mixing of supervised and unsupervised learning.

Unsupervised strategies come with the advantage that little a priori knowledge is necessary for their application. On the other hand, it happens that we wish to see how new results match with our experience. In Chapter 5 (Section 5.5), we discussed an alert system based on SOM and fuzzy clustering. The integration of a priori knowledge is mandatory in this case because we want to issue warnings on phenomena that were recognized as critical in the past. Langer et al. (2011) accomplished this integration by combining the actually recorded data with a large reference dataset. In this context, the reference dataset corresponds to the a priori knowledge, although the learning process formally remains unsupervised.

CHAPTER 7

Software manuals

7.1 Example scripts related to Chapter 2

7.1.1 Linear discrimination, principal components, and marginal distributions

In Chapter 2, we have applied Fisher's linear discrimination method to pairs of body wave and surface magnitudes (mb and MS) measured for earthquakes and nuclear tests. The routines S2.1 and S2.2 illustrate how the discrimination of datasets can be facilitated by using an appropriate rotation of the samples. In the light of this action, we restate the discrimination problem in principal component analysis (PCA), where eigenvalues and corresponding eigenvectors are considered. Eigenvectors and eigenvalues allow us to create a new system of axes, and the original feature vectors are transformed to this new coordinate system. The two small routines illustrate the concept in the 2D-case (two-dimensional feature vectors).

In Fig. 7.1, obtained with the script

S2_1

we start with the frequency distribution of earthquake and nuclear test magnitudes. Although some differences between the two datasets can be recognized, the distinction is blurred as the distributions of earthquake and nuclear test magnitudes overly. Figs. 7.2 and 7.3 show the distributions in two rotated systems of axis, using the components of the eigenvectors for the projection. In one of the two systems, the two datasets can be nicely distinguished. The position of the discrimination function is depicted in Fig. 7.5, which is obtained by rotating the values shown in Fig. 7.4.

The script

S2_2

considers two randomly generated data clouds. For the sake of simplicity, the data samples consist of 2D-feature vectors. X1 and X2 relate to the first cloud, and X3 and X4 relate to the second. The features X2 are derived multiplying features X1 by a constant. Note that X3 has the same dispersion as X1, and X4 has the same dispersion as X2. X3 and X4 are shifted with respect to X1 and X2 by 1.8 units (see Fig. 7.6). We now pool the two datasets and plot the marginal distributions shown in Fig. 7.7. For the sake of distinction,

Advantages and Pitfalls of Pattern Recognition. https://doi.org/10.1016/B978-0-12-811842-9.00007-8

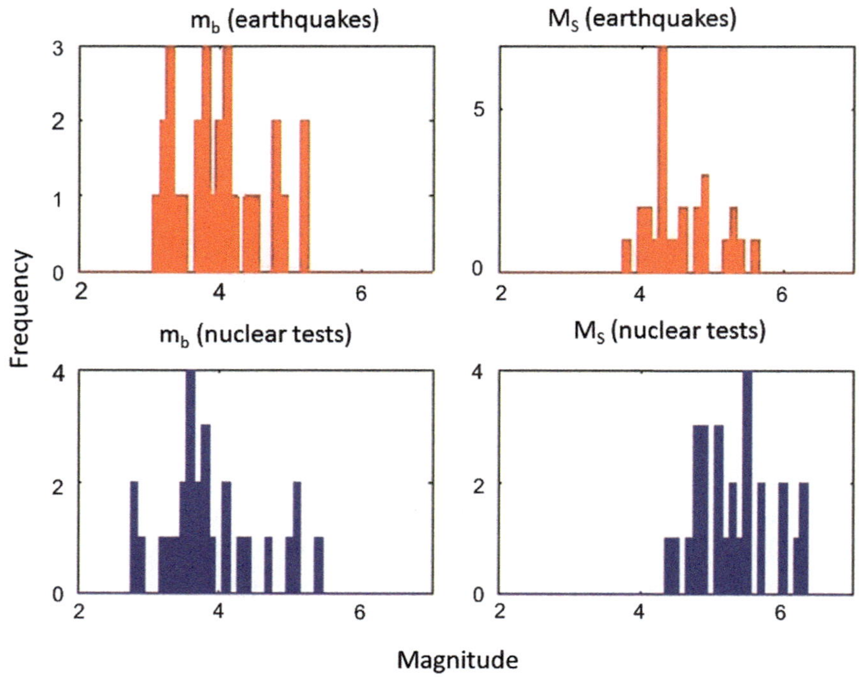

Figure 7.1
Frequency distribution of m_b and MS for earthquakes (in red) and nuclear tests (in blue).

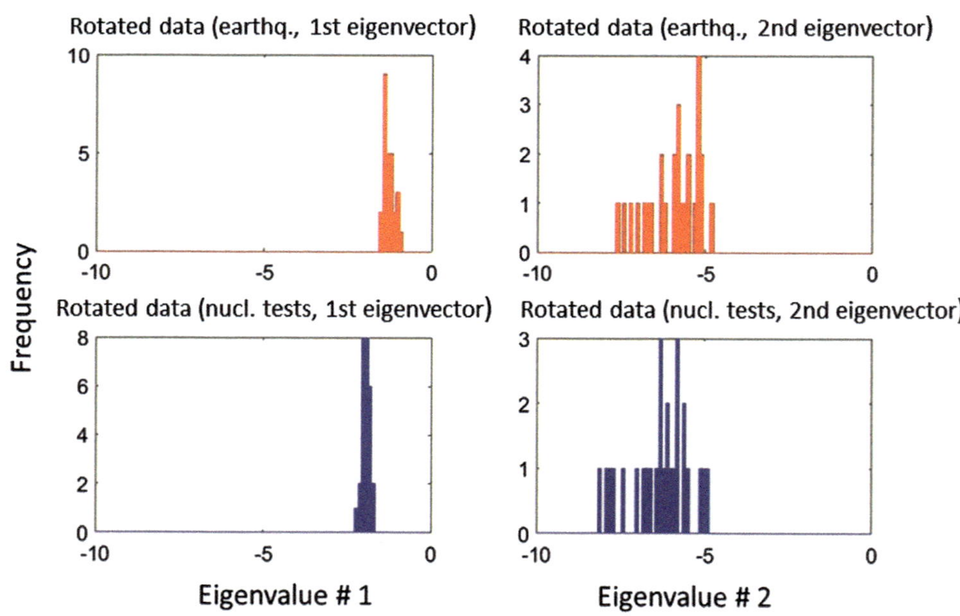

Figure 7.2
Frequency distribution for earthquakes (in red) and nuclear tests (in blue) after the rotation according to the first and the second eigenvector.

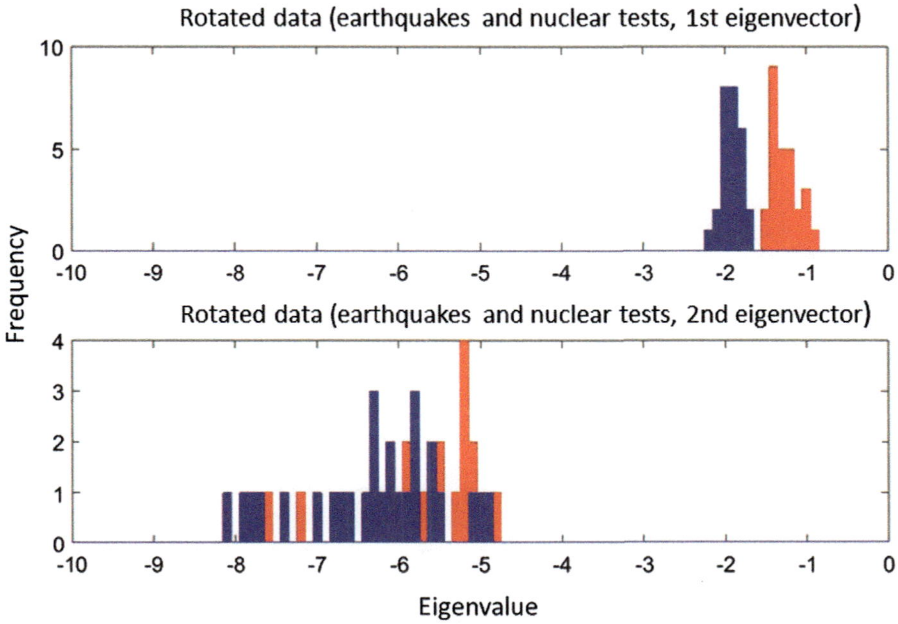

Figure 7.3

Synopsis of Fig. 7.2, again with earthquakes in red and nuclear tests in blue. In the first panel, we note a clear distinction between the two groups.

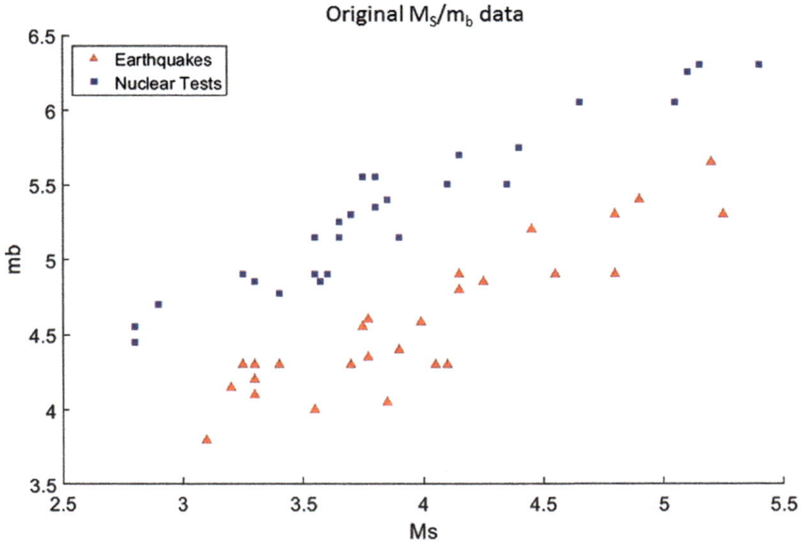

Figure 7.4

Reproduction of Fig. 2.1. M_S—m_b relation for earthquakes (red triangles) and nuclear tests (blue squares).

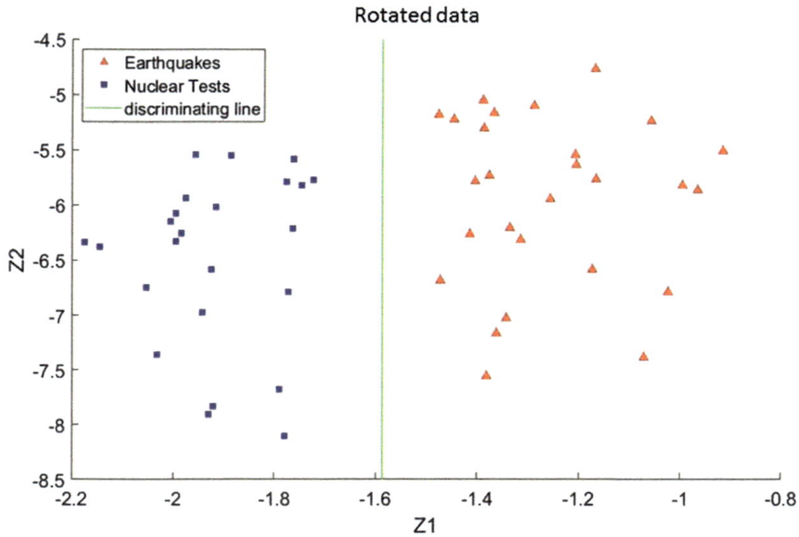

Figure 7.5
$M_S - m_b$ relation for earthquakes (red triangles) and nuclear tests (blue squares) after the rotation of the axes.

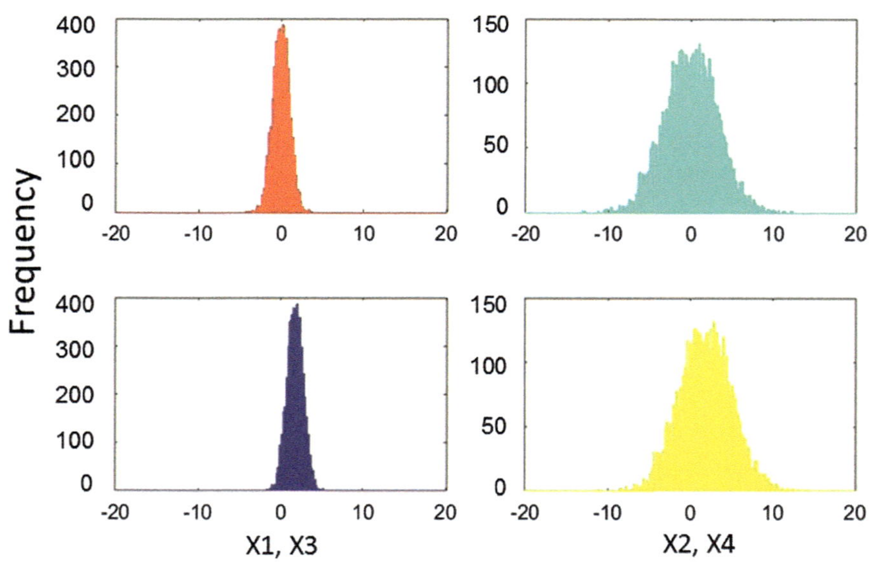

Figure 7.6
Single univariate distributions x1, x2, x3, and x4, obtained as linear combinations, starting from two normal distributions of random numbers: x0 and x1.

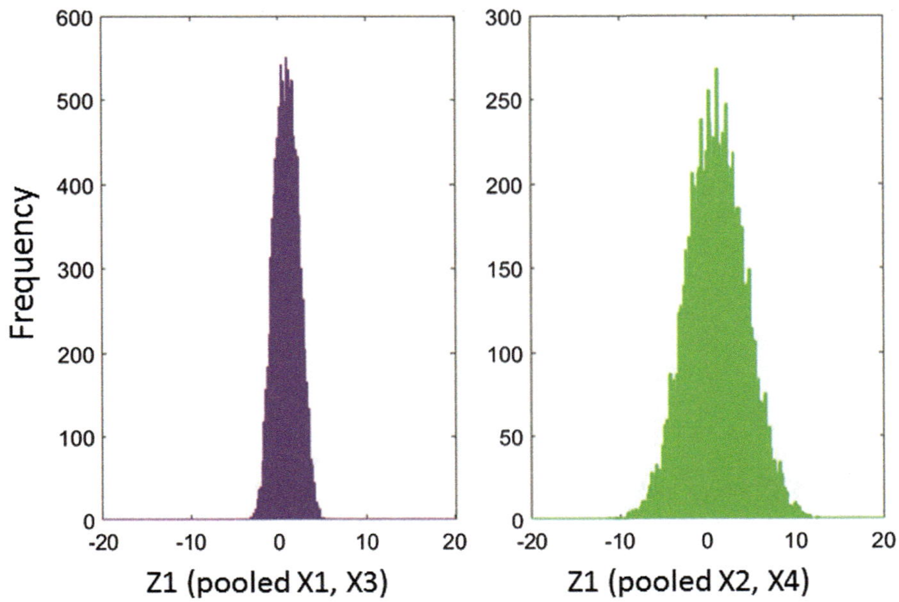

Figure 7.7
Univariate distributions Z1 andZ2 obtained from the merger between X1 and X3 and between X2 and X4, respectively.

we address axes of the two components of the pooled dataset as Z1 and Z2. They have the same orientation as X1 and X2 (or X3 and X4, respectively).

The marginal distributions do not allow a distinction of the two groups (see Fig. 7.7). In Fig. 7.8, we conduct a rotation, again exploiting the eigenvectors obtained from the PCA performed on the pooled covariance matrix of the two groups. N.B.: The pooled covariance matrix is obtained as explained in Chapter 2 and must not be confused with the covariance matrix of the pooled dataset! The reader is invited to use the latter and reproduce Fig. 7.8.

In Fig. 7.9, we depict the samples in the original system of axes (X1, X2) and in the rotated system of axes (here named Y1 and Y2). The distinction between the two groups is evident in both plots of Fig. 7.9, as the two-dimensional feature vectors can be represented in a 2D sheet without loss of information. In case of higher dimensional feature vectors such 2D-plots are insufficient, as they are marginal projections, bringing along all the problems discussed earlier. Pattern recognition aims at resolving these problems.

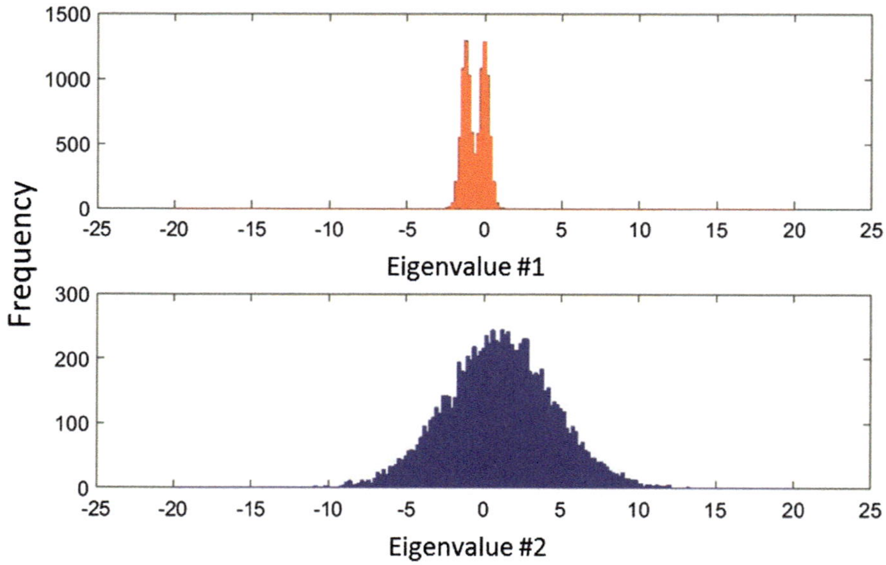

Figure 7.8
Distributions of the pooled and rotated dataset using the two eigenvectors obtained from PCA. Both vectors were calculated for the pooled covariance matrix (not to be confused with the covariance matrix of the pooled dataset!).

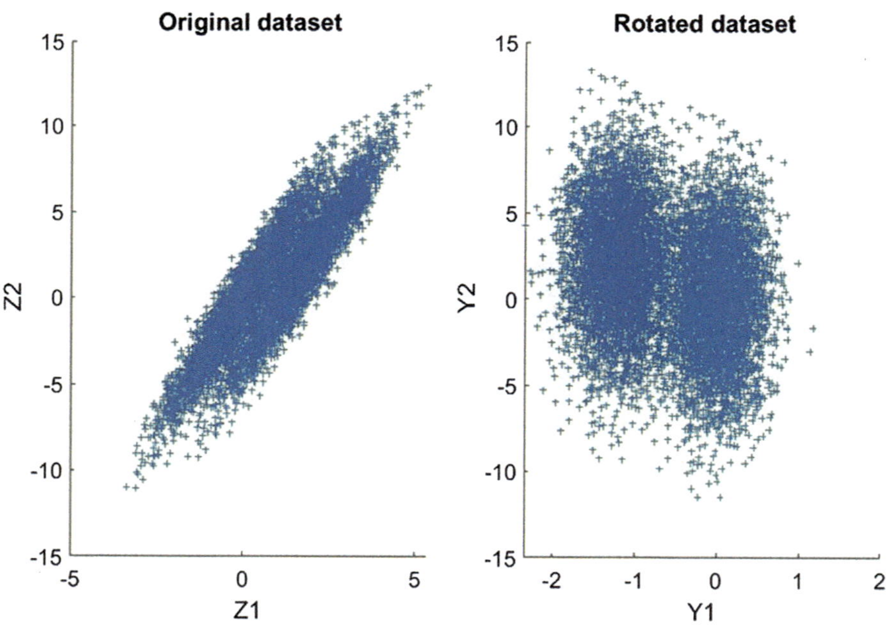

Figure 7.9
Z1–Z2 relation in the original dataspace, on the left, Y1 – Y2 (rotated dataset according to the eigenvectors of the covariance matrix), on the right. For two-dimensional feature vectors, the separability of the two groups is evident in both panels.

7.1.2 The perceptron

The script

S2_3

is a small routine based on the perceptron concept. It exploits some routines published in Theodoridis et al. (2010), which can be downloaded from the website http://booksite.elsevier.com/9780123744869. In its original form, it is an alternative for solving linear discrimination problems. The difference with respect to the afore discussed methods resides in the definition of the measure of separability, which is user-defined in the case of the perceptron. The program loads two files representing the groups A and B. The learning rate *r* of the perceptron is fixed to 0.01. A Boolean class membership is assigned to each pattern, that is, "1" or "−1," depending on its a priori defined class (here earthquakes or nuclear test magnitudes). This simple artificial neural network converges if the classes can be separated by a linear element. For our earthquake/nuclear test example, we get the result shown in Fig. 7.10.

N.B: An MLP software allowing for a hyperbolic-tangent activation function in the hidden layer is presented in the context of the programs related to Chapter 4. The program comes with a GUI, the numerical kernel has been written in C-language that renders the

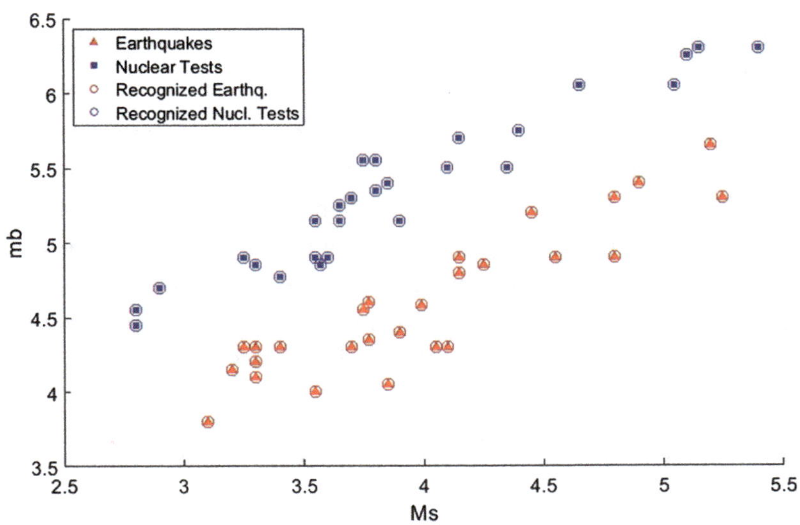

Figure 7.10

Discrimination between earthquakes and nuclear explosions using an artificial neural network based on the perceptron. Original datasets are represented by red triangles and blue squares whereas classification results are shown by means of circles colored according to the class-membership. The diagram demonstrates a perfect separation of the two classes, with a misclassification equal to zero.

application efficient and fast when applied to large datasets. It has a multiple output thus allowing the straightforward treatment of multiclass problem.

7.1.3 Support Vector Machines

In script

S2_4

we demonstrate the application of various support vector machine (SVM) kernels to the earthquake and nuclear test magnitudes example. The SVMs are trained with the original dataset and applied to a second test set, which has the same structure as the training set but values are blurred with some noise. The kernels available in this script are "linear," "polynomial," and "RBF" (*R*adial *B*asis *F*unction). The user can play with controlling parameters, in particular "C" and "tol." Besides, there are the kernel parameters "kpar?," etc. In the "RBF" kernel, we specify only "kpar1" that corresponds to a sigma and controls the resolution of the kernel. In the polynomial option, one can play with "kpar2" that controls the degree of the kernel polynomial. The following plots give a summary how the program may work. First, the program loads the training data, which consists of the members of the groups A and B. In a similar way, a test set is loaded, again containing members of the two groups. Fig. 7.11A and B show the diagrams for the two files.

Note that we have been using the noisy dataset of Chapter 2 for training to highlight chances and pitfalls of the various kernels. We intuitively see that our training set will probably favor a nonlinear kernel.

Comparing the performance of the linear and RBF kernel, we realize that the latter performs better for the training dataset (see Figs. 7.12 and 7.13). Fig. 7.14 is a 3D plot of the scores, and we can recognize that the delimiting level is given by the dark green color that fills the plots in Figs. 7.12 and 7.13. In the test set, we see the risk of overfitting one may run when a nonlinear kernel is used in the training phase. Being the test data linearly separable, the RBF kernel leads to some misclassification in this set. On the other hand, the linear kernel has less misclassified samples in the test set though being less performant during training.

Besides the figures, the script creates ASCII files reporting the support vector for each kernel, and files formatted as [X Y score] to be easily imported to standard graphic software, such as "Surfer."

N.B.: The SVM presented here allows only the resolution of the two-class problem. It exploits routines of Theodoridis et al. (2010), which can be downloaded freely from http://booksite.elsevier.com/9780123744869). For the multiclass SVM and SVM regression, use the SVM_multi_class program, which has been designed on the base of the LIBSVM

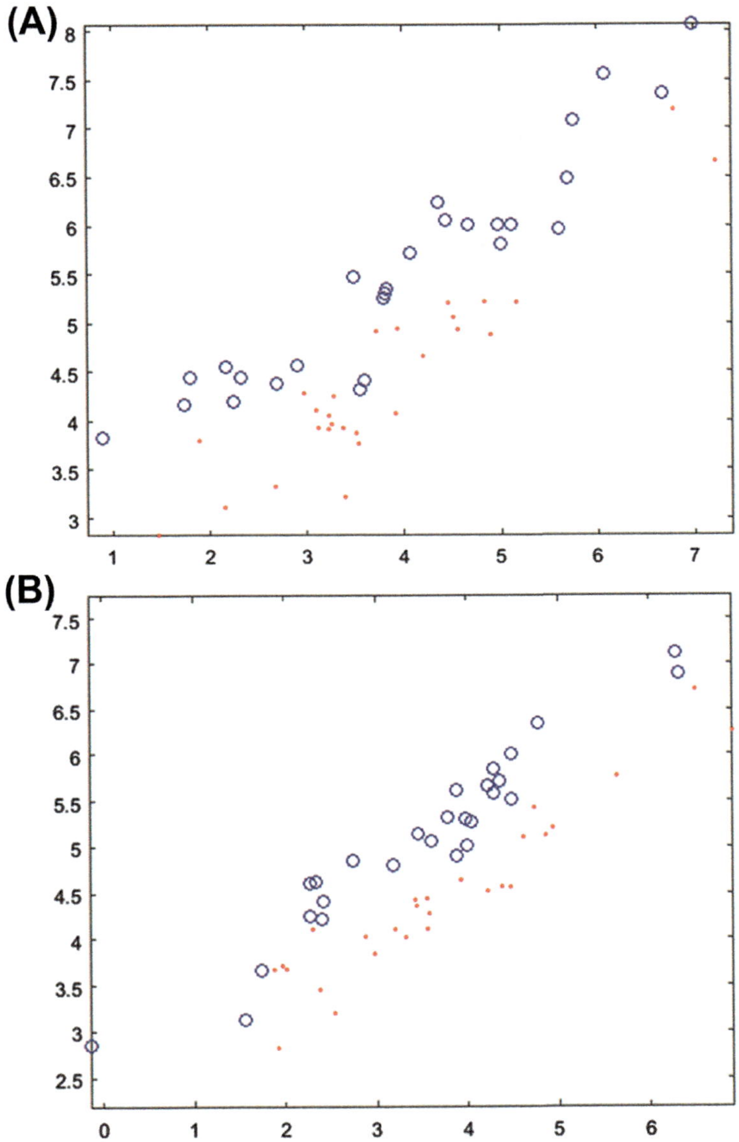

Figure 7.11
Training (A) and test data (B) used in the SVM example code.

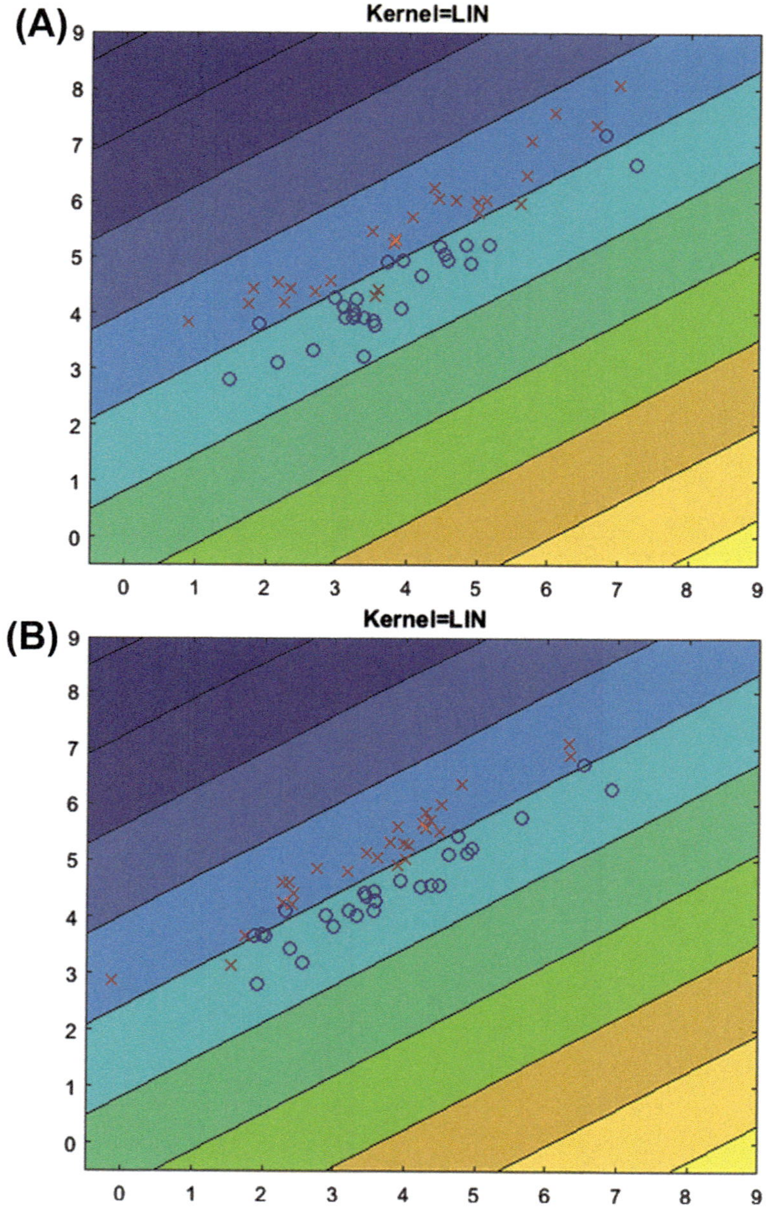

Figure 7.12
Shape of the linear kernel decision function, and position of (A) training set samples, (B) test set samples.

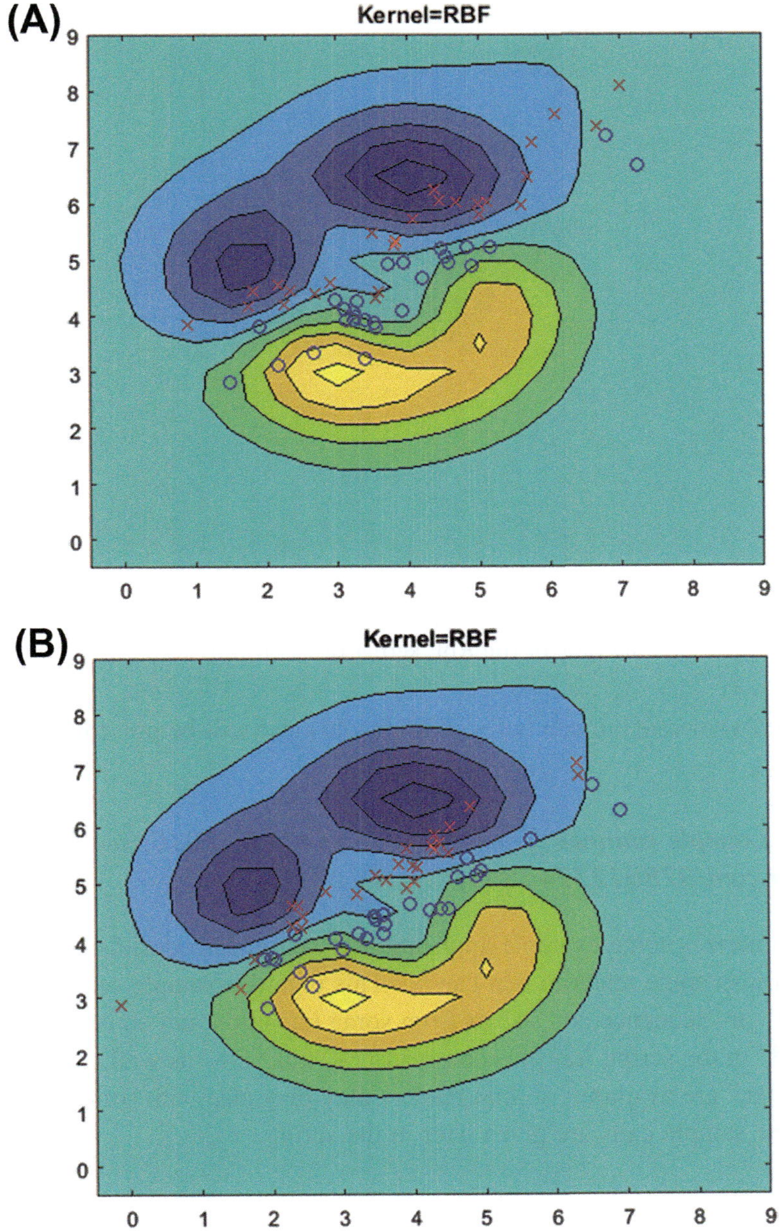

Figure 7.13
Shape of the RBF kernel decision function, and position of (A) training set samples, (B) test set samples.

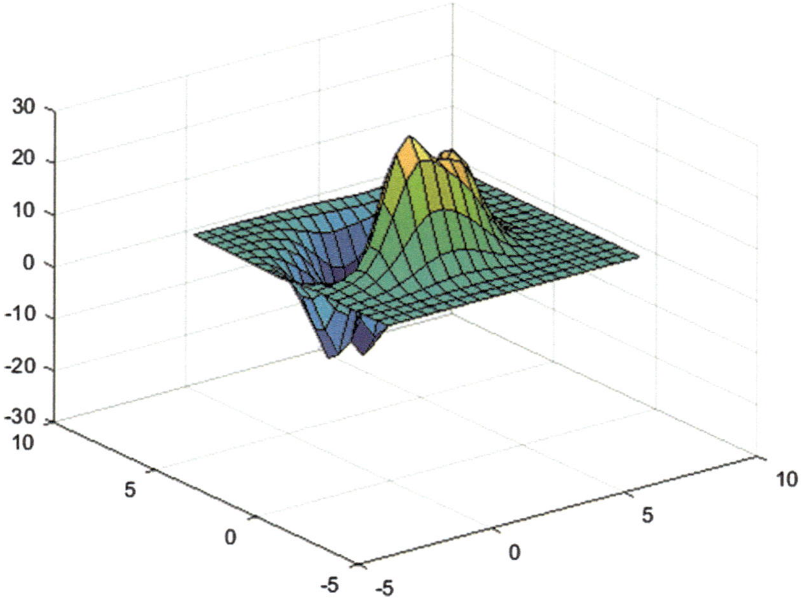

Figure 7.14
3D plot of the RBF kernel decision function.

package (https://www.csie.ntu.edu.tw/~cjlin/libsvm/) and will be presented later in this document.

7.1.4 HMM example routines (from Theodoridis et al., 2010, see http://booksite.elsevier.com/9780123744869)

The script summarizes some example applications given in Theodoridis et al. (2010). Besides, we furnish some material that can be. This regards example sequences such as "sequence2.txt" or "sequence.txt". Apart from some transition and emission matrices given as "hard coded" in the script, transmission matrices "trans.txt," as well as emission matrices "emi.txt," are available, so that readers may play also with some self-defined situations. Some sample calls are given later in the script

S2_5

1. What you need: "BackTraining (*.m-files in Chapter 5 of Theodoridis et al., 2010)

"*.m-files in Chapter 6 of Theo2010"

A sequence of observations, for instance generated by an HMM (*H*idden *M*arkov *M*odel) (it comes along with our examples, and can also be generated using the "hmmgenerate" function of MATLAB).

2. Define example transition and emission matrices (or load them from the disk):

   Suppose

```
trans =
[0.80,0.05,0.15;
0.05,0.90,0.05;
0.1,0.2,0.7];
emis =
[1/6, 1/6, 1/6, 1/6, 1/6, 1/6;
1/10, 1/10, 1/10, 1/10, 1/10, 1/2;
1/12, 1/12, 1/12, 1/12, 1/12, 7/12];
```

   and define the a priori vector pi, for example, `pi = [0.7 0.2 0.1]'`

Such an HMM describes a situation in which three dies are rolled, the first being a fair one (all six outcomes are equally probable), while the two others are biased. The probabilities of outcomes of the three dies are reported in the "emission" matrix. Using the fair die means that we are in state 1, otherwise we are in state 2 or 3. At any roll dice, we have to decide in which state we are. For this purpose, the transition matrix is used.

Now use the "BWDoHMMst" function to calculate the probability that the sequence O was generated by the HMM, that is,

```
[Pr1]=BWDoHMMst(pi,trans_guess, emis_g',0);
```

or

```
[Pr1]=BWDoHMMsc(pi,trans_guess, emis_g',0);
```

which is often preferred as it gives the logarithm of the probability, and does not suffer from round-off problems. The algorithm uses the "Baum—Welch" method, which we have presented in Chapters 2 and 4. Alternatively, one can use the "Viterbi" algorithm, that is,

```
[Pr1, BestPath1]=VitDoHMMst(pi,trans_guess, emis_g',0);
[Pr1, BestPath1]=VitDoHMMsc(pi,trans_guess, emis_g',0);
```

Recall that different from the Baum—Welch algorithm, which considers all paths in the trellis, the Viterbi algorithm considers only those with the highest scores (i.e., the Viterbi path). The `BestPath1` is expressed as a complex variable, by convention being the y-coordinate of the node in the trellis (state number) real, the imaginary part stands for the x-coordinate, that is, the index of the observation. At the end, we obtain the sequence of states by taking the real part only.

A variant of

```
[Pr1, BestPath1]=VitDoHMMsc(pi,trans_guess, emis_g',0);
```

is

```
[Pr1, BestPath1]=VitCoHMMsc(pi,trans_guess, emis_g',O);
```

where the sequence is composed of continuous variable, that is, floats. Note that the entries in the emission matrix receive a different definition and meaning. The emission probabilities are given as a 2 × K matrix, where K is the number of states. It has the form

```
emis=

[b11 b12 b13..b1k
 b21 b22 b23..b2k];
```

The two elements in a column express the probability that a (float) value is emitted in the current state *i,* assuming that the value follows a Gaussian distribution, with averages `m` and standard deviation `s`. The emission matrix therefore is composed of the elements, which specify the Gaussian distributions:

```
[m_1 m_2 m_3..m_k
 s_1 s_2 s_2..s_k]
```

In classification problems, we typically do not know the emission and transition matrices. These have to be learned using a training set of sequences. In the software downloaded from Theodoridis et al. (2010), we find the routines “MultSeqTrainDoHMMBWsc,” which has the following calling sequence

```
[pi_Train, trans_train, emis_train, sum_probs] =
MultSeqTrainDoHMMBWsc(pi_init, trans_guess, emis_guess, Data(1:n), maxEpoch);
```

that is, we use initial guesses for the a priori setting of the state, transition and emission matrices. maxEpoch is an integer giving the number of training cycles; `Data` is a MATLAB™ cell structure, and it contains a one-dimensional sequence of values, such as ‘1 1 1 2 2 1’ in a binary problem. The sequence is a multiple sequence, as it consists of a multitude of sequences—in the example below, they have a length of 3—just in a row. The cell structure of `Data` looks like this:

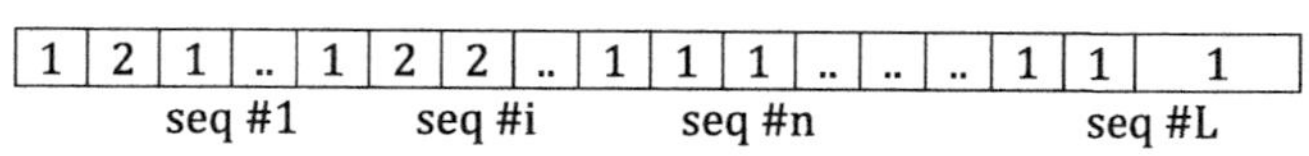

The length of each elementary sequence is defined by the properties of the cell structure. Specifying “`Data(1:n)`,” we specify that we are considering the sequences with index from 1 to *n* out of the L elementary sequences present in the data. Once having learned the parameters of the HMM, that is, `pi_Train, trans_train, emis_train,`

we can apply them to new sequences and check the quality of the HMM, just calculating the probabilities using

```
[Pr1]=BWDoHMMsc(pi,trans_train, emis_train,O);
```

where “`O`” is a sequence of interest.

Remark: A general environment for HMM application is the "HTK Toolbox," which provides a variety of HMM-related routines in C-language. Sources and documentation can be downloaded after registration on http://htk.eng.cam.ac.uk/download.shtml. Both Linux and Microsoft™ versions are available. For complex and big datasets, we address the reader to this software, which has been frequently used in geophysical applications.

7.2 Example scripts and programs related to Chapter 3 (unsupervised learning)

7.2.1 K-means clustering

"K-Means" algorithm is perhaps the most popular partitioning clustering method for its simplicity. Besides, formal criteria regarding the choice of the number of clusters are available. The small script

S3_1

presented here performs a K-means cluster analysis on mb/MS values of earthquakes and nuclear tests. Note that the K-means clustering is also available in the KKAnalysis software. Script S3.1 reads the two files separately and allows comparison of the two files and the resulting clustering (see Fig. 7.15A, B).

Script

S3_2

loads a bidimensional file (here mbms_all2.txt) and carries out a clustering using the "Adaptive Determinant criterion." A similar option is available in the package "KKAnalysis." The number of cluster number is set to "N = 2" (see m-file) and can be changed if desired. It is interesting to compare the results of this script shown in Fig. 7.16 to the application of script S3_4b that uses essentially the same data. Caveat: Avoid clustering of high-dimensional data when the number of samples is small.

7.2.2 Mixed models

The script

S3_3

generates a mixed dataset composed of three Gaussian distributions, each of which given by the mean vectors and corresponding covariance matrices. The program produces a figure (see Fig. 7.17) where each of the Gaussians is shown as a cloud of colored symbols. As a result a bidimensional "test_data.txt" is obtained with a length of 500 samples.

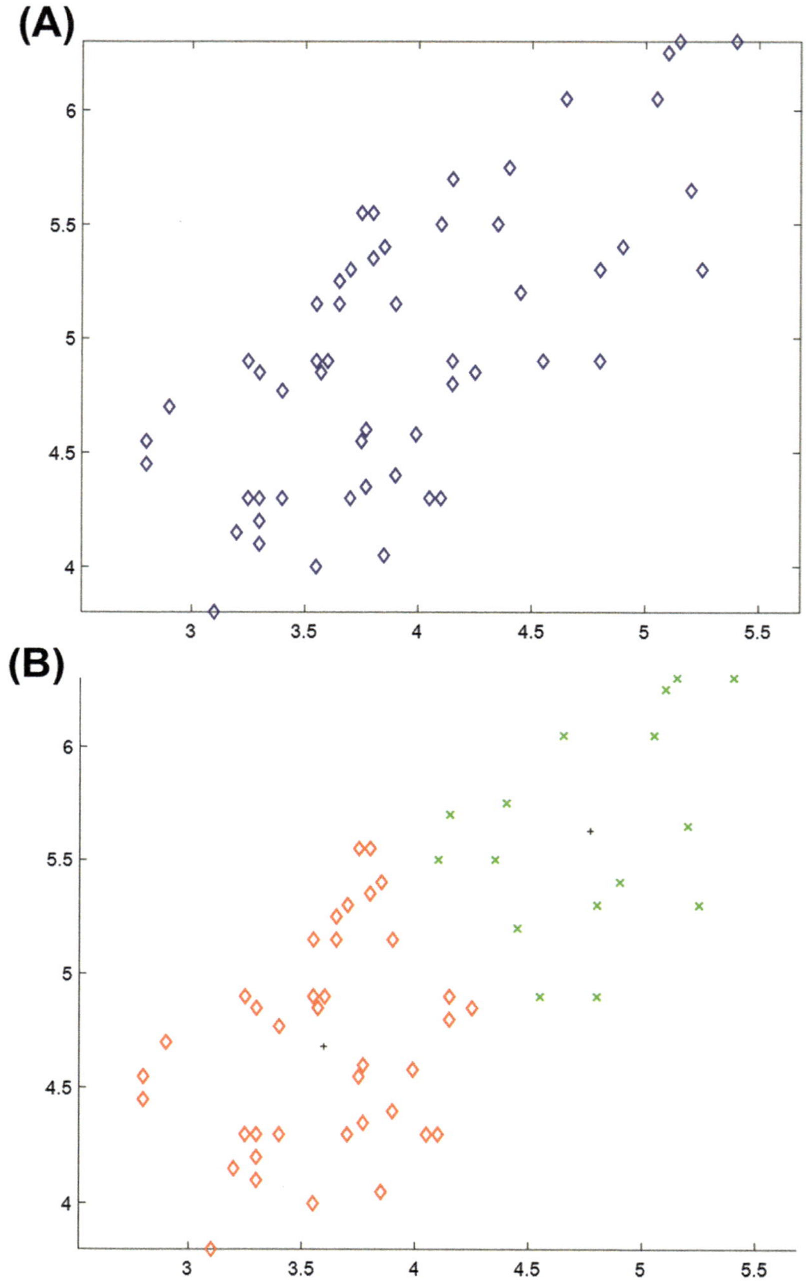

Figure 7.15
(A) Original data (mb/MS of earthquakes and nuclear tests), (B) clusters obtained with K-means clustering.

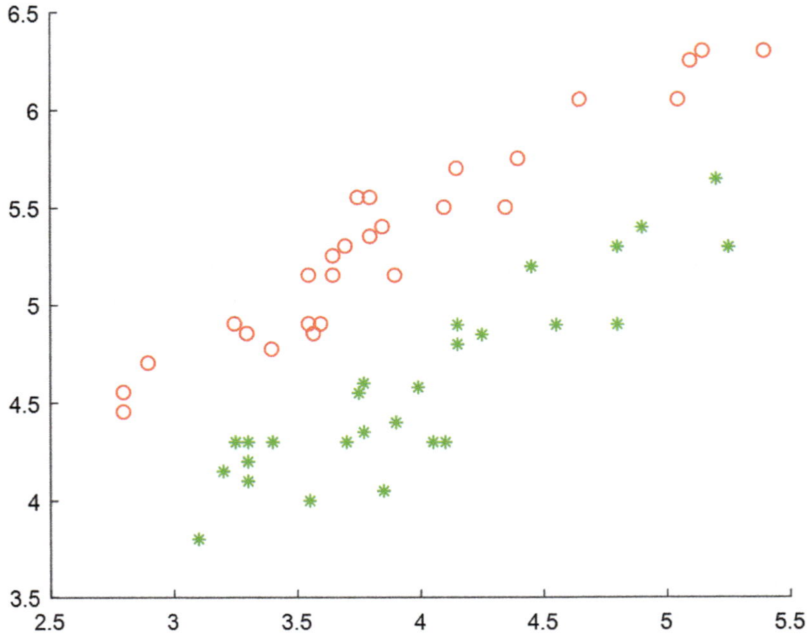

Figure 7.16
Application of the "Adaptive Distance" criterion to clustering of data shown in Fig. 2.1.

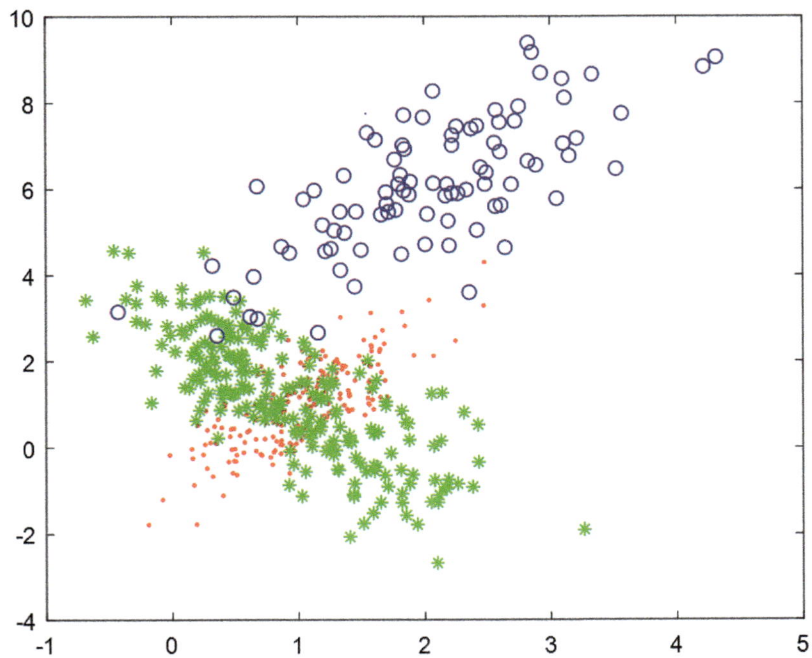

Figure 7.17
A Gaussian mixture composed of three groups.

7.2.3 Expectation maximization clusters

In the script

S3_4a

we apply the generalized mixture decomposition algorithmic scheme (GMDAS, see Chapter 3, Box 3.2) to three bidimensional Gaussian distributions. The script loads three bidimensional files "test_1.txt, test_2.txt, test_3.txt," which were randomly generated a priori. The files can be obtained, for instance, by randomly splitting the "test_data.txt" earlier mentioned into three parts. Initial guesses for the parameters of the Gaussians are the mean vectors and the covariance matrix of each single file. At the end, the user obtains a "Bayesian" clustering, that is, Gaussian models where each sample can be assigned to a group by expectation maximization of the probability of its membership. The program produces a figure (Fig. 7.18) where each sample is assigned to a Gaussian; membership is expressed by colored symbols.

The program exploits the routine "gmdas.m" that can be found in Theodoridis et al. (2010, http://booksite.elsevier.com/9780123744869). Caveat: Avoid clustering of high-dimensional data when the number of samples is small.

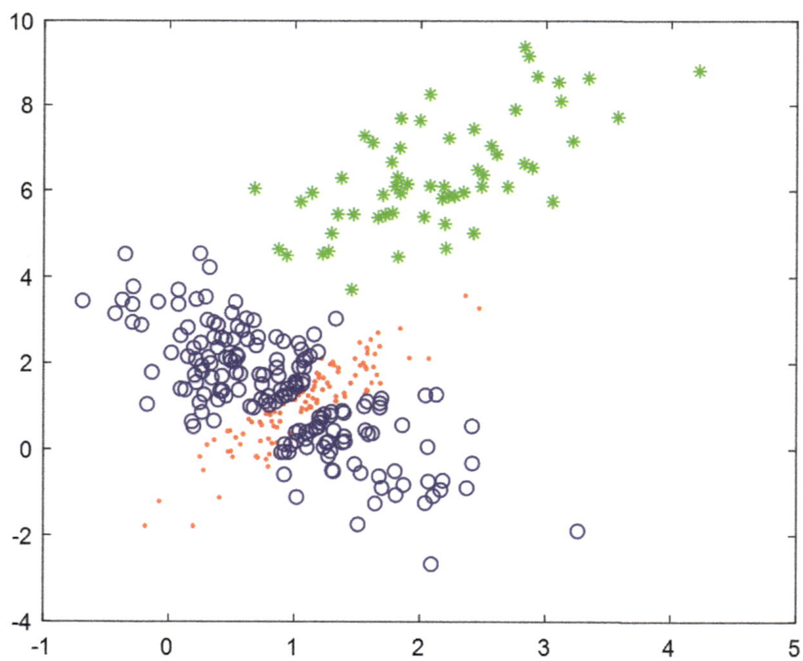

Figure 7.18
Clustering obtained with expectation maximization. The dataset considered corresponds to the samples depicted in Fig. 7.17.

The script

S3_4b

applies the same principles as script S3_4a to two files with mb and MS values of earthquakes and nuclear tests. The initial guesses of the statistical parameters are obtained from the mean vectors and covariance matrices of the original files. The script tries to find a mixed model (two Gaussians) that maximizes the expectation of probability of membership of the samples (see Fig. 7.19).

7.2.4 Fuzzy clustering

The script

S3_5

performs a fuzzy-C-means clustering using the standard MATLAB function "fcm." In this code, clustering is carried out on the file "data_rocks_tot.txt" (a 13-dimensional dataset) with three clusters. In the script, we set the exponent $q = 2$. The bidimensional plot in Fig. 7.20 (using the first and second feature vectors out of the 13) shows the results, assigning each sample to the cluster for which the highest membership is obtained.

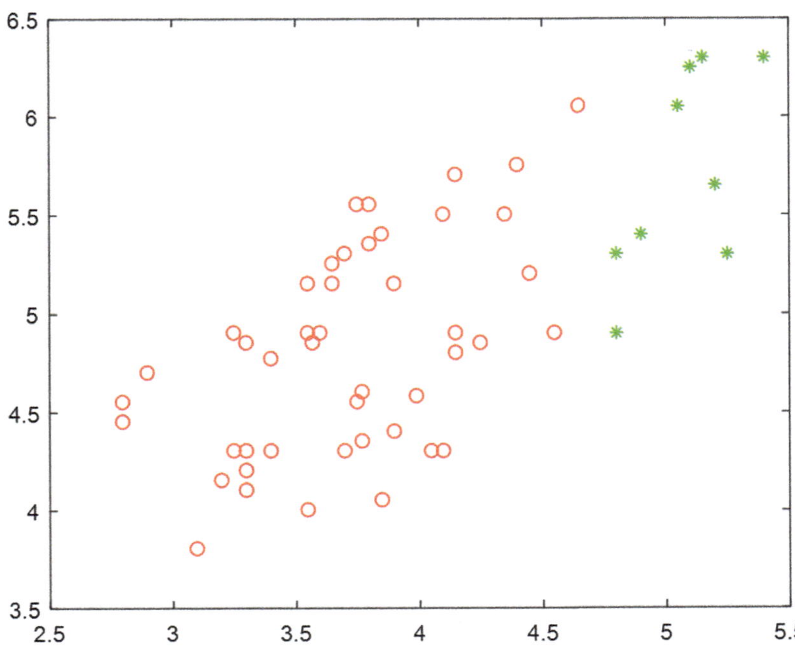

Figure 7.19
Application of expectation maximization clustering to the data shown in Fig. 7.15.

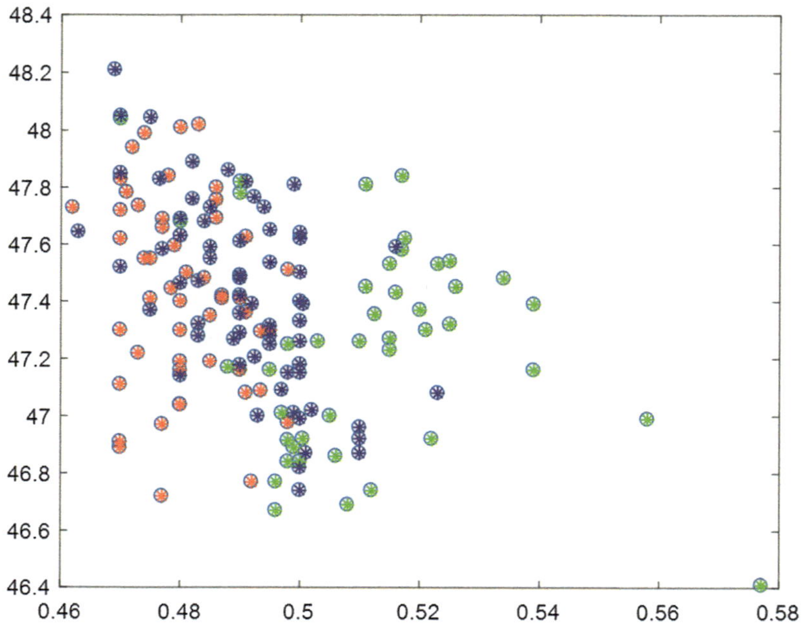

Figure 7.20
Fuzzy clustering of a 13-dimensional dataset regarding geochemical analysis. The prevailing class membership is marked by the colors of the symbols.

The reader is invited to play with the exponent q and compare the fuzzy cluster membership vectors. These values can be accessed in the file "classMemb.txt" that reports the fuzzy cluster membership vectors for all samples.

7.2.5 Hierarchical clustering

The script

S3_6

performs agglomerative clustering using either "single" or "complete linkage." Here, we load the file "Iris_data.txt." The number of clusters can be controlled by setting the parameter "n_clust" (line 120 in the script). The program creates a number of figures, such as the dendrogram (see Fig. 7.21A) and some plots showing the clusters in the original data space (Fig. 7.21B). The vector "belx" reports the cluster membership for each sample. These are written to a file "agglom_clust.txt."

7.2.6 Density-based clustering

The script

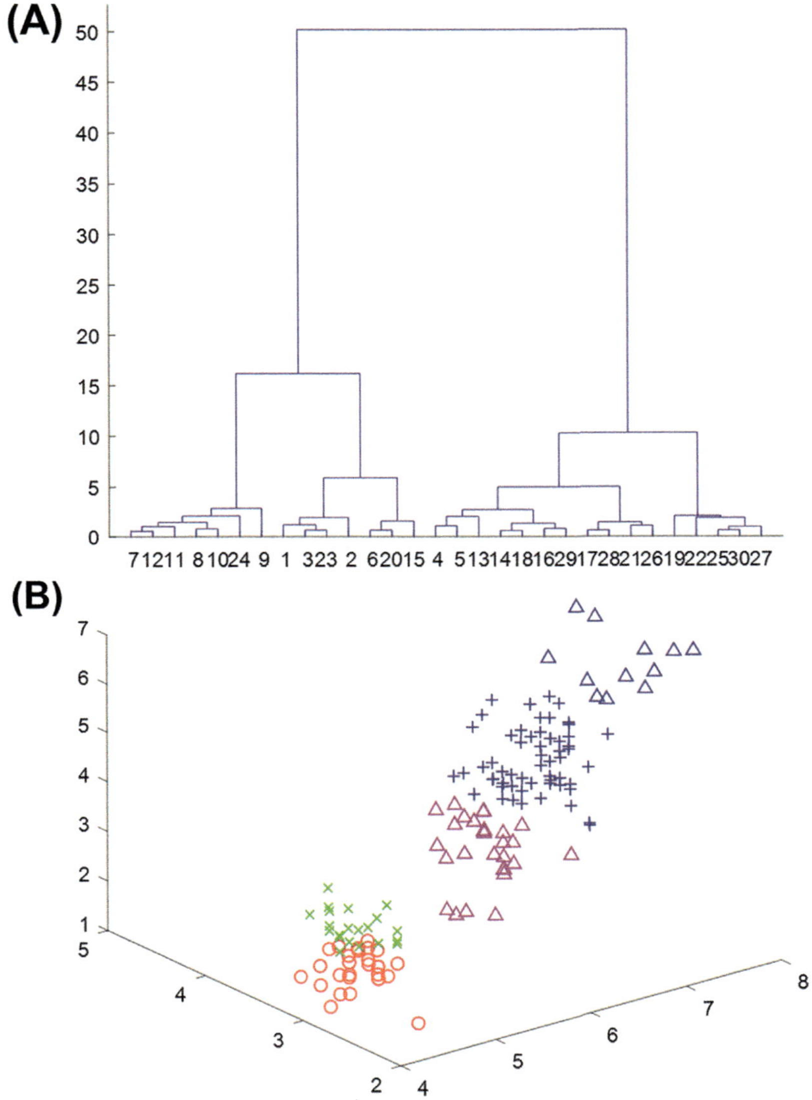

Figure 7.21
(A) Dendrogram obtained with agglomerative clustering of the "Iris" dataset, using the "complete linkage" option. (B) Distribution of clusters in a 3D plot.

S3_7

carries out density-based clustering, here on an example ("mydata") provided by the developers (yarpiz.com). Controlling parameters are the critical distance (epsilon) and the parameter q (in Fig. 7.22A addressed to as "MinPts"). The program identifies two clusters, besides some samples, which are noise as they are not density reachable. The reader can

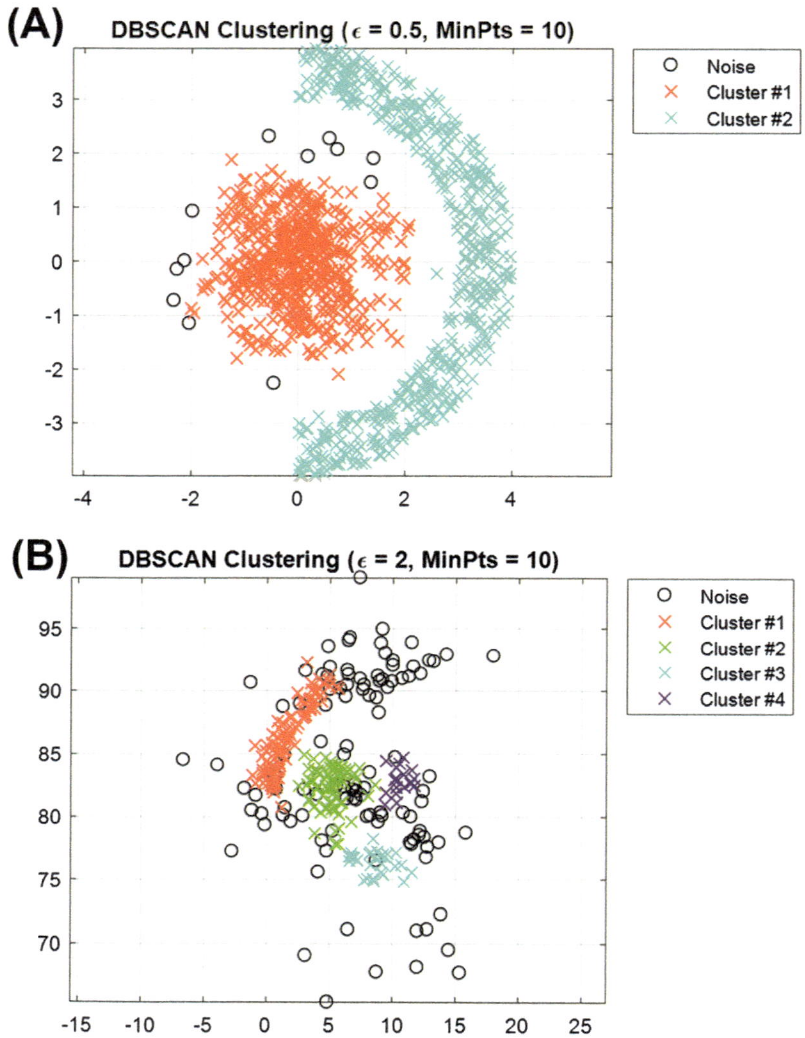

Figure 7.22

(A) Density-based clustering of a test dataset. (B) Density-based clustering of a dataset regarding hypocenter locations (courtesy of T. Tuvè, see also Mostaccio et al., 2013).

also play with a file "LocOct2003_rev." Good results are obtained setting q = 10, and epsilon = 2 (see Fig. 7.22B).

7.2.7 Unsupervised learning toolbox: KKAnalysis

7.2.7.1 Preliminaries

KKAnalysis is a collection of methods for unsupervised classification and clustering. In its original form, the software package was published by Messina and Langer (2011) and widely exploits routines of the SOM Toolbox 2 for MATLAB (Vesanto et al., 2000, see http://www.cis.hut.fi/projects/somtoolbox/). Besides functions related to the SOM, it includes K-means clustering, available in the toolbox, the MATLAB fuzzy clustering as well as adaptive determinant clustering, the latter taken from Späth (1983). All clustering schemes can be applied to the map created by the SOM routines and to the original patterns as well.

KKAnalysis was tested on x86 architecture Microsoft Windows 7™, 8™, 10™.[1] The requirements regarding the hardware are moderate. A reasonable performance was achieved on a machine with a Pentium 4 processor with a 3 GHz clock frequency, and 1 GB RAM.

For the sake of simplicity all alphanumeric files are standard ASCII. Besides, figures can be saved either as MATLAB™ "*.fig* or *.tiff* files. The latter option allows the creation of plots for users who do not have a MATLAB™ environment installed on their computers.

7.2.7.2 Installation

KKAnalysis was developed under MATLAB. This programming language requires a "virtual machine" (using the common Java terminology) called MCR (MATLAB Component/Compiler Runtime) for the execution of deployed applications. If a MATLAB environment is not available, the MCR (the original version, downloaded from http://earthref.org/ERDA/974/, requires v.7.7 or higher for Windows) must be installed to run KKAnalysis. The MATLAB Component Runtime installation on Windows systems can be carried out clicking "MCRInstaller.exe" file and following the wizard. As MCR is included in the MATLAB environment, no extra installation is necessary when a MATLAB is already present on the computer. The MATLAB version should be R2007b or higher for Windows operating systems.

On Windows systems, the installation of the program can be carried out clicking "KKAnalysis_setup.exe" file. Accepting the license conditions the installation performs straightforward. KKAnalysis can be invoked either from the "Programs → KKAnalysis" item or by clicking the shortcut on the desktop (if this option was accepted during the installation process). For the versions downloaded from the companion website of this book see the documention coming along with the programs.

7.2.7.3 Files

7.2.7.3.1 Input files

CRLF the input files of KKAnalysis concern the data used in classification.

The data file must be provided by the users. In its typical format, it consists of rows and columns, where a row contains a feature vector of a pattern. In the example of Table 7.1, the turquoise area highlights such a kind of data file. Sometimes a data file may contain

---

[1] The original version of the program KKAnalyis together with documentation can be found on the site https://earthref.org/ERDA/974/. This older version is similar to the present one, but lacks the sheet options for a Torus and Cyclinder geometry.

Table 7.1: A typical KKAnalysis input file, with: headers row, columns label, and a specific column for the abscissa of the graphical output. The format is plain ASCII.

Header	Kopf	Testa	Hair	Hat
Header	Kopf	Testa	Hair	Hat
red	4.927	4.953	4.568	1
green	4.872	4.905	4.52	3
blue	4.85	4.916	4.594	5
yellow	4.821	4.861	4.516	6
grey	4.959	4.975	4.604	7
black	4.996	5.03	4.686	9

additional information, which is not exploited for classification purposes, although being useful for the user. For instance, there may be descriptive rows at the top, so-called "header." Furthermore, there may be labels for each pattern written in the first columns of the data file (in our example, the labels "red," "green," "blue," etc.). KKAnalysis automatically recognizes these nonnumeric rows and columns and does not consider them in the analysis. If desired, one of the columns (purple, in the example of Table 7.1) may be exploited for the creation of the abscissa in the graphical representation of the results. In this case, make sure that the corresponding column contains only integer values.

7.2.7.3.2 Output files

CRLF files created by KKAnalysis can be distinguished in two types: log files and files containing the results of a session. In the log files, KKAnalysis reports controlling parameters related to the various runs carried out during the session, producing a sort of "execution history." The log file is created by the program at the beginning of each session, that is, every time KKAnalysis is started. Its name is "KKA_Log_YYYYMMDDThhmmss, where YYYYMMDD is the date (year/month/day) and hhmmss is the time (hours/minutes/seconds), both of them referred to the beginning of the session.

The settings used during a specific run can be saved in a configuration file, the name of which is specified by the user. A configuration file created during a previous session can be reloaded to reproduce the corresponding classification results. However, note that KKAnalysis automatically reloads the setting values of the last working session at the beginning of a new one. More details on the configuration parameters are given in the following.

KKAnalysis creates files storing the results of each session in alphanumeric form using ASCII standard. Files "KKA_Results_YYYYMMDDThhmmss" (indicating date and time of execution) generally contain five columns: the index of the pattern (column 1), the cluster membership (column 2), and the RGB color code of the BMU (*B*est *M*atching *U*nit) to which each pattern belongs (columns 3, 4, 5). Another file, named "KKA_Node_Weights_YYYYMMDDThhmmss," has a first column containing the indexes of the map node, then a sequence of columns with their corresponding weights. The number of weight columns corresponds to the number of components used in the classification. As mentioned earlier, KKAnalysis permits saving the graphical output; this is described in more detail in Section 7.2.7.4.2.

7.2.7.4 Getting started

Click on the desktop shortcut "KKAnalysis" (or use the corresponding item in your Program Files menu). KKAnalysis prompts you the subsequent "Welcome" sheet (Fig. 7.23A) offering various options. If you are running the program for the first time, all fields in the sheet will be empty. Otherwise, all fields report the choices of the last session, which you can reuse if you want. In the "welcome" sheet, there is also a "Console" that is used by KKAnalysis to prompt some information during run time, permitting the user to check whether the program is working properly.

7.2.7.4.1 The "Input File" frame

CRLF the "Input File" frame (Fig. 7.23B) contains four subframes: "Path," "Rows," "Columns," and "Show." In the field "Path," you insert the full name (path included) of the raw input data, which should have the format of a matrix, where data are written in rows and columns. The operation of file selection can be carried out either by using the "Browse" button next to the edit box or writing its name by hand.

Once identified the right file, KKAnalysis reads it and reports the structure found, that is, the number of rows and columns encountered. This information is shown in the frames "Rows" and "Columns." In KKAnalysis, the feature vectors of the patterns correspond to the rows in the input file. The number of rows is equivalent to the number of patterns, whereas the number of columns corresponds to the number of components of the feature vector. As mentioned in the section "Input Files," nonnumeric lines and columns inside the file are automatically discarded, and are not used for classification purposes. The values shown in the fields "From" and "To" (placed inside the frames "Rows" and "Columns") are referred to the effective range of data matrix inside the whole file. For example, it can be possible to see a "From" value equal to "4" because the first three lines of the file are row headers. The entries in these fields can be modified to choose the number of the data columns and rows to be used for the construction of the SOM and for clustering. Note that any part of data within the table can be selected, with the only condition that lines and columns neighbor each other.

The "Show" and "X Axis" frame contain useful settings concerning the graphical output. The former allows the user to set a range of patterns to bring into focus. In other words, the analysis is done, for example, on the entire dataset whereas the output files and figures contain only the range of patterns chosen by the user. Finally, the "X Axis" option can be enabled whenever you want to represent the cluster-ID or the color of the BMU not only

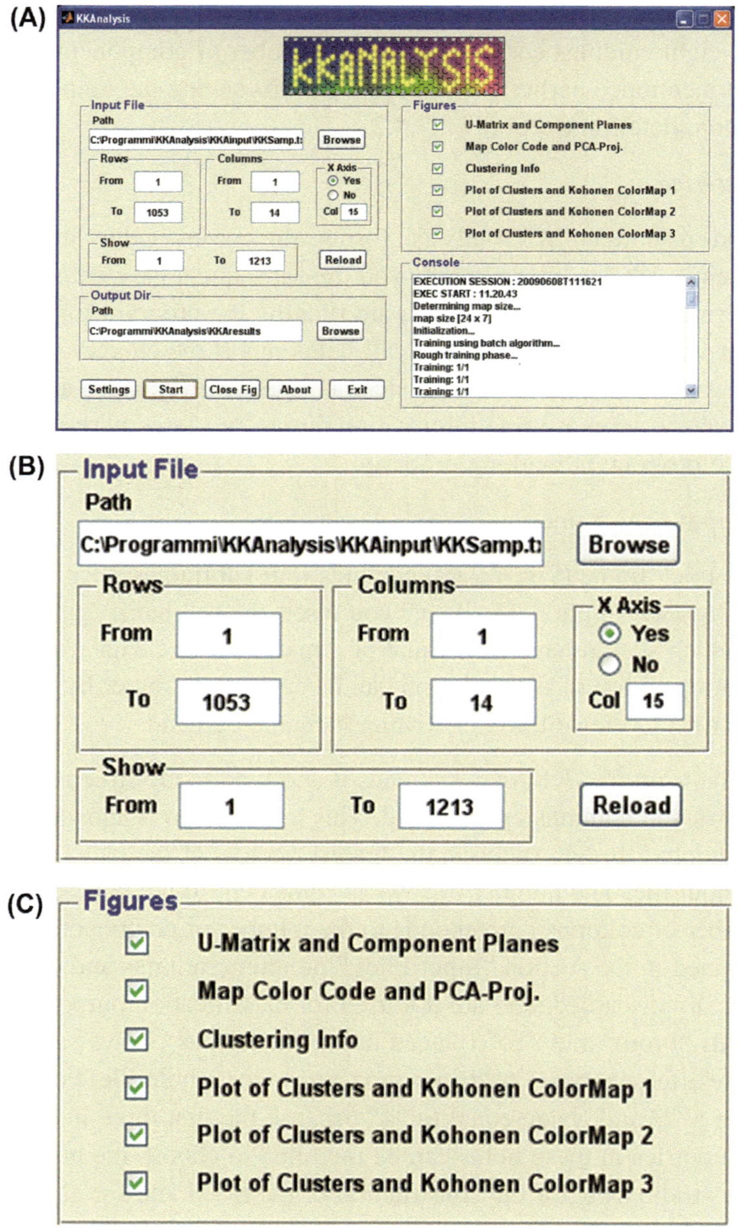

Figure 7.23

(A) Welcome to KKAnalysis! (B) The "Input File" frame. (C) The "Figures" frame.

one by one, but as a function of some additional variable, for an index representing time (e.g., hours, Julian day number, etc.). Note that a column used as "X Axis" must contain integers! Activating this option, KKAnalysis will display the clustering results as a function of the values given in the column specified in the "Col" field.

7.2.7.4.2 The "figures" frame

CRLF besides storing alphanumerical information, such as weights, class membership IDs, etc., KKAnalysis produces a number of graphs helping the user to understand its results (Fig. 7.23C). On the right-hand side of the frame, we may select two graphs related to the SOM, "**U-Matrix and Component Planes**" (Fig. 7.24) and "**Map Color Code and PCA-Proj**" (Fig. 7.25).

The idea of the U-Matrix is to represent the dissimilarities measured between neighboring units of the SOM. To improve the readability of the graph, the units of the SOM themselves make part of the U-matrix as well. At the end, the total number of elements in the U-matrix is given by the number of nodes of the map and the number of distances between neighboring nodes. Suppose a grid with a 24*7 SOM units. That means we measure 23 distances in vertical and 6 distances in horizontal dimensions, giving a total of $24 + 23 = 47$ rows and $7 + 6 = 13$ columns. As stated earlier, the nodes of the SOM themselves are represented in the U-matrix as well. The distance of an object measured with respect to itself does not make sense; therefore, the value reported for the nodes is calculated as average of distances measured to all neighboring nodes.

Distances in the U-matrix are represented by colors. Dark and blue stand for small values, bright to red for large values. An inspection of the U-matrix may reveal an immediate idea about the presence of major clusters among the SOM nodes. These clusters would

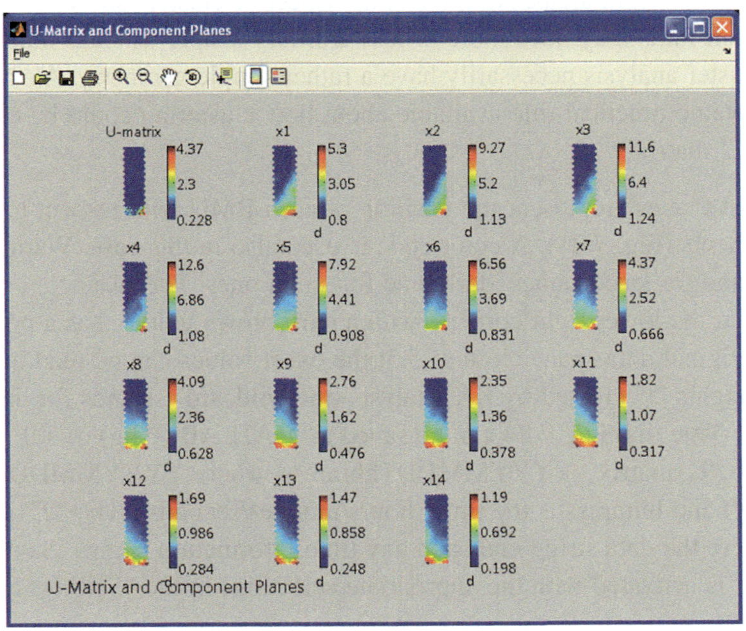

Figure 7.24

U-matrix and component planes. The number of columns of the data file used here is 14.

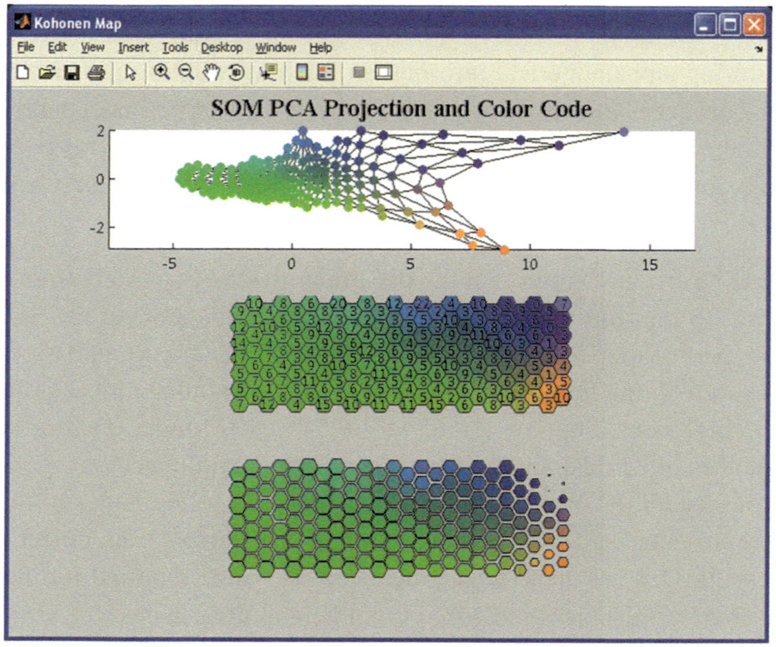

Figure 7.25
Example of SOM using a lattice with a hexagonal topology.

correspond to extended areas with prevailing blue color, meanwhile cluster borders are easily recognized by bright areas. An interesting aspect of clusters showing up in the U-matrix is that their hulls may have more or less arbitrary shape, whereas hulls of conventional cluster analysis necessarily have a rather simple geometry. At the moment, however, we have no practical rule available about how clustering could be established on the base of the U-matrix.

The "Component Planes" represent the position of each BMU with respect to the original variable components (Fig. 7.24). A color code is used also in this case. Warm colors (red–brown) stand for large values, dark blue for small ones. For instance, on the panel "x1," the BMU at the lower right corner—with a dark brown color—has a position on the x1 axis of the original data space close to 5.3; the exact values can be read in the files "KKA_NodeWeights_*" created by KKAnalysis (the wildcard * stands for date and time of file creation). Note that Fig. 7.24 can be saved as MATLAB (*.fig) or TIF file and carries the name "U_matrix_ YYYYMMDDThhmmss, where YYYYMMDD is the date (year/month/day) and hhmmss is the time (hours/minutes/seconds). The "T" in the middle helps to see where the data string ends and day time information begins. Saving of graphical output is activated with the appropriate selections in the "Settings Screen."

Setting the tick on "**Map Color Code and PCA Projection,**" KKAnalysis summarizes basic properties of the SOM created during the training process.

Here, the program has created a SOM made up of $7 \times 24 = 168$ nodes. For the definition of the map geometry and size, KKAnalysis internally performs a principal component analysis of the covariance matrix of the data vector set. The relation length/width of the map approximately corresponds to the ratio of the first two eigenvectors.

The uppermost frame in Fig. 7.25 shows the position of each BMU in a system of axes made up by the eigenvectors corresponding to the two largest eigenvalues. Note, however, that eigenvectors and eigenvalues used for the creation of the first frame are obtained from the SOM values instead of the data vectors. We see that the gross of the nodes scatter in a range from ~ 5 to -5 on the abscissa (that is 10 units in total) and from ~ -1.5 to ~ 1.5 on the ordinate (i.e., ~ 3 units). By taking the ratio $10/3 = 3.3$, we understand that the geometrical shape spanned by the BMUs, that is, $24/7 = 3.4$, corresponds well to the one made up by the eigenvalue ratio obtained from the covariance matrix of the data.

The second frame shows the color coding of the BMUs. As mentioned earlier, the colors of the BMUs are obtained from a 2D PCA of the SOM values, firstly extracting the two largest eigenvalues plus corresponding eigenvectors of the covariance matrix to define a new representation space. In this new space, the axes are identified with the principal colors, such as red and green. The third color (blue) scales with respect to a linear function of one of the two principal components. Then, the position of a node can be easily inferred from its color, which is a mixture of the three principal colors. The numbers given for each node correspond to the number of patterns for which this node was identified as BMU. In our example, the green node on the upper left corner of the frame was identified as BMU for nine patterns. Some nodes turn out to be "losers" as they were never BMUs; consequently, the middle panel in Fig. 7.25 reports a "0" within these nodes.

SOM nodes can be understood as microclusters, each attracting and representing a number of patterns. In the third frame of Fig. 7.25, KKAnalysis shows how compact these microclusters are. For the definition of compactness, KKAnalysis uses the information stored in the U-matrix, that is, the average of distances separating a node from the neighboring ones. These large symbols represent compact clusters not sharply distinguished from the surrounding ones; small ones indicate clusters with a high degree of heterogeneity. Loser nodes, for which no measure of separation can be defined, are simply marked by a dot.

Note that Fig. 7.25 can be saved as MATLAB (*.fig) or TIF file and has the name "ColCode_ YYYYMMDDThhmmss, where YYYYMMDD is the date (year/month/day) and hhmmss is the time (hours/minutes/seconds). The "T" in the middle helps to see where the data string ends and the day time information begins. Saving of graphical output is activated with the appropriate selections in the "Settings Screen."

The "**Clustering Info**" figure is related to options of KKAnalysis for cluster analysis sensu stricto, that is, K-means, fuzzy C-means, and clustering with the adaptive determinant criterion for the metrics of distance (the various available clustering criteria are shown in Fig. 7.32). It provides some basic parameters to check the quality of achieved results, helping the user to compare various set ups and to identify the most suitable configuration. As we shall discuss later, KKAnalysis incorporates three strategies of cluster analysis. The first, default option is the traditional "*K-Means*."

In Fig. 7.26A, we see how many patterns were assigned to the various clusters ("**Number of Patterns in Clusters**"). Here a partition with three clusters is considered. Different from the nodes of the SOM (for which we claimed the property of "topological fidelity"), the cluster IDs do not have a numerical or topological meaning. Instead of using cluster "1," "2," "3," we could have also used "A," "B," "C." KKAnalysis assigns by convention a "1" to the biggest cluster, "2" to the second biggest, and so on.

As mentioned earlier, in partitioning nonhierarchical clustering algorithms, the number of clusters must be chosen a priori by the user. A popular criterion for this choice is the "**Davies–Bouldin Index**" (DBI). KKAnalysis offers its use as an option (see the paragraph regarding the KKAnalysis configuration, i.e., "Settings"). The DBI balances the summed internal dispersion against distances measured among cluster centroids (external dispersion). In doing so, we account for the trade-off between compactness of clusters (which is optimum when dispersion is minimum) and number of clusters, which is also preferred to be small. The most suitable partition for which the DBI is minimum is identified. Note that when the DBI option is used, all related results make reference to this partition.

As we shall learn later—in the context of KKAnalysis configuration—clustering can be carried out both on the original data as well as SOM nodes. Using the K-means clustering on the original data, one obtains a **Clustering Info** figure such as shown in Fig. 7.26A. For K-means on SOM values, the **Clustering Info** figure looks like the example shown in Fig. 7.26B.

The cluster ID here is given by assigning a color to the nodes of the map. Numbers given in the single nodes correspond to the number of patterns for which a node is a BMU. A representation like Fig. 7.26B is only available for K-means clustering on SOM data.

A further option for clustering in KKAnalysis is fuzzy C-means. Fuzzy clustering is similar to classical, crisp K-means clustering beside the fact that the class membership of a pattern is given by a vector of scores, stating the degree to which a pattern belongs to a certain class or cluster. The upper frame ("**Number of Patterns in Clusters**") of Fig. 7.27 resembles to the one seen for the K-means clustering, with the difference that a partition with three clusters was adopted. The frame reports how patterns would be grouped considering only the highest score of class membership. The second frame "**Fuzzy C-Means Clustering Quality**" depicts the degree of fuzziness of the clustering. Here, 18% have a maximum score of 0.9 or greater; 49% have maximum scores between 0.7% and

(A)

(B)

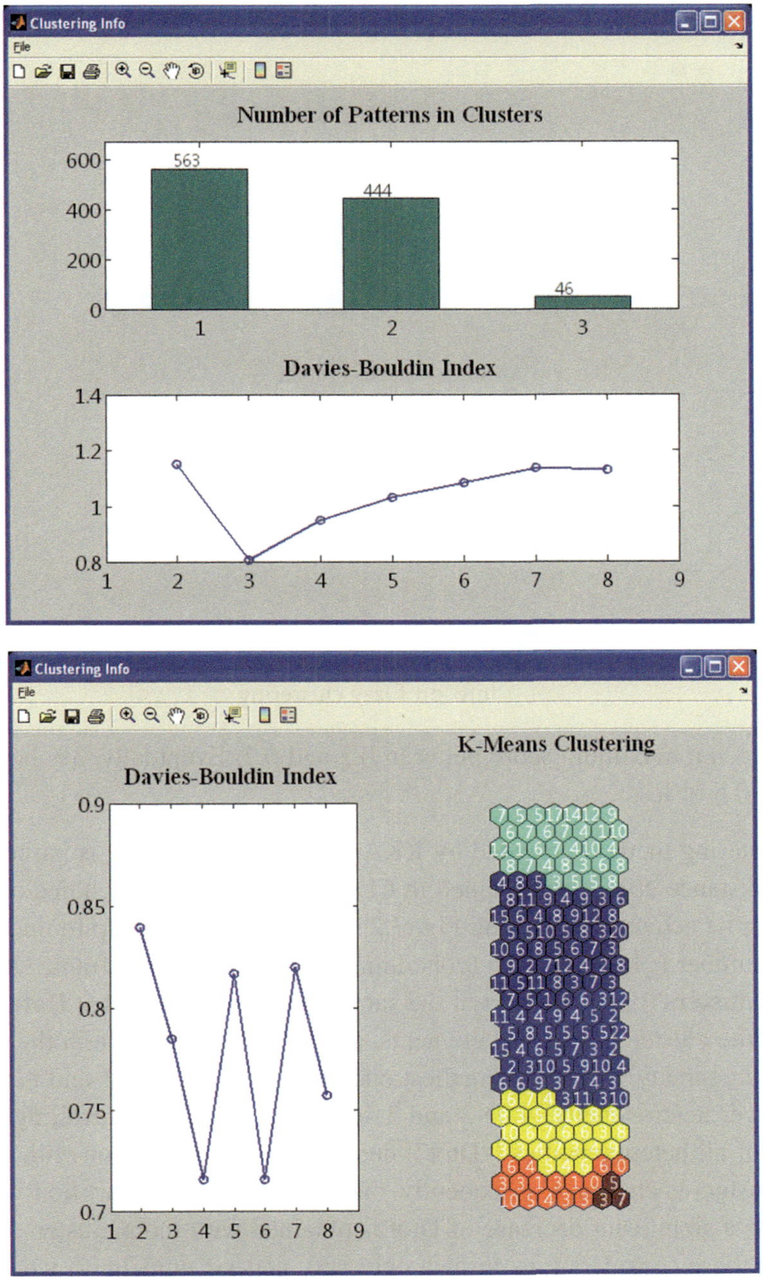

Figure 7.26

(A) Results of K-means clustering performed on original data. Based on the Davies–Bouldin index, a partition with three clusters is identified as the most suitable. (B) Clustering Info figure for K-means clustering on SOM data.

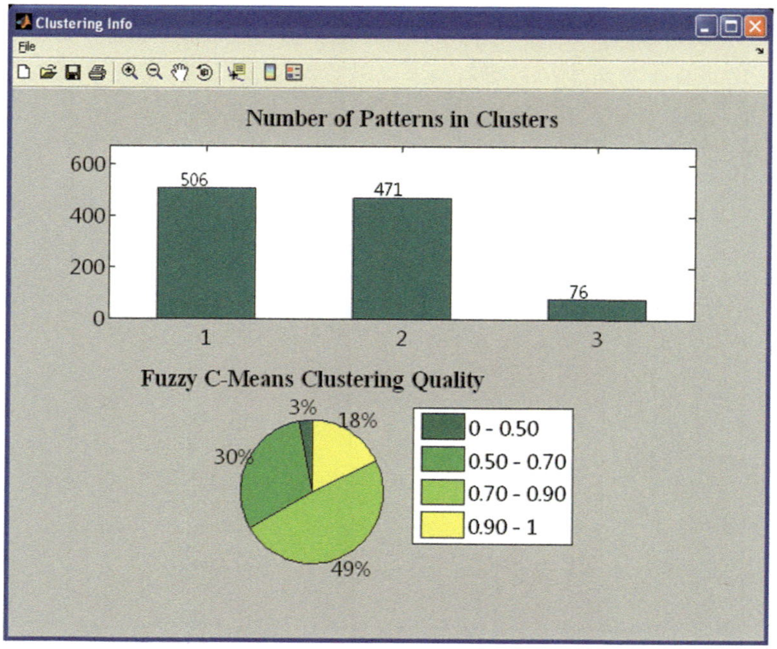

Figure 7.27
Info on fuzzy clustering.

0.9%; 30% reach a maximum score between 0.5 and 0.7. Eventually, 3% have a maximum score of only 0.5 or less.

A further clustering method provided by KKAnalysis, named "CA," is based on the use of the adaptive distance criterion explained in Chapter 3. The optimal choice of the number of clusters can be achieved observing Fig. 7.28. This figure may help to understand whether the number is large enough (containing many patterns, see frame "**Number of Patterns in Clusters**") and compact at the same time (frame "**Cluster Heterogeneity**"). For instance, the cluster #3 has 152 elements, and is the smallest one in the partition. Its degree of heterogeneity is also the smallest one (1.9e-2). Clusters #1 and #2 have somewhat larger heterogeneity (4.0e-2 and 3.4e-2), but are by far prevailing. It is expected that the sum of all heterogeneities ("Dtot") decreases when a partition with a higher number of clusters is chosen. Consequently, one tends to prefer a partition with few clusters unless a significant decrease of Dtot is obtained with more clusters. Note, however, that these considerations furnish only very generic guidelines, whereas the final decision on the suitable partition may include other considerations going beyond these numbers.

Note that Figs. 7.27 and 7.28, can be saved as MATLAB (*.fig) or TIF file and have the name "ClustInfo_YYYYMMDDThhmmss, where YYYYMMDD is the date (year/month/day) and hhmmss is the time (hours/minutes/seconds). The "T" in the middle helps to see

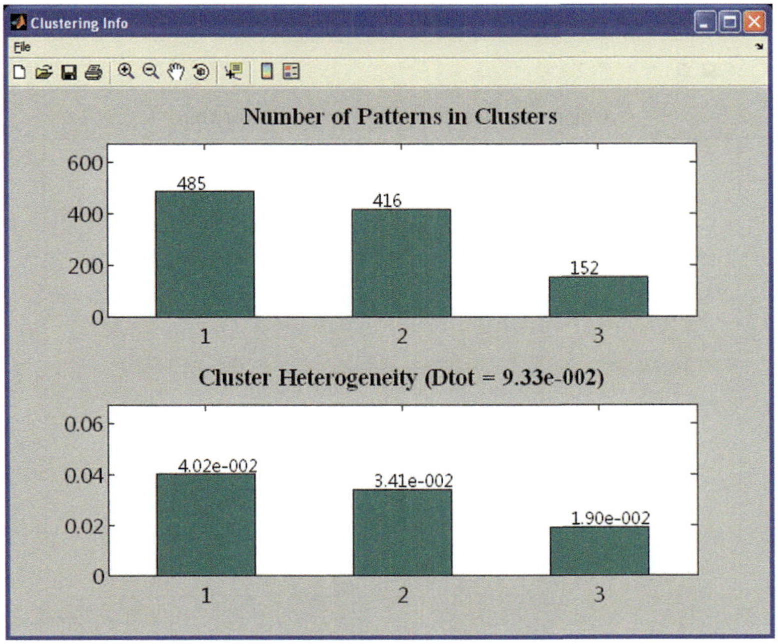

Figure 7.28
Clustering Info figure for adaptive determinant cluster analysis.

where the data string ends and the day time information begins. Saving of graphical output is activated with the appropriate selections in the "Settings Screen."

On the lower part of the "**Figure**" frame (see Fig. 7.23C), the user has three options to visualize—pattern by pattern—the classification results obtained from the SOM and the various clustering methods. For the generation of Fig. 7.29, we have applied cluster analysis using the adaptive determinant criterion. The corresponding figures obtained with K-means clustering have the same layout and are therefore not shown. To obtain the graph depicted in Fig. 7.29, we have applied the option "**Plot of Clusters and Kohonen ColorMap 1.**" The cluster membership of each pattern can be read from the vertical position of its colored shaped marker, whereas the color of each triangle corresponds to the one of the BMU the pattern belongs to.

In the "**Plot of Clusters and Kohonen ColorMap 2**" and "**Plot of Clusters and Kohonen ColorMap 3"** the same information is represented in an alternative manner. For the sake of conciseness, we omit the details here.

The content of Fig. 7.29 and its variants are slightly modified when the clustering is carried out with the fuzzy C-means option. The light gray bars indicate the "preferred" class membership. In addition, dark bars represent the maximum score, that is, the

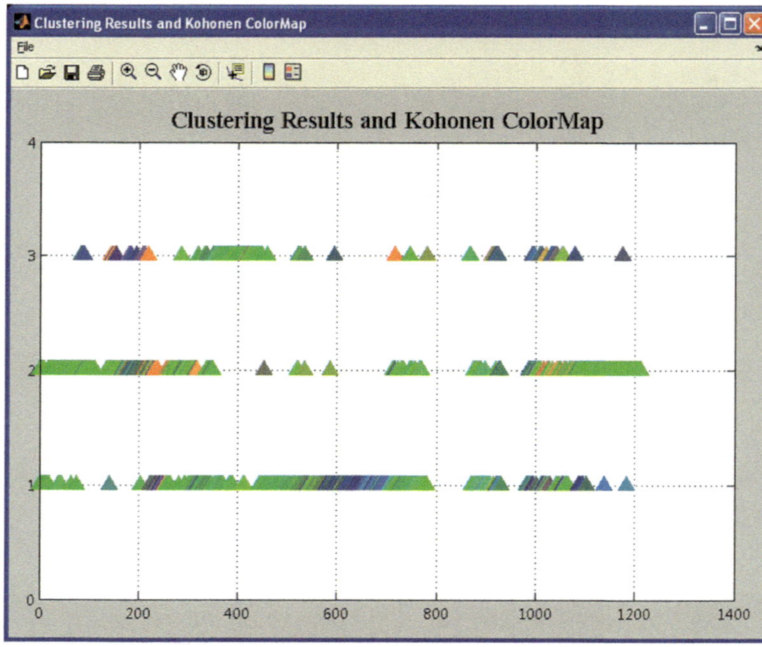

Figure 7.29
The first of three alternatives for representing the classification results (called as "ColorMap 1").

maximum of all encountered class membership values for a pattern. It determines to which cluster the pattern should be preferably assigned (see Fig. 7.30). The coverage of light gray with a dark gray bar expresses the degree of fuzziness of the class membership. If the coverage is complete, then the assignment of the class membership is crisp, that is, with no fuzziness. Conversely, fuzziness is high where large parts of the light gray bars remain visible. In the graphs, the patterns between #0 and #100, for instance, are assigned with a good score to the cluster 2, that is, their membership is rather crisp. The degree of fuzziness is substantial for some patterns belonging to the cluster 3. For example, patterns between #660 and #710 have a maximum score of class membership slightly above 0.5, that is, they show a high degree of fuzziness.

The colored stripe placed underneath the abscissa in Fig. 7.30 reports the full class membership vector encountered for all patterns. The cluster IDs are represented with colors. Each bar of this graph is composed of as many subbars as the number of clusters. Obviously, the sum of all class membership values is 1, and the length of the subbars is proportional to the membership rate of the patterns to the clusters. For instance, in Fig. 7.30, the largest subbars of the first 100 patterns are blue colored, and these patterns are assigned to cluster 1.

The synoptic graphs (see Figs. 7.29 and 7.30) can be zoomed either by editing the plots with MATLAB or by setting the interval in the "Show" panel in the "Welcome Screen" (Fig. 7.23A). To prepare Fig. 7.31, we have applied the fuzzy C-means with three clusters,

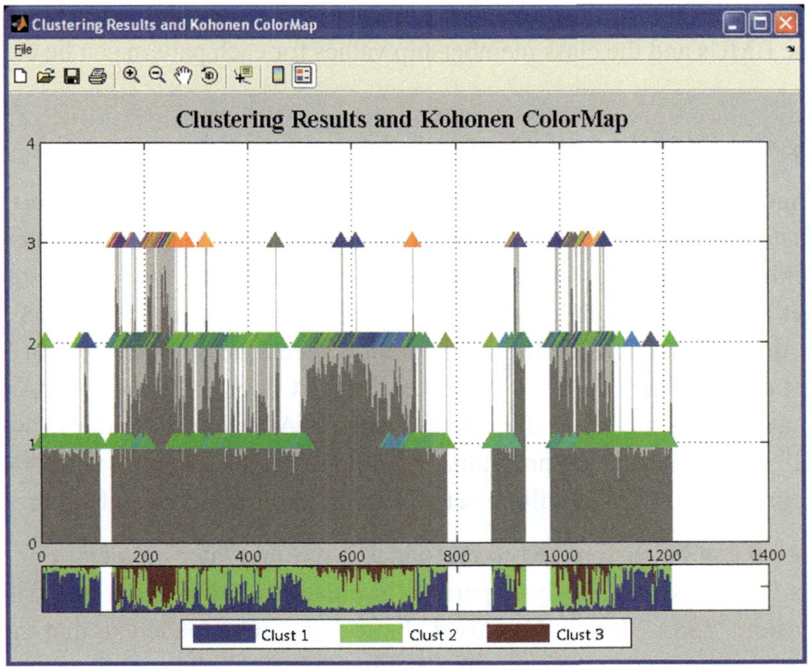

Figure 7.30
"Clustering Results and Kohonen ColorMap 2" obtained with fuzzy C-means clustering.

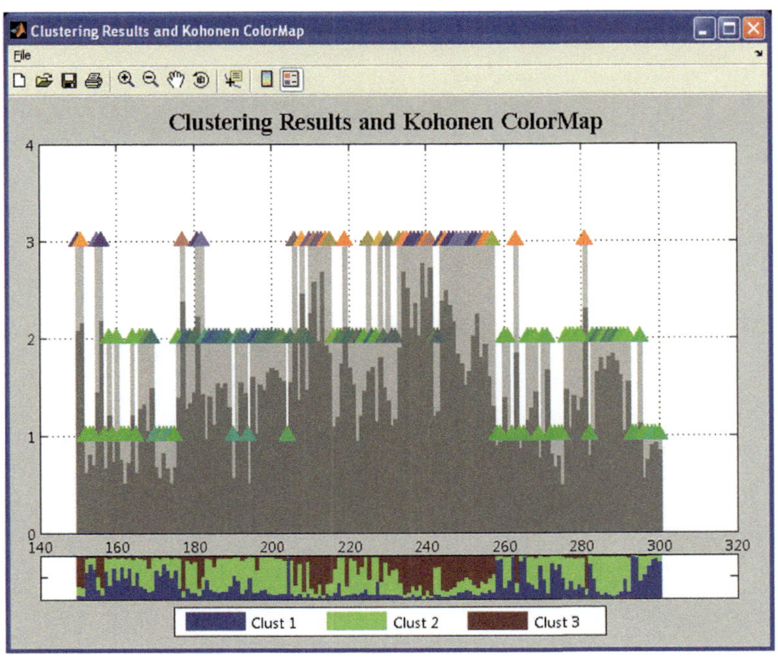

Figure 7.31
An example of result using the zoom function of KKAnalysis.

indicating "150" and "300" as the first and the last pattern to show, respectively. In Fig. 7.31, the color of the BMUs and the class membership values for each pattern can be clearly recognized.

7.2.7.5 Configuring KKAnalysis—the "settings"

KKAnalysis has been designed to be applicable to a wide variety of fields. This flexibility entails the existence of a considerable number of parameters that govern the tasks of the program. Clicking the "Settings" button on the "Welcome sheet," KKAnalysis prompts the "Settings sheet" (Fig. 7.32) with a number of configuration options shown later. Before discussing all the various items on the configuration sheet, please note the four buttons at the lower right corner: "Load," "Save As," "OK," and "Cancel." Running KKAnalysis for the first time, all options are set to their default values. After choosing the various options, click the "OK" button to start the program. Note that changes accepted with "OK" will be valid for the current session and will be set as default choices in the following session.

The configuration can be customized to reproduce a run in a later moment. For this purpose, press "Save As." KKAnalysis permits to create a configuration file, which should have the extension "*.kfg." Using the "Load" button, it can be loaded so that the program recovers the parameters of the desired run. Clicking "Cancel," the user stops editing the

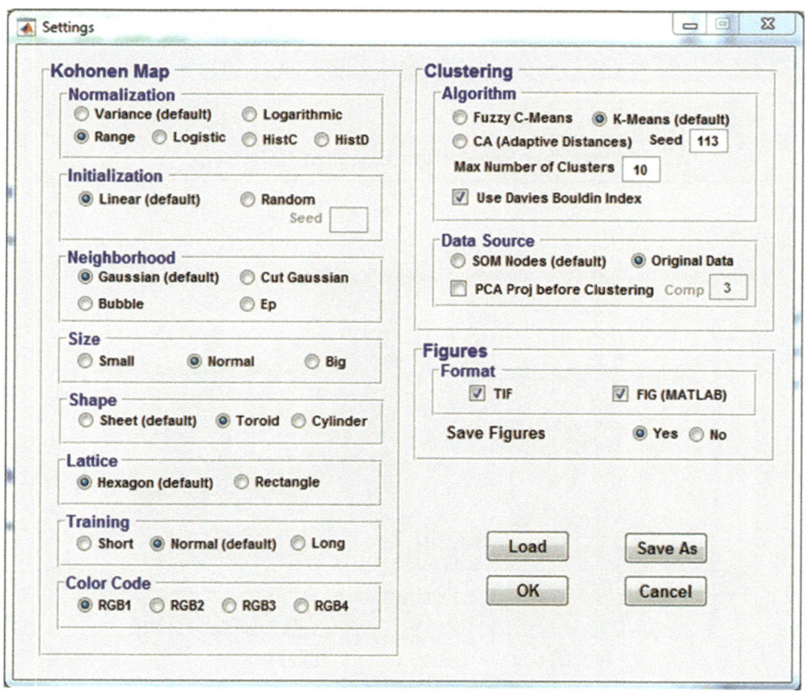

Figure 7.32
The "Settings" sheet of KKAnalysis.

configuration and returns to the "Welcome window," ignoring all changes of the configuration that were not previously saved.

The options in the "**Normalization**" frame give the possibility to keep the data in a certain range. KKAnalysis uses in this context the procedures provided by the SOM Toolbox 2. All normalization operations are carried out considering each component (feature) separately. In detail,

– "Variance" is the default option. The normalized values x' are obtained with the relation: $x' = (x\text{-}mean(x))/stdev(x))$ where $mean(x)$ is the average and $stdev(x)$ is the standard deviation of x.
– "Logarithmic" is the preferred option for data with a high dynamic range. Normalization is carried out using the formula $x' = \log(x - \min(x) + 1)$, $\min(x)$ being the minimum of all values in the dataset.
– "Range" scales values so that x' ranges from 0 to 1, that is, $x' = (x\text{-}\min(x)/(\max(x)\text{-}\min(x))$, $\max(x)$ being the largest value in the dataset.
– "Logistic" uses the famous sigmoidal function: $x'' = 1/(1 + \exp(-x'))$. x' is the normalized value of x with respect to the standard deviation, just like in the "Variance" normalization.
– "HistD" and "HistC" transform the metric values of x into ordinal ones, sorting them with respect to their amount and keeping only the index (or rank). In "HistD" sorting and ranking is carried out as is, whereas in "HistC" the ranking is carried with respect to bins, which is faster as the number of items to be sorted is lower. As in "HistC," the number of bins is obtained from the up-rounded square root of the number of patterns; consequently, the distribution of data within the bins will not be perfectly uniform. The data falling within a certain bin are linearly scaled with respect to the lower and upper rank boundary of the bin. For example, data in the bin 3 will be linearly transformed such that they cover the range between 3 and 4.

The "**Initialization**" frame has two options: "linear" and "random."

– The linear option is based on the eigenvalues of the covariance matrix. Having a bidimensional SOM, only the two largest eigenvalues are considered. The initial weight of each node of the SOM is obtained interpolating the two largest eigenvectors, which represent length and width of the SOM.
– The "random" option bypasses problems when the calculation of the eigenvalues of the covariance matrix is ill-conditioned. In this case, the initial values x_i for the i-th feature are set to $rand(1)*(\max(x_i)\text{-}min(x_i))$.

The "**Neighborhood**" frame offers various choices on how the influence area during the SOM training is defined, and how the upgrade of nodes neighboring a BMU is handled. The distance dependence of the upgrade of a node is described by a function φ. The latter can be as follows: "Gaussian" (this implies that all nodes are considered, even having small and gradually decreasing importance); "Cut Gaussian" (i.e., tails of the Gaussian representing distant nodes are omitted); "Bubble" (a rectangular or box-car function, 1 inside the radius of influence, 0 outside); "Ep" (with importance decreasing proportionally to the square of the distance).

An important issue is the "**Map Size.**" KKAnalysis follows the strategy applied in the SOM toolbox 2, where the design of the map is based on the covariance matrix of the data and its eigenvectors. The number of nodes for a "Normal" sized map is obtained heuristically according to *n_nodes* = 5 * *n_patterns*^0.54321. The ratio of the two largest eigenvalues is used to determine the length and the width of the map. The actual side lengths are chosen in a way that the resulting number of nodes is as close as possible to the number of nodes calculated with the heuristic formula. The choice "Large" gives a map with double-sided length (i.e., 4 times the number of nodes of the "Normal"); conversely, "Small" produces a map where the side length is only 50% of a "Normal."

KKAnalysis offers various options regarding the **Shape** of the map. Beside the classical sheet, which applies well in many problems, the "Toroid" or "Cylindrical" geometry options may be preferred when, for instance, angular data are considered. Furthermore, the two geometries may be preferred in case of normalized feature vectors, when we focus on the shape of an object rather than its size.

The "**Lattice**" can be made up either by "Hexagon(s)" or "Rectangle(s)" nodes. The choice affects the definition of neighborhood of nodes. In a hexagon lattice, each node has six neighbors with exactly the same distance, in a rectangle lattice only four. From a geometrical point of view, hexagons are an optimum polygon, as one can cover a mesh without leaving gaps, minimizing the sum of side length. The "Hexagon" option is, therefore, proposed as default choice.

The "**Training**" is carried out using the batch training algorithm proposed in the SOM Toolbox 2. Batch training means that an upgrade of the node weights is made only once during a cycle, that is, after comparing all patterns with the nodes of the map. Conversely, in sequential training—not adopted here—weights are updated every time a pattern is presented to the map. Benchmark tests revealed that batch training is the fastest among other methods. The training length depends on the ratio of the number of data and nodes of the SOM, called as *mpd*. Suppose a map with 100 nodes and 450 patterns in input; then *mpd* = 0.22. KKAnalysis carries out *ceil*(10 * *mpd*) = *ceil* (2.2) = 3 steps in 'rough tuning," and *ceil* (40*mpd*) = 9 steps in "fine tuning" mode, adjusting the learning rate first rapidly, then more slowly. Choosing "short" training, the number of steps is four times smaller, for "long" training four times larger. The learning rate is adjusted as a function of time following a relation $\lambda(t) = \lambda_0/(1 + 100\, t/T)$, T being the training length and λ_0 the initial learning rate. It is equal to 0.5 in the "rough" tuning phase and 0.05 in the "fine" tuning phase.

The **Color Code** option allows the user to choose among different configurations of heuristic color code. It permits to show the same results with different color maps. It can be possible that some details are visible in a better way using a color code rather than another. RGB1 is the default. The first eigenvalue and eigenvector are assigned to "Red," the second to "Green," whereas "Blue" corresponds to a linear combination of the first two eigenvalues. In the options RGB2, RGB3, and RGB4, the assignment of colors to eigenvalues and vectors is changed in a round-robin manner. For the example shown in this document, we have selected the "RGB4" option.

In the frame "**Clustering Algorithm,**" the user can choose among the classical "K-Means", the "CA (Adaptive Distances)" based on the adaptive determinant criterion, and the "Fuzzy C-Means" option. All algorithms are partitioning ones; therefore, the number of clusters has to be specified a priori in the field "Maximum Number of Clusters." Note, that the term "Maximum" is related to the K-means option. In this case, setting the tick in the field "Use Davies–Bouldin Index," the clustering is carried out from two to the number given in the field "Maximum Number of Clusters." When the Davies–Bouldin Index is not used, KKAnalysis creates only the partition corresponding to the number of clusters selected in the field "Maximum Number of Clusters".

In "**Clustering Source,**" the user specifies whether clustering is carried out on the raw data or using the SOM values. With the latter option, the scatter of patterns considered in clustering may considerably reduce, which can facilitate the task. Note that with the adaptive distance option there is the risk of ill-conditioned inverse of the dispersion matrix, especially when the number of components is high and only a limited amount of patterns is available.

It is possible to simplify the clustering problem reducing the dimensionality of data, performing a principal component analysis. For this option, set a tick on "PCA Proj. before clustering" and specify the number of principal components to be considered (field "Comp"). Of course, the number of principal components should be less or equal to the original number of components of the dataset or of the SOM.

In the frame "**Figures,**" the user can choose between the possibility of saving the produced figures or not. In the former case, there are two available formats: TIF format guarantees an optimum quality of the picture; FIG format allows the user to open and edit the picture using MATLAB. Note that figure files saved in the MATLAB format are much smaller than the corresponding TIF images.

7.3 Programs related to applications (Chapter 4)

The codes we are presenting here exploit methods and techniques explained earlier, but come in a fashion to be applied to large multivariate sets. In particular, these codes regard the multilayer perceptron (MLP) and the SVM.

7.3.1 Back propagation neural network (BPNN)

The neural network presented here is a MLP with one hidden layer of nodes, in which a hyperbolic-tangent activation function is applied. This function is similar to the sigmoidal function mentioned in Chapter 2 and allows us in principle to solve classification problems of arbitrary complexity. Recall that the MLP resolves the classification problem in a nonlinear prediction. Thus, its application to both inversion and regression is straightforward, as there is no need to adjust the architecture of the MLP to perform such tasks.

The GUI (Fig. 7.33) asks to load two files: the training data and the test data. Both datasets have the same format. The feature vectors are given in rows, similar to the standard used in KKAnalysis. The targets are given in tail to the feature values. Clicking "Load," the user can browse the filesystem and select the desired file. The user is prompted the preliminary length of the feature vector (field "Input Layer Size"), which by default corresponds to the total length of a row -1. The last item is reserved to the target, at minimum 1. If there is a multiple output, for instance four targets, the user can adjust it

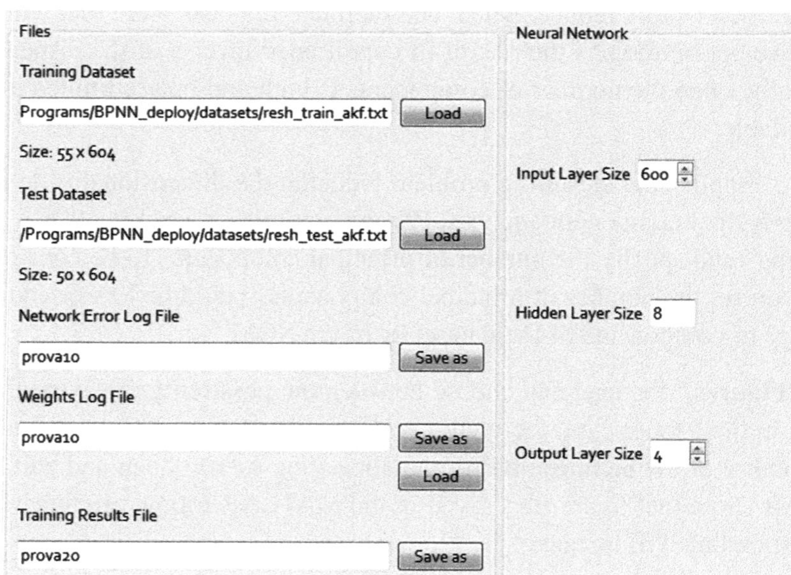

Figure 7.33
General input for the BPNN code.

in the field "Output Layer Size." Consequently, in the "Input Layer Size," the length is the total length of the row − number of targets (here equal to four).

Defining the "Hidden Layer Size" (the number of nodes in the hidden layer of the perceptron), we complete the definition of the network architecture. The remaining fields are related to operation parameters. The user must specify the "Network Error Log File" (reporting the global training and test errors at each training cycle), the "Weights Log File" (saving the weights encountered after a number of iterations), and the "Training Results file" where the errors for each sample after n (n = "Intermediate Saving Step") iterations are stored.

Setting the "Random Generator Seed" (Fig. 7.34), we randomly initialize the weights. It can be necessary to repeat the training changing the seed, as the back propagation does not guarantee a global optimum. The "MSE" stop value puts a limit on when the iteration should be interrupted. A defined end iteration is also warranted to be defined with the "Max Number of Cycles."

The learning rate is controlled by two parameters, that is, the "Init. Learning Rate" and the "Init. Momentum." The first parameter specifies the adjustment of the weights along the direction of the steepest descent. The momentum represents a memory parameter, stating that the adjustment should account the direction and amount of the rate of adjustments in the previous steps. In this way, small local minima can be escaped, as the weight adjustment maintains the earlier direction, even though a small inversion of the gradient is encountered. By activating the "Autotuning," the learning rate gradually increases as long as the training error "L.M.S.E." decreases; otherwise, the learning rate decreases. The moment term behaves similarly.

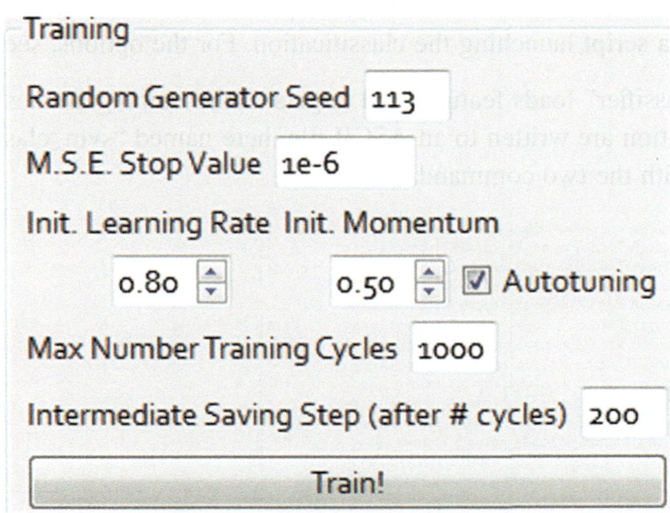

Figure 7.34
Operational parameters for BPNN.

During training, the program provides some output concerning the accuracy reached at each cycle. The reported parameters are as follows: the learning rate ("eps"), the momentum ("mom"), and the global errors for training and test set ("L.M.S.E." and "T.M.S.E.," respectively). The graphical representation (Fig. 7.35) is certainly helpful when training takes many cycles. "Bumpy" learning curves are often a bad sign, questioning some choices with respect to the size of the training dataset and the number of hidden nodes.

7.3.2 SVM library

The tool presented here is based on a library for SVMs (http://www.csie.ntu.edu.tw/~cjlin/libsvm). It is organized in the following steps:

(1) Load data and targets (Training and Test)
(2) SVM training and model creation
(3) SVM prediction of test dataset
(4) Writing output ASCII files

In the code coming along with this book, we have added a GUI that facilitates the use of the SVM scripts. We start reading training and test data. These two datasets contain alphanumeric values, with rows forming the feature vectors. A second pair of files adds the corresponding labels/targets of the patterns. Using SVM as classifier, the targets can be strings; using SVM in regression, the targets correspond to the desired output of the regression function. In that case, the values are supposed to be numerical. In the example already considered in Chapter 4, the features regard the geochemical composition of igneous rock samples; the labels correspond to their a priori classification following the scheme in the "Streckeisen" diagram.

For SVM training, the library offers a variety of options. These can be specified with command lines in a script launching the classification. For the options, see Table 7.2:

The script "svmclassifier" loads features and targets of the training and test data. Results of the SVM prediction are written to an ASCII file, here named "svm_class.txt." The script invokes training with the two command lines.

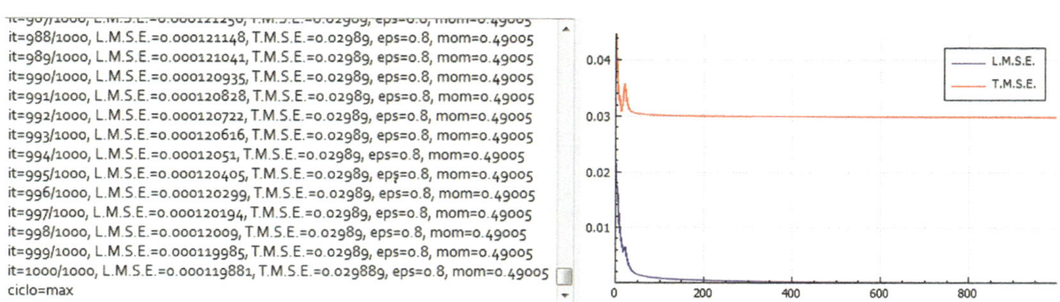

Figure 7.35
Graphical output during a run of BPNN.

Table 7.2: Options for SVM classification.

```
% The 'svmtrain' function returns a model that can be used for future predictions.
% LIBSVM allows one to set several options. Here the flag to be used:
%
% -s svm_type: set type of SVM (default 0)
%        0 -- C-SVC                  (multi-class classification)
%        1 -- nu-SVC                 (multi-class classification)
%        2 -- one-class SVM
%        3 -- epsilon-SVR   (regression)
%        4 -- nu-SVR        (regression)
% -t kernel_type: set type of kernel function (default 2)
%        0 -- linear: u'*v
%        1 -- polynomial: (gamma*u'*v + coef0)^degree
%        2 -- radial basis function: exp(-gamma*|u-v|^2)
%        3 -- sigmoid: tanh(gamma*u'*v + coef0)
%        4 -- precomputed kernel (kernel values in training_instance_matrix)

%%%%% Parameters related to kernel type %%%%%
% -d degree: set degree in kernel function (default 3)
% -g gamma: set gamma in kernel function (default 1/num_features)
% -r coef0: set coef0 in kernel function (default 0)

%%%%% Parameters related to kernel type %%%%%
% -d degree: set degree in kernel function (default 3)
% -g gamma: set gamma in kernel function (default 1/num_features)
% -r coef0: set coef0 in kernel function (default 0)

%%%%% Parameters related to svm type %%%%%
% -c cost: set the parameter C of C-SVC, epsilon-SVR, and nu-SVR (default 1)
% -n nu: set the parameter nu of nu-SVC, one-class SVM, and nu-SVR (default 0.5)
% -p epsilon: set the epsilon in loss function of epsilon-SVR (default 0.1)

%%%%% Other parameters %%%%%
% -b probability_estimates: whether to train a SVC or SVR model for probability estimates, 0 or 1 (default 0)
% -m cachesize: set cache memory size in MB (default 100)
% -e epsilon: set tolerance of termination criterion (default 0.001)
% -h shrinking: whether to use the shrinking heuristics, 0 or 1 (default 1)
% -wi weight: set the parameter C of class i to weight*C, for C-SVC (default 1)
% -v n: n-fold cross validation mode
% -q: quiet mode (no outputs)

% Options have to be written inside 'tr_opts' variable (ex. '-s 3 -t 3 -b 1')
% default values are: '-s 0 -t 2 -d 3 -r 0 -c 1 -n 0.5 -p 0.1'
```

```
tr_opts = '-s 0 -b 1';
model = svmtrain(tr_lab_dou, tr_data, tr_opts);
```

where `tr_lab_dou` are the labels, and `tr_data` are the values of the training set (Fig. 7.36).

That means, we use an "RBF" kernel (default) and transform the scores into probabilities. We recover the test results using the SVM for prediction. The corresponding commands in the script are as follows:

```
ts_opts = '-b 1';
[pr_lab_dou, acc, dec_vals] = svmpredict(ts_lab_dou, ts_data, model, ts_opts);
```

where `ts_lab_dou` are the labels, and `ts_data` are the values of the test set.

For the results of the training set, use

```
[pr_lab_dou, acc, dec_vals] = svmpredict(tr_lab_dou, tr_data, model, ts_opts);
```

that is, carry out the SVM prediction of the training data. The overall accuracy for the training data is ∽83%.

The script for SVM regression is similar. After loading training and test data, the script invokes training as follows:

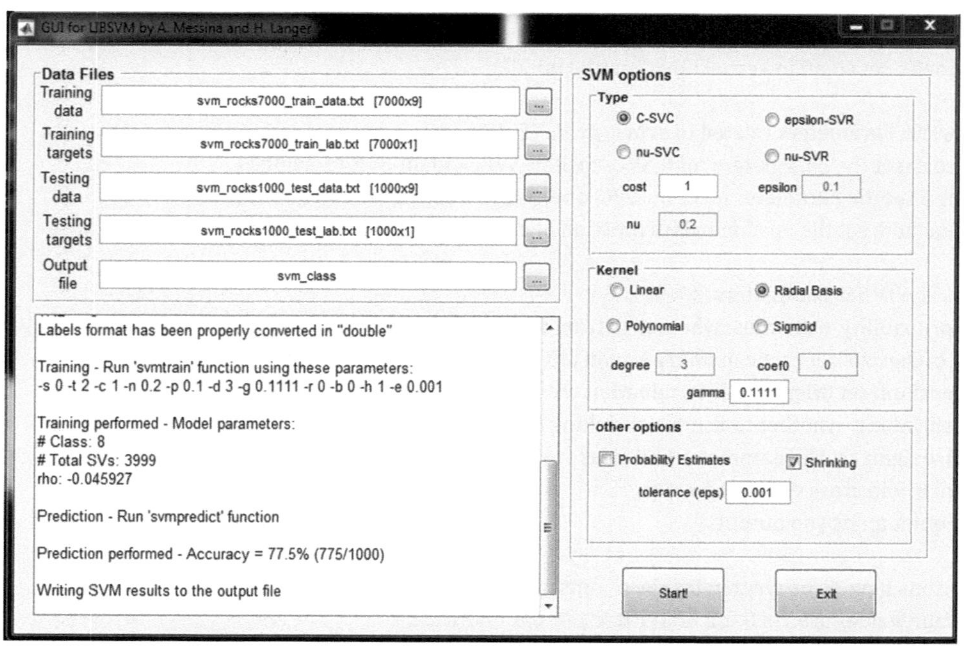

Figure 7.36
SVM-GUI, settings for SVM use as classifier.

tr_opts = '-s 3 -b 1 -t 2';

model = svmtrain(y_trdata, x_trdata, tr_opts)

and produces the results with the commands.

ts_opts = '-b 1';

[pr_lab_dou, acc, dec_vals] = svmpredict(y_tsdata, x_tsdata, model, ts_opts)

Instead of using the earlier-mentioned command lines, the user can use the GUI deployed in the directory ".libsvmLIB_SVM_GUIcode" and click on "SVM_GUI.m" (or "SVM_GUI.exe" when using the compiled version). The GUI asks for training and test files as well as corresponding targets, which can be string-type labels when SVM is used as classifier.

The labels are temporarily converted to numbers to count the columns. In the results file "svm_class.txt," the program reports the probability that a pattern belongs to a specific class (see Table 7.3). Fig. 3.4 gives an example for a session where SVMs are used as classifiers. Four files must be specified: the training set with the numerical feature, the corresponding targets, then a test set, again with the numerical features, and the targets, which are compared to the computed output to assess the success. The predictions for the test are written to the output. During runtime, the program prompts some messages, such as the conversion of string targets to integer class IDs, the command line invoked by the GUI, and finally some information concerning the overall success. In the panel on the right, the user specifies the operation characteristics. There is a choice between SVC (use SVM as classifier) and SVR (for regression). In the classification option, we can further choose among "C-CVC" and so-called nu ν-SVC (here called as "nu-SVC"). Recall that in the first case one specifies the weight of patterns falling inside the separating margins. Using the ν-"nu" option, the parameter C is replaced by a parameter $\nu \in [0, 1]$, which defines the lower and upper bound of the number of examples that are support vectors and that lie on the wrong side of the hyperplane, respectively (Chen et al., 2005). We further specify the kernel function. Activating "shrinking" (set by default), specific operations are enabled simplifying the optimization problem. Furthermore, we can activate the transformation of the scores into a probability measure, as mentioned earlier. As a result, we obtain Table 7.3.

Note that we have applied the option of transforming the scores into probabilities following the formalism given in Chapter 4, Eqs. (4.4a) and (4.4b). For the example test set, the program reports an overall success of 77% (Fig. 7.36).

The SVM regression can be carried out by activating the "epsilon-SV" or "nu-SVR" options. The "epsilon" version considers the width of the regression tube as explained in Section 4.7. The GUI asks for a training file containing the features and a target file (in regression, the values that should be predicted). The program performs a test on test features and compares the predictions to corresponding target values.

Table 7.3: Summary of SVM multiclass classification results (test).

Pred. Lab	MaxProb.	Prob.AND	Prob.BAS	Prob.DAC	Prob.RYD	Prob.RYL	Prob.TAN	Prob.TRA	Prob.TRB
RYL	0.8740	0.0026	0.0016	0.0063	0.0990	0.8740	0.0081	0.0076	0.0009
TRA	0.8945	0.0203	0.0035	0.0110	0.0136	0.0224	0.0301	0.8945	0.0045
DAC	0.4619	0.0628	0.0037	0.4619	0.0355	0.0079	0.3805	0.0417	0.0061
AND	0.8491	0.8491	0.0292	0.0062	0.0133	0.0020	0.0919	0.0037	0.0046
RYL	0.8665	0.0053	0.0092	0.0118	0.0711	0.8665	0.0123	0.0212	0.0026
RYL	0.8753	0.0024	0.0020	0.0074	0.0998	0.8753	0.0055	0.0060	0.0017
DAC	0.8192	0.0375	0.0007	0.8192	0.1134	0.0124	0.0033	0.0130	0.0005
TAN	0.7222	0.2084	0.0273	0.0082	0.0068	0.0075	0.7222	0.0096	0.0100
TRA	0.4159	0.0907	0.1168	0.0215	0.0191	0.0357	0.2507	0.4159	0.0497
AND	0.8120	0.8120	0.0364	0.0103	0.0038	0.0025	0.1281	0.0025	0.0045
DAC	0.6807	0.1579	0.0095	0.6807	0.0922	0.0318	0.0110	0.0130	0.0039
TRB	0.6267	0.0113	0.0399	0.0037	0.0030	0.0033	0.2986	0.0136	0.6267
TRA	0.7898	0.0361	0.0027	0.0026	0.0043	0.0055	0.1483	0.7898	0.0106
AND	0.9677	0.9677	0.0120	0.0032	0.0020	0.0012	0.0086	0.0027	0.0026
TRA	0.9199	0.0213	0.0017	0.0061	0.0035	0.0142	0.0309	0.9199	0.0025
BAS	0.8260	0.0429	0.8260	0.0080	0.0083	0.0120	0.0472	0.0177	0.0379
RYL	0.8873	0.0033	0.0025	0.0061	0.0858	0.8873	0.0072	0.0065	0.0013

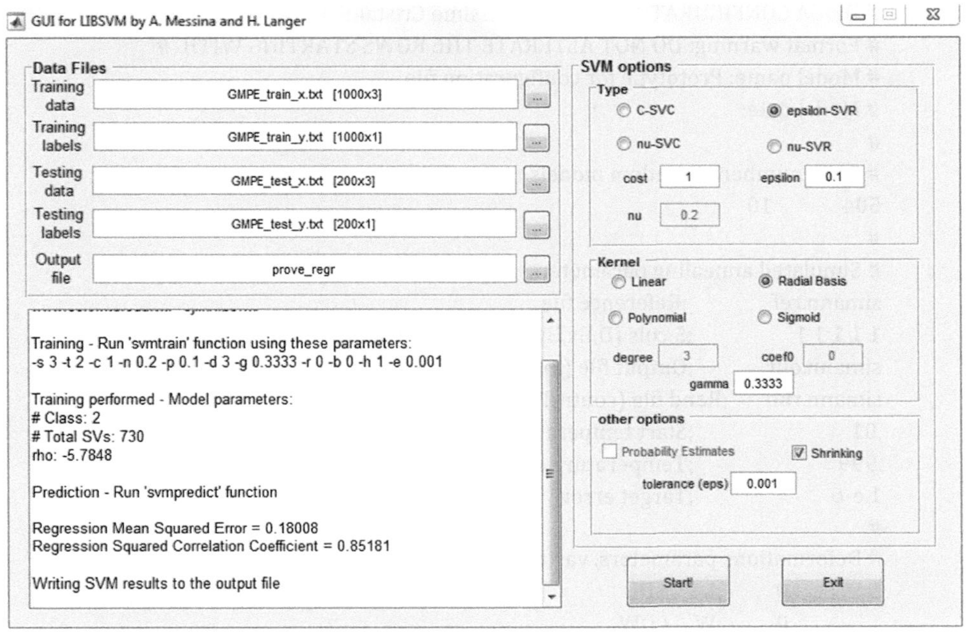

Figure 7.37
GUI for SVM regression.

7.4 Miscellaneous

For the sake of the reader's convenience, we also provide some codes and scripts mentioned in Chapters 4 and 5, not directly related to pattern recognition but propaedeutic for the applications discussed in these chapters.

7.4.1 DMGA—generating ground deformation, magnetic and gravity data

The inversion of geophysical model parameters using MLP is based on the generation of training and test datasets using synthetic simulations. In this case, our target vector $\boldsymbol{Y}$ is given by the geophysical model parameters, for which we simulate the data vector $\boldsymbol{X}$. To allow the readers to conduct their own experiments, we provide the software DMGA. The program follows essentially Nunnari et al. (2001). Here we explain a few, essential steps using the software. Starting DMGA, the user sees the prompt

MAIN MENU
1. Load input file
6. Generate prototype for input file
99. End program

For new users, the best option is "6," that is, generate a prototype for input file. This file looks like this

```
# DMGA CONFIGURATION FILE by Massimo Cristaldi, H. Langer, 1996,1997
# Format warning: DO NOT ALTERATE THE ROWS STARTING WITH '#'
# Model name: Prototype for configuration file.
# Model date:
#
# Seed, number of random models
504        10
#
# Simulated annealing parameters
simann.ref          ;Reference file
1 1 1 1 1           ;5 cols (D,Bx,By,Bz,Gz) 1=activated, = excluded
simann.out          ;Output file (model parameters found during optimization)
simann.rnd     ;Rand file (control purpose only)
.01                 ;Start temperature
.999                ;Temperature decrease factor
1.e-6               ;Target error
#
# Deformations parameters, variability
1        0      ;L    / DL
1        0      ;W    / DW
30       0      ;dip  / Ddip
0        0      ;xx   / Dxx
0        0      ;yy   / Dyy
0        0      ;zz   / Dzz
10       0      ;azimuth / Dazimuth
1        0      ;aper   / Daper
.25      0      ;poiss  / Dpoiss
#
# Magnetism parameters, variability
0.1      0      ;s1          / Ds1
0.1      0      ;s2          / Ds2
12.56e-7    0      ;s        / Dm
1        0      ;S0          /  DS0
#
# Density contrast, variability
0.1      0      ;Rho         / DRho
#
# Station informations
1               ;nStat
#
# Station coordinates
-4.0     4.0       ; X, Y
```

The program generates data vectors using random variations of the model parameters. We set a seed (of integer type) to initialize the random generator, and specify how many data vectors may be created; these can be used in the application of the MLP (e.g., the back propagation neural network (BPNN) program mentioned earlier) in inversion. The simulated annealing parameters control the options for the global optimization. The "simman.ref" (or some other file specified by the user) contains the reference data vector that the optimization tries to match; "simman.out" is the file in which the inverted model parameters are written; "simman.rnd" is a control file for checking the random number (not of specific interest here). The next two lines regard the control parameters for the simulated annealing and when to stop the optimization (i.e., "Target error").

The following lines regard the model parameters of a dike, together with the range of variation during the random generation of the model parameters. We specify the average length L with range DL, and the average width W with range DW. The values xx, yy, and zz indicate the location of the dike center, while the values in the second column are their range of variation. We then need azimuth, crack opening, and Poisson's ratio. The next four parameters regard the parameters in the Murakami model for the electrokinetic effect; the second column always reports the range of variation

s1 = streaming coefficient upper side
s2 = streaming coefficient lower side (given in S[iemens]/m)
m = magnetic permeability (N[ewton]/A[mpere]^2, or H[enry]/m. Typically m = 1.26e-6 H/m.
S = source (in V)

Finally, we specify the density contrast of the intruding material with respect to the surrounding rock, indicate the number of stations, and add the x and y coordinates of each site.

As an exercise, one may use the parameter file given in Chapter 4, Table 4.6. We recommend the use of SI units (the gravitation constant in the program is specified in SI); an example is reported in the following (Table 7.4).

Table 7.4: Example of parameter settings for random simulations in DMGA.

Crack length (m)	5000	±4000
Crack width (m)	2000	±2000
Longitude (UTM) (m)	0	±10,000
Latitude (UTM) (m)	0	±10,000
Depth(m)	5000	±4000
Azimut (deg)	0	±90
Dip (deg)	90	±50
Crack opening (m)	2	±1
Poisson's ratio	0.25	±0.0
Density contrast (kg/m^3)	300	±100

Specify the coordinates of the network x and y in meters. "16" gives the number of stations.

```
# Station information
16
#
# Station coordinates (m)
5000  5000
5000  -5000
-5000  5000
-5000  -5000
10000 5000
10000-5000
-10000 5000
-10000 -5000
5000  10000
5000 -10000
-5000 10000
-5000 -10000
10000 10000
10000 -10000
-10000 10000
-10000 -10000
```

Now we load an input file and get the menu below. In addition, typing "1" we can load an existing input file. We can use options from "2" to "5" to modify our input, and then save the modified model parameters. "7" is important for the generation of a reference data vector for option "10" (simulated annealing). Invoking option "10," the user will be asked for such a file. Together with the data vector, it contains the physical model parameters for its generation.

```
MAIN MENU

1.    Load input file
2.    Modify model parameters
3.    Modify station coordinates
4.    Add/delete station
5.    Write input file
6.    Generate prototype of input file
7.    Calculate deformation, magnetic, (gravity) field
8.    Use random generator
9.    Modify Simulated Annealing parameters
10.   Simulated Annealing
11.   Generate Plot Files
99.   End Program.
```

Option "8" is used for the generation of a set of data vectors, which can be directly used for network training and testing. We create a large number of data vectors by setting a large number of random simulations. It is a good idea to split the file in hindsight into test and training set to warrant that all the feature vectors are normalized in the same way. Output data and parameters are written all in a column; therefore, use the script "reshape" to transform it into a format that can be imported directly to BPNN. Two versions of the output files are provided. The file without extension reports normalized values in a range [−1.0, 1.0] both for the simulated geophysical data as well as the model parameters. At the tail of the file, the user will find maximum ranges of the model parameters encountered during the simulations. Discard them when passing the data to BPNN. Note that the normalization of the geophysical data is nonlinear (see Chapter 4, Appendix 4.1). For the sake of documentation, a second *.tmp file is created reporting the absolute values of both synthetic geophysical values and the model parameters.

Option "11" is used for purposes of plotting. The user is asked to define a 2D field, specifying the number of nodes and spacing in x (horizontal, or "easting") and y direction (vertical, or "northing"). The plot files can be directly imported to standard graphical software, for example, the Microsoft "Surfer" programs.

7.4.2 Treating fault plane solution data

Fault plane solution data discussed in Chapter 5 are given in a format shown below:

40.96 19.67 10.00 4 70 −167 270 78 -21 2.330 23 R199701121210A.

40.82 19.67 10.00 2 65 −177 270 87 -25 1.460 23 R199701191942A.

41.40 14.63 10.00 280 27−110 122 65 -80 7.700 22 R199703192310A.

33.96 8.29 10.00 66 44 89 247 46 91 3.770 23 R199703201802A.

The first three columns give the hypocenter coordinates, columns 4−9 report strike, dip, and rake of the two planes, columns 10 and 11 give the scalar moment (dyn cm) (mantissa and exponent as a power of 10). Finally, the string starting with "R.." gives the origin time of the event. Here, we are interested only in the columns 4−9, columns 1−3 may be of interest for the creation of the plots showing the geographical distribution of the beach balls. As explained in Chapter 5, moment tensor components are preferred to the angles. The small routine "mom" can be used for the transformation of the angles into moment tensors. The user can enter the angles manually or create a list like the one shown below:

282 23 119

95 37 63

89 48 60

..

Type “mom <list >out”, which produces a file “out” containing the conversions from angles to moment tensor components.

Strike, Dip, Lambda −999 to end
dMxx, dMxy, dMxz
 −6.790006E-01 4.510440E-02 6.870698E-01
dMyy, dMyz, dMzz
 4.985183E-02 -3.101986E-01 6.291487E-01
Strike, Dip, Lambda −999 to end
dMxx, dMxy, dMxz
 −8.025405E-01 -3.434316E-01 -2.130597E-01
For the use in KKAanlysis, the comment lines “Strike, Dip, Lambda …,” “dMxx.”, “dMyy …” must be deleted.

In Chapter 5, KKAnalysis was applied to moment tensor components. The plots showing the geographical position of the fault planes solutions were created using the “General Mapping Tools” (downloaded from https://www.soest.hawaii.edu/gmt/), in particular the “psmeca” command. Note that “psmeca” uses the fault plane solutions given as angles. A file reporting the necessary information looks like this.

```
17.1    43.55  20   115  17   39   347  79   104  1  0  0  0
18.75   42.26  8    141  28   80   332  63   95   1  0  0  0
13.87   43.58  6    161  27   117  311  66   77   1  0  0  0
13.69   43.58  10   144  30   93   321  60   89   1  0  0  0
18.75   42.16  10   125  36   70   329  56   104  1  0  0  0
```

Columns 1−3 are the hypocenter coordinates (longitude E, latitude N, depth), and columns 4−9 are the angles.

The plot can be obtained, for instance, by typing

```
psmeca RGB_blue90.txt −N -Jm0.5 -R5.0/25.0/30.00/48.0 -Sc2.5 -W1/0/0/0 -Gblack
-Fa0.05c/cc -P −O >> test_7.ps
```

which considers a file “RG-blue90.txt”; “-Jm0.5” means a standard meridional projection around a central meridian of 0.5; “-R5.0..” defines west/east longitude and south/north latitude shown in the map; “Sc2.5” controls the size of the beach balls; “-W..” controls the option of the pen; “-Gblack” means that the compressive fields of the beach-balls are filled in black; with “Fa0.05c/cc we define how P- and T-axes are represented. In addition, “-O” means that we append to an existing postscript image, “-N” means that symbols outside the area (defined by the “-R..” frame) are not skipped. Further information can be found in the documentation of psmeca.

For showing the coastlines, we used first the command

```
pscoast.exe -Jm0.5 -R5.0/25.0/30.0/48.0 -B0.5 -Dh -G230/188/143 -S205/238/252
-Wblack -K > test_7.ps
```

before the psmeca command. "-S.." controls the filling of the wet areas, "-G.." the filling of land. "-K" allows the user to add more information to the plot, for instance, applying "psmeca" discussed earlier.

Bibliography

Abramowitz, M., Stegun, I.A., 1972. Handbook of Mathematical Functions with Formulas, Graphs, and Mathematical Tables. Dover, New York, 9th printing.

Akaike, H., 1973. Information theory and an extension of the maximum likelihood principle. In: Proc. of the 2nd International Symposium on Information Theory, vol. 1. Akademiai Kiado, Budapest, pp. 267–281.

Aki, K., Richards, P., 1980. Quantitative Seismology – Theory and Methods. Freeman and Company, San Francisco, 932 pp.

Aki, K., Fehler, M., Das, S., 1977. Source mechanism of volcanic tremor: Fluid-driven crack models and their application to the 1963 Kilauea eruption. Journal of Volcanology and Geothermal Research 259–287.

Aliotta, M., Cannata, A., Cassisi, C., Giugno, R., Montalto, P., Pulvirenti, A., 2011. DBStrata: a system for density-based clustering and outlier detection based on stratification. In: Proceeding SISAP '11 – Proceedings of the Fourth International Conference on Similarity Search and Applications, pp. 107–108. https://doi.org/10.1145/1995421 1995432. Lipari, Italy, June 30–July 01, 2011.

Alparone, S., Andronico, D., Lodato, L., Sgroi, T., 2003. Relationship between tremor and volcanic activity during the Southeast Crater eruption on Mount Etna in early 2000. Journal of Geophysical Research 108 (B5), 2241. https://doi.org/10.1029/2002JB001866.

Ambraseys, N.N., Simpson, K.A., Bommer, J.J., 1996. Prediction of horizontal response spectra in Europe. Earthquake Engineering and Structural Dynamics 25, 371–400.

Anderberg, M.R., 1973. Cluster Analysis for Applications. Academic Press, New York, 359 pp.

Ankerst, M., Breunig, M.M., Kriegel, H.P., Sander, J., 1999. OPTICS: Ordering Points To Identify the Clustering Structure. In: ACM SIGMOD International Conference on Management of Data. ACM Press, pp. 49–60.

Anzidei, M., Lambeck, K., Antonioli, F., Furlani, S., Mastronuzzi, G., Serpelloni, E., Vannucci, G., 2014. Coastal structure, sea-level changes and vertical motion of the land in the Mediterranean. Geological Society, London, Special Publications 388, 453–479.

Archie, G.E., 1942. The electrical resistivity log as an aid in determining some reservoir characteristics. Petroleum Technology 1, 55–67.

Aspinall, W.P., Loughlin, S.C., Michael, F.V., Miller, A.D., Norton, G.E., Rowley, K.C., Sparks, R.S.J., Young, S.R., 2002. The Montserrat volcano observatory: its evolution, organization, role and activities. In: Druitt, T.H., Kookelaar, B.P. (Eds.), The Eruption of Soufrière Hills Volcano, Montserrat, from 1995–1999, Geological Society, vol. 21, pp. 71–91. Memoirs, London.

Aspinall, W., Carniel, R., Jaquet, O., Woo, G., Hincks, T., 2006. Using hidden multi-state Markov models with multi-parameter volcanic data to provide empirical evidence for alert level decision-support. Journal of Volcanology and Geothermal Research 153 (1–2), 112–124.

Azzaro, R., Branca, S., Gwinner, K., Coltelli, M., 2012. The volcano-tectonic map of Etna volcano, 1:100,000 scale: morphotectonic analysis from high-resolution DEM integrated with geologic, active faulting and seismotectonic data. Italian Journal of Geosciences 131 (1), 153–170.

Baker, J., 1975. Dragon system – Overview. IEEE Transactions on Acoustics, Speech, & Signal Processing AS23 (1), 24–29.

Barnhardt, C.L., 1957. The American College Dictionary. Random House, 1431 pp.

Barron, A., 1993. Universal approximation bounds for superposition of a sigmoidal function. IEEE Transactions on Information Theory 39, 930–945.

Baum, L., Sell, G., 1968. Growth transformations for functions on manifolds. Pacific Journal of Mathematics 27 (2), 211–227.

Bebbington, M., Lai, C.D., 1996. Statistical analysis of New Zealand volcanic occurrence data. Journal of Volcanology and Geothermal Research 74, 101–110.

Bebbington, M.S., 2007. Identifying volcanic regimes using hidden markov models. Geophysical Journal International 171 (2), 921–942.

Bellman, R.E., 1961. Adaptive Control Processes: A Guided Tour. Princeton University Press, New Jersey, 94 pp.

Benitez, M.C., Ramirez, J., Segura, J.C., Ibanez, J.M., Almendros, J., Garcia-Yeguas, A., Cortes, G., 2007. Continuous HMM-based seismic-event classification at Deception Island, Antarctica. IEEE Transactions on Geoscience and Remote Sensing 45 (1), 138–146.

Beyreuther, M., Wassermann, J., 2008. Continuous earthquake detection and classification using discrete hidden Markov models. Geophysical Journal International 175 (3), 1055–1066.

Bezdek, J.C., 1981. Pattern Recognition with Fuzzy Objective Functions. Plenum Press (Springer), New York, 256 pp. https://doi.org/10.1007/978-1-4757-045-1.

Bishop, M., Thompson, E., 1986. Maximum-likelihood alignment of DNA-sequences. Journal of Molecular Biology 190 (2), 159–165.

Bishop, C.M., 1995. Neural Networks for Pattern Recognition. Oxford University Press, Oxford, 482 pp.

Blandford, R., 1977. Discrimination between earthquakes and Underground Explosions. Annual Review of Earth and Planetary Sciences 5, 111–122.

Blaser, L., Ohrnberger, M., Riggelsen, C., Scherbaum, F., 2009. Bayesian Belief Network for Tsunami Warning Decision Support. In: Sossai, C., Chemello, G. (Eds.), ECSQARU 2009. Springer-Verlag, Berlin, pp. 757–768.

Blaser, L., Ohrnberger, M., Riggelsen, C., Babeyko, A., Scherbaum, F., 2011. Bayesian networks for tsunami early warning. Geophysical Journal International 185 (3), 1431–1443.

Blaser, L., Ohrnberger, M., Krüger, F., Scherbaum, F., 2012. Probabilistic tsunami threat assessment of 10 recent earthquakes offshore Sumatra. Geophysical Journal International 188 (3), 1273–1284.

Bonaccorso, A., Calvari, S., Coltelli, M., Del Negro, C., Falsaperla, S. (Eds.), 2004. Mt. Etna: Volcano Laboratory, Geophys. Monogr. Ser., vol. 143. AGU, Washington, D.C. https://doi.org/10.1029/GM143, 369 pp.

Boore, D.M., Atkinson, G., 2008. Ground-motion prediction equations for the average horizontal component of PGA, PGV, and 5%-damped PSA at spectral periods between 0.01 s and 10.0 s. Earthquake Spectra 24, 99–138.

Borradaile, G., 2003. Statistics of Earth Science Data. Their Distribution in Time, Space and Orientation. Spriner, Berlin-Heidelberg, 351 pp.

Bradley, J.V., 1968. Distribution-free Statistical Tests. Prentice-Hall, 388 pp.

Budetta, G., Carbone, D., 1995. Improvement and assessment of gravity networks on Etna, 2nd Interim Report. In: ETNATECH – the Shallow Magmatic Plumbing System at Etna: Method and Technique Development for Spatial and Temporal Evolution, pp. 14–21.

Bulow, R.C., Johnson, C.L., Bills, B.G., Shearer, P.M., 2007. Temporal and spatial properties of some deep moonquake clusters. Journal of Geophysical Research 112, E09003. https://doi.org/10.1029/2006JE002847.

Bussian, A.E., 1983. Electromagnetic conductance in porous media. Geophysics 48, 1258–1268.

Campbell, C., Ying, Y., 2011. Learning with Support Vector Machines. Morgan and Claypool Publishers, ISBN 978-1-60845-616-1, 95 pp.

Cannata, A., Montalto, P., Privitera, E., Russo, G., 2009. Characterization and location of infrasonic sources in active volcanoes: Mt. Etna, September–November 2007. Journal of Geophysical Research 114. https://doi.org/10.1029/2008JB006007.

Cannata, A., Montalto, P., Aliotta, M., Cassisi, C., Pulvirenti, A., Privitera, E., Patanè, D., 2011. Clustering and classification of infrasonic events at Mount Etna using pattern recognition techniques. Geophysical Journal International 185, 253–264. https://doi.org/10.1111/j.1365-246X.2011.04951.x.

Cannon, A.J., 2012. Köppen versus the computer: comparing Köppen-Geiger and multivariate regression tree climate in terms of climate homogeneity. Hydrology and Earth System Sciences 16, 217–229. https://doi.org/10.5194/hess-16-217-2012.

Carcione, J.M., Ursin, B., Nordskag, J.I., 2007. Cross-property relations between electric conductivity and seismic velocities of rocks. Geophysics 72, 193–204. https://doi.org/10.11901/1.2762224.

Cassisi, C., Ferro, A., Giugno, R., Pigola, G., Pulvirenti, A., 2013. Enhancing density-based clustering: parameter reduction and outlier detection. Information Systems 38 (3), 317–330.

Cassisi, C., 2013. Geophysical Time Series Data Mining (Ph.D. thesis). Università degli Studi di Catania, Catania, Italy.

Cesca, S., Şen, A.T., Dahm, T., 2014. Seismicity monitoring by cluster analysis of moment tensors. Geophysical Journal International 196, 1813–1826. https://doi.org/10.1093/gji/ggt492.

Charniak, E., 1991. Bayesian Networks without Tears: making Bayesian networks more accessible to the probabilistically unsophisticated. Journal AI Magazine 12 (4), 50–63.

Chen, D., Chen, W.E., 2013. Using the Köppen Classification to Quantify Climate Variation and Change: An Example for 1901-2010, vol. 6. Environmental Development, pp. 69–79. https://doi.org/10.1016/j.envdev.2013.03.007.

Chen, H., Lin, C., Schöllkopf, B., 2005. A tutorial on ν-support vector machines. Applied Stochastic Models in Business and Industry 25, 111–136. https://doi.org/10.1002/asmb.537.

Chiarabba, C., De Gori, P., Patanè, D., 2004. The Mt. Etna plumbing system: the contribution of seismic tomography. In: Bonaccorso, A., Calvari, S., Coltelli, M., Del Negro, C., Falsperla, S. (Eds.), Mt Etna Volcano American Geophysical Monograph, vol. 143, pp. 191–204.

Chouet, B., 1985. Excitation of a buried magmatic pipe: a seismic source model for volcanic tremor. Journal of Geophysical Research: Solid Earth 90 (B2), 1881–1893.

Cohen, J., 1960. A coefficient of agreement for nominal scales. Educational and Psychological Measurement 20, 37–46.

Cook, R.K., 1962. Strange Sounds in the Atmosphere, Part 1, pp. 12–16. Sound 1.

Corsaro, R.A., Cristofolini, R., Lo Giudice, A., Nunnari, G., Occhipinti, L., 1996. Recognition of Etnean succession units by artificial neural network. Acta Vulcanologica 8 (2), 41–46.

Corsaro, R.A., Falsaperla, S., Langer, H., 2013. Geochemical pattern classification of recent volcanic products from Mt. Etna, Italy, based on Kohonen maps and fuzzy clustering. International Journal of Earth Sciences 102, 1151–1164. https://doi.org/10.1007/s00531-012-0851-7.

Cover, T.M., 1965. Geometrical and statistical properties of systems of linear inequalities with applications in pattern recognition. IEEE Transactions on Electronic Computers 14 (3), 326–334.

Crampin, S., Fyfe, C.J., 1974. Automatic analysis of tape recordings from seismic networks. Geophysical Journal of the Royal Astronomical Society 39, 155–168.

Curilem, M., Vergara, J., San Martin, C., Fuentealba, G., Cardona, C., Huenupan, F., Chacón, M., Khan, S., Hussein, W., Becerra, N., 2014. Pattern recognition applied to seismic signals of the Llaima volcano (Chile): an analysis of the events' features. Journal of Volcanology and Geothermal Research 282, 134–177.

Cybenko, G., 1989. Approximation by superpositions of a sigmoidal function. In: Math of Control Signals & Systems, vol. 2. Springer, New York, pp. 303–314.

D'Agostino, M., Di Grazia, G., Ferrari, F., Langer, H., Messina, A., Reitano, D., Spampinato, S., 2013. In: Zobin, V. (Ed.), Volcano Monitoring and Early Warning on Mt Etna Based on Volcanic Tremor – Methods and Technical Aspects. Nova Publishers, New York, U.S.A.

Dainty, A.M., Toksöz, M.N., 1981. Seismic codas on the Earth and Moon: a comparison. Physics of the Earth and Planetary Interiors 26, 250–260.

Dammeier, F., Moore, F.J., Hammer, C., Haslinger, F., Loew, S., 2016. Earth Surface seismic data using hidden Markov models. Journal of Geophysical Research: Earth Surface 121, 351–371. https://doi.org/10.1002/2015JF003647.

Daubechie, I., 1991. The Lectures on Wavelets. SIAM, Philadelphia.

Davies, D.L., Bouldin, D.W., 1979. A cluster separation measure. IEEE Transactions on Pattern Analysis and Machine Learning 1 (2), 224–227.

Davis, J.C., 1986. Statistics and Data Analysis in Geology. John Wiley & Sons, New York, 646 pp.

De Candolle, A., 1874. Constitution dans le règne végétal des groupes physiologigues applicables à la geographie botanique ancienne et moderne. Archives des Sciences Physiques et Naturelles 50, 5–42.

Del Negro, C., Ferrucci, F., Napoli, R., 1997. The permanent network for magnetic surveillance of Mt Etna. Changes in the geomagnetic total intensity observed in 1995. Acta Vulcanologica 9, 1–7.

Dempster, A.P., Laird, N.M., Rubin, D.B., 1977. Maximum likelihood from incomplete data via the EM algorithm. Journal of the Royal Statistical Society 39, 509–512.

Derras, B., Bard, P.Y., Cotton, F., Bekkouche, A., 2012. Adapting the neural network approach tp PGA prediction: an example bases on the KiK-net data. Bulletin of the Seismological Society of America 102, 1446–1461. https://doi.org/10.1785/0120110088.

Di Grazia, G., Falsaperla, S., Langer, H., 2006. Volcanic tremor location during the 2004 Mt. Etna lava effusion. Geophysical Research Letters 33, L04304. https://doi.org/10.1029/2005GL025177.

Diehl, T., Kissling, E., 2000. Users guide for Consistent Phase picking at local to regional scales. In: ETH (Swiss Federal Institute of Technological Zurich), Lamont-Doherty Earth Observatory, Columbia, U.S.A, 36 pp.

Druitt, T.H., Kookelaar, B.P., 2002. The Eruption of Soufrie're Hills Volcano, Montserrat, from 1995–1999, Geological Society, vol. 21. Memoirs, London.

Duda, O.D., Hart, P.E., Stork, D.G., 2001. Pattern Classification. John Wiley and Sons, 654 pp.

Dunn, J.C., 1973. A fuzzy relative of the ISODATA process and its use in detecting compact well-separated clusters. Journal of Cybernetics 3, 32–57.

Durbin, R., Eddy, S.R., Krogh, A., Mitchison, G., 1999. Biological Sequence Analysis: Probabilistic Models of Proteins and Nucleic Acids. Cambridge University Press.

Ehrismann, W., Müller, G., Rosenbach, O., Sperlich, N., 1966. Topographic reduction of gravity measurements by the aid of digital computers. Boll Geofisica Teorica ed Applicata 8 (29), 3–20.

Evers, E.G., Haak, H.W., 2010. The characteristics of infrasound, its propagation and some early history. In: Le Pichon, A., Blanc, E., Hauchecorne, A. (Eds.), Infrasound Monitoring from Atmospheric Studies. Springer, pp. 3–28. https://doi.org/10.1007/978-1-4020-9508-5.

Falsaperla, S., Montalto, A., Spampinato, S., 1989. Analysis of seismic data concerning explosive sequences on Stromboli volcano in 1989. Boll. GNV 1, 249–258.

Falsaperla, S., Graziani, S., Nunnari, G., Spampinato, S., 1996. Automatic classification of volcanic earthquakes by using multi-layered neural networks. Natural Hazards 13, 205. https://doi.org/10.1007/BF00215816.

Falsaperla, S., Langer, H., Spampinato, S., 1998. Statistical analyses and characteristics of volcanic tremor on Stromboli volcano. Bulletin of Volcanology 60/2, 75–88.

Falsaperla, S., Alparone, S., D'Amico, S., Di Grazia, G., Ferrari, F., Langer, H., Sgroi, T., Spampinato, S., 2005. Volcanic tremor at Mt. Etna, Italy, preceding and accompanying the eruption of July–August, 2001. Pure and Applied Geophysics 162, 2111–2132. https://doi.org/10.1007/s00024–005-2710-y.

Falsaperla, S., Behncke, B., Langer, H., Neri, M., Salerno, G.G., Giammanco, S., Pecora, E., Biale, E., 2014. "Failed" eruptions revealed by pattern classification analysis of gas emission and volcanic tremor data at Mt. Etna, Italy. International Journal of Earth Sciences. https://doi.org/10.1007/s00531-013-0964-7.

Falsaperla, S., Neri, M., Di Grazia, G., Langer, H., Spampinato, S., 2017. What happens to in-soil Radon activity during a long-lasting eruption? Insights from Etna by a multidisciplinary data analysis. Geochemistry, Geophysics, Geosystems 18. https://doi.org/10.1002/2017GC006825.

Falsaperla, S., Neri, M., Di Grazia, G., Langer, H., Spampinato, S., 2018. Radon tells unexpected Tales of Mt Etna's unrest. EOS Transactions American Geophysical Union 99. https://doi.org/10.1029/2018EO094693.

Fawcett, T., 2006. An Introduction to ROC analysis. Pattern Recognition Letters 861–874.

Ferrick, M.G., Qamar, A., St Lawrence, W.F., 1982. Source mechanism of volcanic tremor. Journal of Geophysical Research: Solid Earth 87 (B10), 8675–8683.

Fisher, R.A., 1936. The use of multiple measurements in taxonomic problems. Annals of Eugenics 7, 179–188 reprinted in Contributions to Mathematical Statistics, John Wiley, New York (1950).

Fittermann, D.V., 1979. Theory of electrokinetic-magnetic anomalies in a faulted half-space. Journal of Geophysical Research 84, 6031–6040.

Fittermann, D.V., 1981. Correction to "Theory of electrokinetic-magnetic anomalies in a faulted half-space". Journal of Geophysical Research 86, 9585–9588.

Flack, V.F., Afifi, A.A., Lachenbruch, P.A., Schouten, H.J.A., 1988. Sample size determination from the Two Rater Kappa Statistic. Psychometrika 53, 321–325.

Fleiss, J.L., Cohen, J., Everitt, B.S., 1969. Large sample standard errors of kappa and weighted kappa. Psychological Bulletin 72/5, 323–327.

Foulger, G. (2018). http://www.foulgerconsulting.com/Services.html.

Freeman, J.A., Skapura, D.M., 1992. Neural Networks – Algorithms, Applications and Programming Techniques. Addison-Wesley Publishing Company, 401 pp.

Frohlich, C., Nakamura, Y., 2006. Possible extra-solar-system cause for certain lunar seismic events. Icarus 185, 21–28. https://doi.org/10.1016/j.icarus.2006.07.002.

Geman, S., Bienenstock, E., Doursat, R., 1992. Neural networks and the bias-variance dilemma. Neural Computation 4, 1–58.

Georges, T.M., Beasley, W.H., 1977. Infrasound refraction by winds. Journal of the Acoustical Society of America 61, 28–34.

Gold, T., Soter, S., 1970. Apollo 12 seismic signal: Indication of a deep layer of powder. Science 169, 1071–1075.

Goldberg, D.E., 1989. Genetic Algorithms in Search, Optimization, and Machine Learning. Addison-Wesley, Reading, MA.

Hamilton, E., 1978. Sound velocity-density relations in sea-floor sediments and rocks. Journal of the Acoustical Society of America 63, 366–377.

Hamilton, E., 1979. Vp/Vs and Poisson ratios in marine sediments and rocks. Journal of the Acoustical Society of America 66, 909–922.

Hammer, C., Beyreuther, M., Ohrnberger, M., 2012. A seismic event spotting system for volcano fast-response systems. Bulletin of the Seismological Society of America 102, 948–960. https://doi.org/10.1785/0120110167.

Hammer, C., Ohrnberger, M., Schlindwein, V., 2015. Pattern of cryospheric seismic events observed at Ekström ice shelf, Antarctica. Geophysical Research Letters 42 (10), 3936–3943. https://doi.org/10.1002/2015GL064029.

Hamming, R.W., 1983. Digital Filters, second ed. Prentice-Hall, Englewood Cliffs, New Jersey, pp. 257.

Han, J., Kamber, M., Pei, J., 2011. Data Mining – Concepts and Techniques. Morgan and Kaufmann Waltham, MA, U.S.A., 703 pp.

Hastie, T., Tibshirani, R., Friedman, J., 2002. The Elements of Statistical Learning. Springer, New York, 533 pp.

Heidbach, O., Custodio, S., Kingdon, A., Mariucci, M.T., Montone, P., Müller, B., Pierdominici, S., Rajabi, M., Reinecker, J., Reiter, K., Tingay, M., Williams, J., Ziegler, M., 2016. Stress Map of the Mediterranean and Central Europe 2016. GFZ Data Services. http://doi.org/10.5880/WSM.Europe2016.

Hinks, T., 2008. Probabilistic Volcanic Hazard and Risk Assessment (Ph.D. thesis). University of Bristol.

Hinneburg, A., Keim, D., 1998. An efficient approach to clustering large multimedia databases with noise. Proceedings of the 4th ACM-SIGKDD 58–65. New York.

Hochspringen, M.E., Kriegel, H.P., Sander, J., Xu, X., 1996. A density-based algorithm for discovering clusters in large spatial databases with noise. In: Simoudis, E., Han, J., Fayyad, U.M. (Eds.), Proceedings of the Second International Conference on Knowledge Discovery and Data Mining (KDD-96). AAAI Press, ISBN 1-57735-004-9, pp. 226–231.

Holland, J.H., 1975. Adaptation in Natural and Artificial Systems, second ed. University of Michigan Press, MIT Press. 1992.

Holland, J.H., 1992. Genetic algorithms. Scientific American 66–72.

Hornik, K., Stinchcombe, M., White, H., 1989. Multilayer feedforward networks are universal approximators. Neural Networks Computers 2, 359–366.

Hossin, M., Sulaiman, M.N., 2015. A review on evaluation metrics for data classification evaluations. International Journal of Data Mining and Knowledge Management Process 5, 1–11. https://doi.org/10.5121/ijdkp.2015.5201.

Hotelling, H., 1933. Analysis of a complex of statistical variables into principal components. Journal of Educational Psychology 24, 417-441+498-520.

Hsu, C.W., Chang, C.C., Lin, C.J., 2016. A Practical Guide to Support Vector Classification. National Taiwan University, Taipei 106, Taiwan. http://www.csie.ntu.edu.tw/~cjlin.

Huang, N.E., Wu, Z., 2008. A review on Hilbert-Huang transform: method and its application to geophysical studies. Review of Geophysics 46, RG2006. https://doi.org/10.1029/200/2007/RG00228.

Huang, X., Jack, M., Ariki, Y., 1990. Hidden Markov Models for Speech Recognition. Edinburgh University Press.

Huang, J., Zhu, Q., Yang, L., Cheng, D., Wu, Q., 2017. QCC: A Novel Clustering Algorithm Based on Quasi-Cluster Centers.

Hyvärien, A., Karhunen, J., Oja, E., 2001. Independent Component Analysis. John Wiley & Sons, New York, 481 pp.

Iten, K., 2016. http://www.iten-online.ch/klima/klimatabellen.htm.

Jaeger, J.C., Cook, N.G.W., 1979. Fundamentals of Rock Mechanics, third ed. Chapman and Hall, London, New York. 593 pp.

Jagric, T., Zunk, M., 2013. Neural network world: Optimized Spiral Spherical SOM (OSS-SOM). Neural Network World 5/13, 411–426.

Jelinek, F., Bahl, L., Mercer, R., 1975. Design of a linguistic statistical decoder for recognition of continuous speech. IEEE Transactions on Information Theory 21 (3), 250–256.

Jiang, H., 2010. Discriminative training of HMMs for automatic speech recognition: a survey. Computer Speech and Language 24 (4), 589–608.

Johnson, J.B., Ripepe, M., 2011. Volcano infrasound: a review. Journal of Volcanology and Geothermal Research 206, 61–69. https://doi.org/10.1016/j.jvolgeores.2011.06.006.

Joshi, J.C., Srivastava, S., 2014. A hidden markov model for avalanche forecasting on Chowkibal-Tangdhar road axis in Indian Himalayas. Journal of Earth System Science 123 (8), 1771–1779.

Joshi, J.C., Tankeshwar, K., Srivastava, S., 2017. Hidden Markov model for quantitative prediction of snowfall and analysis of hazardous snowfall events over Indian Himalaya. Journal of Earth System Science 126 (3), 33.

Kagan, Y.Y., 2007. Simplified algorithms for calculating double-couple rotation. Geophysical Journal International 171, 411–418. https://doi.org/10.1111/j.1365-246X.2007.03538.x.

Kanasewich, E.R., 1981. Time Sequence Analysis in Geophysics. University of Alberta, 480 pp.

Karush, W., 1939. Minima of Functions of Several Variables with Inequalities as Side Constraints (M.Sc. Dissertation). Dept. of Mathematics, University of Chicago, Chicago, Illinois.

Kaufmanm, L., Rousseeuw, P.J., 1990. Finding Groups in Data: An Introduction to Cluster Analysis. John Wiley & Sons, 368 pp.

Kirkpatrick, S., Gelett, C.D., Vecchi, P.M., 1983. Optimization by Simulated Annealing. Science 220, 621–630.

Knapmeyer, M., Weber, R.C., 2015. Seismicity and interior structure of the Moon. In: Tong, V., Garcia, R. (Eds.), Extraterrestrial Seismology. Cambridge Univ. Press, Cambridge, U. K.

Knapmeyer-Endrun, B., Hammer, C., 2015. Identification of new events in the Apollo lunar seismic data by Hidden Markov model—based event detection and classification. Journal of Geophysical Research — Planets 120 (10), 1620—1645. https://doi.org/10.1002/2015JE004862.

Köhler, A., Ohrnberger, M., Scherbaum, F., 2009. Unsupervised feature selection and general pattern discovery using self-organizing maps for gaining insights into the nature of seismic wavefields. Computational Geosciences 35 (9), 1757—1767.

Kohonen, T., 1984. Self-organization and Associative memory, Springer Series in Information Sciences, first ed., vol. 8. Springer-Verlag, New York.

Kohonen, T., 2001. Self-Organizing Maps, third ed. Springer, Berlin. 501 pp.

Köppen, W., 1900. Versuch einer Klassifikation der Klimate, vorzugsweise nach ihren Beziehungen zur Pflanzenwelt. Geographische Zeitschrift 6, 593—611.

Köppen, W., 1936. Das geografische System der Klimate. In: Köppen, W., Geiger, R. (Eds.), Handbuch der Klimatologie, vol. I part C 46 pp., Bornträger, Berlin.

Koski, A., 1996. Modelling ECG signals with hidden Markov models. Artificial Intelligence in Medicine 8 (5), 453—471.

Kottek, M., Grieser, J., Beck, C., Rudolf, B., Rubel, F., 2006. World map of the Köppen-Geiger climate classification updated. Meteorologische Zeitschrift 15, 259—263.

Kreyszig, E., 1982. Statistische Methoden und ihre Anwendungen. 7. Auflage. Vandenhoeck & Ruprecht, Göttingen (Germany), 451 pp.

Krzanowski, W.J., 1988. Principles of Multivariate Analysis — A User's Perspective. Oxford Science Publications, Oxford (U.K.), 563 pp.

Kuehn, N.M., Riggelsen, C., Scherbaum, F., 2011. Modeling the joint probability of earthquake, site, and ground-motion parameters using Bayesian networks. Bulletin of the Seismological Society of America 101, 235—249.

Kuhn, H.W., Tucker, A.W., 1951. Nonlinear programming. In: Proceedings of 2nd Berkeley Symposium. University of California Press, Berkeley, pp. 481—492. MR 0047303.

Kumazawa, M., Imanishi, Y., Fukao, Y., Furumoto, M., Yamamoto, A., 1990. A theory of spectral analysis based on the characteristic property of a linear dynamic system. Geophysical Journal International 101, 613—630.

Lacassie, J.P., Ruiz del Solar, J., Roser, B., Hervé, F., 2006. Visualization of volcanic rock geochemical data and classification with artificial neural networks. Mathematical Geology 38 (6), 697—710.

Lafferty, S., McCallum, A., Pereira, F.C.N., 2001. Conditional random fields: probabilistic models for segmenting and labeling sequence data. In: Proceedings of the 18th International Conference on Machine Learning 2001 (ICML 2001). Morgan Kaufmann Publishers Inc., San Francisco, CA, U.S.A., pp. 282—289

Lammlein, D.R., Latham, G.V., Dorman, J., Nakamura, Y., Ewing, M., 1974. Lunar seismicity, structure and tectonics. Review of Geophysics 12, 1—21.

Lammlein, D.R., 1977. Lunar seismicity and tectonics. Physics of the Earth and Planetary Interiors 14, 224—273.

Langer, H., Nunnari, G., Occhipinti, L., 1996. Estimation of seismic waveform governing parameters with neural networks. Journal of Geophysics Research 101, 20109—20118.

Langer, H., Falsaperla, S., Powell, T., Thompson, G., 2006. Automatic classification and a-posteriori analysis of seismic event identification at Soufriere Hills Volcano, Montserrat. Journal of Volcanology and Geothermal Research 153 (1—2), 357—369.

Langer, H., Falsaperla, S., Masotti, M., Campanini, R., Spampinato, S., Messina, A., 2009. Synopsis of supervised and unsupervised automatic classification techniques applied to volcanic tremor data at Mt Etna, Italy. Geophysical Journal International 178 (2), 1132—1144. https://doi.org/10.1111/j.1365-246X.2009.04179.x.

Langer, H., Falsaperla, S., Messina, A., Spampinato, S., Behncke, B., 2011. Detecting imminent eruptive activity at Mt. Etna, Italy, in 2007–2008 through pattern classification of volcanic tremor data. Journal of Volcanology and Geothermal Research 200, 1–17. https://doi.org/10.1016/j.jvolgeores. 2010.11.019.

Langer, H., Tusa, G., Scarfì, L., Azzaro, R., 2016. Ground-motion scenarios on Mt. Etna inferred from empirical relations and synthetic simulations. Bulletin of Earthquake Engineering 14, 1917–1943.

Laws, K., 1980. Rapid texture identification. In: Image Processing for Missile Guidance, vol. 238. SPIE, pp. 376–380.

Le Maitre, R.W., 2002. Igneous rocks: a classification and glossary of terms. In: Recommendations of the IUGS Subcommission on the Systematics of Igneous Rocks, second ed. Cambridge University Press, Cambridge, p. 236.

Lighthill, M.J., 1978. Waves in Fluids. Cambridge University Press, New York, 504 pp.

Lognonné, P., Johnson, C., 2007. Planetary seismology. In: Schubert, G. (Ed.), Treatise on Geophysics Volume 10: Planets and Moons. Elsevier, Oxford, UK, pp. 65–120.

Lognonné, P., 2005. Planetary seismology. Annual Review of Earth and Planetary Sciences 33, 571–604. https://doi.org/10.1109/CVPR.2003.1211373.

Mahalanobis, P.C., 1925. Analysis of race mixture in Bengal. Journal of the Asiatic Society of Bengal (New Series) 23, 301–330.

Malone, S., Endo, E., Weaver, C., Ramey, J., 1981. Seismic monitoring for eruption prediction. In: Lipman, R.W., Mullineaux, D.R. (Eds.), The 1980 Eruptions of Mount St. Helens. Washington, U.S. Geological Survey Professional Paper 1250.

Mandelbrot, B., 1983. The Fractal Geometry of Nature. Freeman, San Francisco, 480 pp.

Marshall, P.D., Basham, P.W., 1972. Discrimination between earthquakes and underground explosions employing an improved Ms scale. Geophysical Journal of the Royal Astronomical Society 28, 431–458.

Marzocchi, W., Sandri, L., Selva, J., 2008. BET EF: a probabilistic tool for long- and short-term eruption forecasting. Bulletin of Volcanology 70 (5), 623–632.

Mattera, D., Palmieri, F., Haykin, S., 1999. An explicite algorithm for training support vector machines. IEEE Signal Processing Letters 6, 243–246.

McHugh, M., 2012. Interrater reliability; the kappa statistics. Biochemia Medica (Zagreb) 22, 276–282.

McNutt, S.R., Roman, D.C., 2015. Volcanic seismicity. In: Sigurdsson, H., Houghton, B., McNutt, S., Rymer, H., Stix, J. (Eds.), The Encyclopedia of Volcanoes, second ed. Academic Press, San Diego, Calif, pp. 1011–1034 https://doi.org/10.1016/B978-0-12-385938-9.00059-6.

McNutt, S.R., 1992. Volcanic tremor. In: Encyclopedia of Earth System Science, vol. 4. Academic Press, San Diego, pp. 417–425.

McNutt, S.R., 1996. Seismic monitoring and eruption forecasting of volcanoes: a review of the state-of-the-art and case histories. In: Scarpa, R., Tilling, R. (Eds.), Monitoring and Mitigation of Volcano Hazards. Springer-Verlag, Berlin/Heidelberg, pp. 99–146.

McNutt, S.R., 2000. Volcanic seismicity. In: Sigurdsson, H., Houghton, B., Rymer, H., Stix, J., McNutt, S. (Eds.), Encyclopedia of Volcanoes. Academic, San Diego, Calif, pp. 1095–1119.

Meilă, M., 2007. Comparing clusterings – an information based distance. Journal of Multivariate Analysis 98, 873–895. https://doi.org/10.1016/j.jmva.2006.11.013.

Messina, A., Langer, H., 2011. Pattern recognition of volcanic tremor data on Mt. Etna (Italy) with KKAnalysis—a software program for unsupervised classification. Computers and Geosciences. https://doi.org/10.1016/j.cageo.2011.03.015.

Metz, C.E., 1978. Basic principles of ROC analysis. Seminars in Nuclear Medicine 8, 283–298.

Meyer, Y., 1993. Wavelets, Algorithms and Applications. SIAM, Philadelphia.

Miller, A.D., Foulger, G.R., Julian, B.R., 1998a. Non-double couple earthquakes 2. Observations. Reviews of Geophysics 36, 551–568. https://doi.org/10.1029/98RG00717.

Miller, A., Stewart, R., White, R., Luckett, R., Baptie, B., Aspinall, W., Latchman, J., Lynch, L., Voight, B., 1998b. Seismicity associated with dome growth and collapse at the Soufrière Hills Volcano, Montserrat. Geophysical Research Letters 25 (18), 3401–3404.

Minshull, T.A., Goulty, N.R., 1988. The influence of tidal stresses on deep moonquake activity. Physics of the Earth and Planetary Interiors 52, 41–55.

Minsky, M.L., Papert, S.A., 1969. Perceptrons. MIT Press, Cambridge, MA., U.S.A (1990).

Moran, S.C., Matoza, R.S., Garcés, M.A., Hedlin, M.A.H., Bowers, D., Scott, W.E., Sherrod, D.R., Vallance, J.W., 2008. Seismic and acoustic recordings of an unusually large rockfall at Mount St. Helens, Washington. Geophysical Research Letters 35 (L19302).

Morrissey, M.M., Chouet, B.A., 1997. Burst conditions of explosive volcanic eruptions recorded on microbaragraphs. Science 275 (5304), 1290–1293.

Mostaccio, A., Tuvè, T., Patanè, D., Barberi, G., Zuccarello, L., 2013. Improving seismic surveillance at Mt. Etna volcano by probabilistic earthquake location in a 3D model. Bulletin of the Seismological Society of America 103/4, 2447–2459. https://doi.org/10.1785/0120110202.

Mulargia, F., Gasperini, P., Tinti, S., 1987. Identifying different regimes in eruptive activity – an application to Etna volcano. Journal of Volcanology and Geothermal Research 34 (1–2), 89–106.

Murakami, H., 1989. Geomagnetic fields produced by electrokinetic sources. Journal of Geomagnetism and Geoelectricity 41, 221–247.

Murphy, K.P., 2002. Dynamic Bayesian Networks: Representation, Inference and Learning (Ph.D. thesis). AAI3082340.

Nakamura, Y., Dorman, J., Duennebier, F., Ewing, M., Lammlein, D., Latham, G., 1974. High-frequency Lunar Teleseismic Events. In: Proc. 5th Lunar Science Conference, vol. 3. Pergamon Press, New York, pp. 2883–2890.

Nakamura, Y., Latham, G.V., Dorman, H.J., 1982. Apollo lunar seismic experiment-Final Summary. Journal of Geophysical Research 87 (S01), A117–A123. https://doi.org/10.1029/JB087iS01A117.

Nakamura, Y., 1977. HFT events: shallow moonquakes? Physics of the Earth and Planetary Interiors 14, 217–223.

Nakamura, Y., 1978. A1 moonquakes: source distribution and mechanism. In: Proceedings of the 9th Lunar and Planetary Science Conference. Pergamon Press, New York, pp. 3589–3607.

Nakamura, Y., 2003. New identification of deep moonquakes in the Apollo lunar seismic data. Physics of the Earth and Planetary Interiors 139, 197–205. https://doi.org/10.1016/j.pepi.2003.07.017.

Nakamura, Y., 2005. Farside deep moonquakes and deep interior of the Moon. Journal of Geophysical Research 110, E01001. https://doi.org/10.1029/2004JE002332.

Nefian, A., Hayes, M.H.I., 1998. Hidden Markov models for face recognition. In: Proceedings of the 1998 IEEE International Conference on Acoustics, Speech and Signal Processing, vol. 5, pp. 2721–2724.

Neidel, N., Tarner, T., 1971. Semblance and other coherency measures for multichannel data. Geophysics 36, 482–497.

Neri, M., Mazzarini, F., Tarquini, S., Bisson, M., Isola, I., Behncke, B., Pareschi, T., 2008. The changing face of Mount Etna's summit area documented with Lidar technology. Geophysical Research Letters 35, L09305. https://doi.org/10.1029/2008GL033740.

Neuberg, J., Luckett, R., Baptie, B., Olsen, K., 2000. Models of tremor and low-frequency earthquake swarms on Montserrat. Journal of Volcanology and Geothermal Research 101 (1–2), 83–104.

Neural Networks Research Centre, Helsinki University of Technology, 1997. World Poverty Map. http://www.cis.hut.fi. http://www.cis.hut.fi/research/som-research/worldmap.html.

Nishio, H., 2005. Spherical SOM for non-euclidean Metric Space (Ph.D. thesis). Nara Institute of Science. NAIST-IS-DD, 51 pp.

Nunnari, G., Bertucco, L., Ferrucci, F., 2001. A neural approach to the integrated inversion of geophysical data of different types. IEEE Transactions on Geoscience and Remote Sensing 39, 736–748.

Oberst, J., Nakamura, Y., 1991. A search for clustering among the meteroid impacts detected by the Apollo lunar seismic network. Icarus 91, 315–325.

Oberst, J., 1987. Unusually high stress drops associated with shallow moonquakes. Journal of Geophysical Research 92, 1397–1405.

Ohrnberger, M., 2001. Continuous Automatic Classification of Seismic Signals of Volcanic Origin at Mt. Merapi, Java, Indonesia (Ph.D. thesis). University of Potsdam.

Okada, Y., 1985. Surface deformation due to shear and tensile faults in a half-space. Bulletin of the Seismological Society of America 75, 1135−1154.

Oppenheim, A.V., Schafer, R.W., 1999. Discrete-Time Signal Processing, second ed. Prentice Hall. 870 pp.

Pachter, L., Sturmfels, B., 2005. Algebraic Statistics for Computational Biology. Cambridge University Press.

Patané, D., De Gori, P., Chiarabba, C., Bonaccorso, A., 2003. Magma ascent and the pressurization of Mt. Etna's volcanic system. Science 299, 2061−2063.

Peel, M.C., McMahon, T., Finlaysin, B.L., 2007. Updated world map of the Köppen-Geiger climate classification. Journal of Hydrology and Earth System Sciences 11, 1633−1644. https://doi.org/10.5194/hess-11-1633-2007.

Peterson, J., Hutt, C.R., 2014. World-Wide Standardized Seismograph Network—A Data Users Guide. U.S. Geological Survey Open-File Report 2014−1218, 74 pp. https://doi.org/10.3133/ofr20141218.

Platt, J., 1999. Fast training of Support Vector Machines using sequential minimum optimization. In: Schöllkopf, B., Burges, J.C., Smola, A.J. (Eds.), Advances in Kernel Methods: Support Vector Learning. MIR Press, MA, U.S.A, pp. 185−208.

Pondrelli, S., Salimbeni, S., Ekström, G., Morelli, A., Gasperini, P., Vannucci, G., 2006. The Italian CMT dataset from 1977 to the present. Physics of the Earth and Planetary Interiors 286−303. https://doi.org/10.1016/j.pepi.2006.07.008,159/3-4.

Prentice, K.C., 1990. Bioclimatic distribution of vegetation for general circulation model studies. Journal of Geophysical Research 95/D8, 11811−11830.

Rabiner, L., 1989. A tutorial in hidden Markov-models and selected applications in speech recognition. Proceedings of the IEEE 77 (2), 257−286.

Raymer, L.L., Hunt, E.R., Gardner, J.S., 1980. An Improved Transit Time to Porosity Transform. 21st Annual Logging Symposium. Transactions of the Society of Professional Well Log Analysis, Paper 456.

Riggelsen, C., Ohrnberger, M., 2014. A machine learning approach for improving the detection capabilities at 3C seismic stations. Pure and Applied Geophysics 171, 395.

Riggelsen, C., Ohrnberger, M., Scherbaum, F., 2007. Dynamic Bayesian networks for real-time classification of seismic signals. In: Kok, J., Koronacki, J., Lopez de Mantaras, R., Matwin, S., Mladenic, D., Skowron, A. (Eds.), Knowledge Discovery in Databases: PKDD 2007, Lecture Notes in Computer Science, vol. 4702. Springer, Berlin/Heidelberg, pp. 565−572.

Riggelsen, C., Ohrnberger, M., Scherbaum, F., 2007b. In: Kok, J.N., et al. (Eds.), Dynamic Bayesian Networks for Real-Time Classification of Seismic Signals. Springer-Verlag Berlin Heidelberg, pp. 565−572. PKDD 2007, LNAI 4702, 2007.

Riggelsen, C., 2008. Learning Bayesian networks: a MAP criterion for joint selection of model structure and parameter. In: Eighth IEEE International Conference on Data Mining, 15−19 December 2008, Pisa, Italy, pp. 522−529.

Ritter, H., 1999. Self-organizing maps on non-Euclidean spaces. In: Oja, E., Kaski, S. (Eds.), Kohonen Maps. Elsevier, Amsterdam, pp. 97−110.

Rosenblatt, F., 1958. The Perceptron: a probabilistic model for information storage and organization in the brain. Cornell Aeronautical Laboratory, Psychological Review 65, 386−408.

Röth, G., Tarantola, A., 1994. Neural network and inversion of seismic data. Journal of Geophysical Research 99, 6753−6768.

Runge, J., Petoukhov, V., Kurths, J., 2014. Quantifying the Strength and Delay of climatic Interactions: the Ambiguities of Cross correlation and a novel measure based on graphical models. Journal of Climate 27 (2), 720−739.

Russell, S., Binder, J., Koller, D., Kanazawa, K., 1995. Local learning in probabilistic networks with hidden variables. In: Proc. 1995 Joint Int. Conf. Artificial Intelligence (IJCA'95), 1146,1152. Montreal, Quebec, Canada.

Sabetta, F., Pugliese, A., 1987. Attenuation of peak horizontal acceleration and velocity from Italian strong-motion records. Bulletin of the Seismological Society of America 77, 1491–1513.

Sander, J., Ester, M., Kriegel, H.P., Xu, X., 1998. Density-based clustering in spatial databases: the algorithm GDBSCAN and its applications. In: Data Mining and Knowledge Discovery. 2. Auflage. Band 2. Springer, Berlin. https://doi.org/10.1023/A:1009745219419.

Sanderson, M., 1999. The classification of climates from Pythagoras to Koeppen. Bulletin of the American Meteorological Society 80/4, 669–673.

Sato, K., Sakakibara, Y., 2005. RNA secondary structural alignment with conditional random fields. Bioinformatics 21 (2), 237–242.

Scarfi, L., Langer, H., Garcia-Fernandez, M., Jimenez, M.J., 2016. Path effects and local site amplifications: two case studies on Mt Etna (Italy) and Vega Baja (Spain). Bulletin of Earthquake Engineering 14, 2117–2127. https://doi.org/10.1007/s10518-016-9883-x.

Schick, R., 1988. Volcanic tremor-source mechanisms and correlation with eruptive activity. Natural Hazards 125–144.

Schöllkopf, B., Smola, A., 2002. Support Vector Machines, Regularization, Optimization and beyond. MIT Pres, 626 pp.

Schön, J., 1983. Petrophysik. Enke Verlag, Stuttgart (Germany), 405 pp.

Sen, M.K., Stoffa, P.L., 1991. Nonlinear one-dimensional seismic waveform inversion using simulated annealing. Geophysics 56, 1624–1638.

Serpelloni, E., Vannucci, G., Pondrelli, S., Argnani, A., Casula, G., Anzidei, M., Baldi, P., Gasperini, P., 2007. Kinematics of the Western Africa-Eurasia plate boundary from focal mechanisms and GPS data. Geophysical Journal International 169 (3), 1180–1200. https://doi.org/10.1111/j.1365-246X.2007.03367.x.

Settles, B., 2005. Abner: an open source tool for automatically tagging genes, proteins and other entity names in text. Bioinformatics 21 (14), 3191–3192.

Shapiro, L., Stockman, G., 2000. Computer Vision. http://nana.lecturer.pens.ac.id/.

Skorohod, B.A., 2017. Diffuse Algorithms for Neural and Neuro-Fuzzy Systems : With Application on Control Engineering and Signal Processing. Elsevier, 220 pp.

Smola, A.J., Schöllkopf, B., 2004. A tutorial on support vector regression. Statistics and Computing 14, 199-222.

Spampinato, S., Langer, H., Messina, A., Falsaperla, S., 2019. Short-term detection of volcanic unrest at Mt. Etna by means of a multi-station warning system. Scientific Reports. https://doi.org/10.1038/s41598-019-42930-3.

Späth, H., 1983. Cluster-Formation und Analyse, Theorie, FORTRAN Programme. Beispiele, Oldenbourg, München.

Spichak, V., Goidina, A.G., 2016. Neural network estimate of seismic velocities and resistivity of rocks from electromagnetic and seismic sounding data. Izvestiya – Physics of the Solid Earth 52, 371–381. https://doi.org/10.1134/S1069351316030125.

Stein, B., Meyer zu Eissen, S., Wißbrock, F., 2003. In: Hanza, M.H. (Ed.), 3rd IASTED Int. Conference on Artificial Intelligence and Applications (AIA 03), ISBN 0-88986-390-3, pp. 216–221. Benalmádena, Spain.

Stewart, S.W., 1977. Real time detection and location of local seismic events in Central California. Bulletin of the Seismological Society of America 67, 433–452.

Streckeisen, A.L., 1974. Classification and nomenclature of plutonic rocks. Recommendations of the IUGS subcommission on the Systematics of Igneous Rocks. Geologische Rundschau. Internationale Zeitschrift für Geologie. Stuttgart 63, 773–785.

Streckeisen, A.L., 1978. IUGS subcommission on the Systematics of Igneous Rocks. Classification and nomenclature of volcanic rocks, lamprophyres, carbonatites and melilite rocks. Recommendations and suggestions. Neues Jahrbuch für Mineralogie – Abhandlungen 141, 1–14.

Sutton, C., McCallum, A., 2007. An Introduction to Conditional Random Fields for Relational Learning. MIT Press. Chapter: Introduction to Statistical Relational Learning.

Symons, G.J., 1888. The Eruption of Krakatoa and Subsequent Phenomena. Trübner, London.
Theodoridis, S., Koutroumbas, K., 2009. Pattern Recognition, fourth ed. Academic Press, Burlington San Diego, U.S.A. 961 pp.
Theodoridis, S., Pikrakis, A., Koutroumbas, K., Cavouras, D., 2010. Introduction to Pattern Recognition – A MATLAB™ Approach. Academic Press, Burlington, San Diego, U.S.A, 219 pp.
Tibshirani, R., Walther, G., Hastie, T., 2001. Estimating the number of clusters in a data set via the gap statistics. Journal of the Royal Statistical Society 63, 411–423.
Tokarev, P.I., 1983. The Great Tolbachik Fissure Eruption: Geological and Geophysical Data 1975–1976. Cambridge University Press.
Tusa, G., Langer, H., 2016. Prediction of ground motion parameters for the volcanic area of Mount Etna. Journal of Seismology 20, 1–42.
Vapnik, V.N., Kotz, S., 2006. Estimation of Dependences Based on Empirical Data. Springer. ISBN 0-387-30865-2, 510 pp.
Vapnik, V.N., 1999. The Nature of Statistical Learning Theory, second ed. Springer-Verlag, ISBN 0-387-98780-0. 319 pp.
Varley, N., Johnson, J., Ruiz, M., Reyes, G., Martin, K., 2006. Statistics in volcanology. In: Mader, H., Connor, C., Coles, S. (Eds.), Applying Statistical Analysis to Understand the Dynamics of Volcanic Explosions. Geological Society of London.
Verbeek, R.D.M., 1885. Krakatau (Uitgegeven Op Last Van Zijne Excellentie Den Gouverneur-Generaal Van Nederlandsch-Indië). Landsdrukkerij, Batavia.
Vesanto, J., Himberg, J., Alhoniemi, E., Parhankangas, J., 2000. SOM Toolbox for Matlab 5. Report A57. http://www.cis.hut.fi/projects/somtoolbox.
Viera, A.J., Garrett, J.M., 2005. Understanding interobserver agreement. The kappa statistics. Family Medicine 37, 360–363.
Vogel, K., Riggelsen, C., Korup, O., Scherbaum, F., 2014. Bayesian network learning for natural hazard analyses. Natural Hazards and Earth System Sciences 14, 2605–2626.
Weber, R.C., Bills, B.G., Johnson, C.L., 2010. A simple physical model for deep moonquake occurrence times. Physics of the Earth and Planetary Interiors 182, 152–160. https://doi.org/10.1016/j.pepi.2010.07.009.
Weisstein, E.W., 2018. Laplace Distribution. From MathWorld – A Wolfram Web Resource. http://mathworld.wolfram.com/LaplaceDistribution.html.
Werbos, P.J., 1974. Beyond Regression: New Tools for Prediction and Analysis in Behavioural Sciences (Master thesis). Harvard University.
Wickman, F., 1966. Repose period patterns of volcanoes. I. Volcanic eruptions regarded as random phenomena. Arkiv för kemi, mineralogi och geologi 4 (4), 291–301.
Wu, Y., Takatsuka, M., 2006. Spherical self-organizing map using efficient indexed geodetic data structure. Neural Networks 19, 901–910.
Young, S., Sparks, S., Robertson, R., Lynch, L., Aspinall, W., 1997. Eruption of Soufriere Hills volcano in Montserrat continues. Eos 78 (38), 401–416.
Zadeh, L.A., 1965. Fuzzy sets. Information and Control 8, 338–353.
Zerenner, T., Friederichs, P., Lehnertz, K., Hense, A., 2014. A Gaussian graphical model approach to climate networks. Chaos: An Interdisciplinary Journal of Nonlinear Science 24 (2), 023103.
Zscheischler, J., Mahecha, M.D., Harmeling, S., 2012. Climate classifications: the value of unsupervised clustering. Procedia Computer Science 9, 897–906. https://doi.org/10.1016/j.procs.2012.04.096.

Index

'*Note*: Page numbers followed by "f" indicate figures and "t" indicate tables.'

A

Accuracy, 51–52, 127, 134–135, 156–158, 162, 173–174, 179–180, 185, 198, 242–243, 304
Activation function, 40–41, 41f, 46, 48–49, 49f, 51, 52f, 56, 76–77, 267–268, 300
Adaptive distance criterion, 97, 216f
Aeolian Archipelago, 132
Agglomerative clustering, 106–107, 280, 281f
Agglomerative scheme, 108–109
AIC. *See* Akaike criterion (AIC)
Akaike criterion (AIC), 162
Angular data, 222, 298
ANN. *See* Artificial neural network (ANN)
ANOVA, 91–92, 119–120, 119t, 190, 193–194
Apollo, 176–179
A-posteriori analysis, 136, 238–239, 255
A-priori information, 26, 33, 127, 141, 182, 189, 251
Area under the curve (AUC), 241–242
Arenal volcano, 129
Artificial neural network (ANN), 40–41, 257–258, 267, 267f
AUC. *See* Area under the curve (AUC)
Autocorrelation function (ACF), 128, 134, 135f, 142, 246–247

B

Backpropagation, 51–52, 76–78
Back propagation neural network (BPNN), 134–135, 300–302, 300f–302f, 310–311
Barron, 249–250, 254–255
Baum-Welch algorithm, 168
Bayes, 65
Bayesian clustering, 278
Bayesian event tree (BET), 181
Bayesian networks (BN), 34, 69–73, 128, 179, 251, 258–259
Beach ball, 225f, 228–229, 228f–229f, 311–312
Beaufort scale, 4
Best matching unit (BMU), 117–119, 118f, 121, 123, 123f, 211–212, 216f, 227–229, 284–286, 288–289, 297–298
BET. *See* Bayesian event tree (BET)
Bias, 5, 42, 47f, 48, 49f, 127–187, 138–139, 240–241
Binary classification, 240
Binomial distribution, 137–139
Bivariate Gaussian, 10–11, 11f, 29f
Black box, 127–128, 159, 166–167
BMU. *See* Best matching unit (BMU)
BN. *See* Bayesian networks (BN)
Body wave magnitude. *See* m_b
Bootstrap, 138–139, 205–206
BPNN. *See* Back propagation neural network (BPNN)
BWDoHMMsc, 273–274
BWDoHMMst, 273

C

Categorical data, 4, 88–89, 258–259
Cauchy probability distribution, 29
C-CVC, 305
Cell structure, 274
Centroid, 25, 28, 29f, 37, 63, 98, 102–103, 109, 189–190, 210–211, 226–227, 240–241, 256–257, 256f
City block distance. *See* Manhattan distance
C-language, 267–268, 275
Climate classification, 203–204
Climate zones, 203–213
Clustering, 4, 49, 90–114, 189–190, 194, 197, 200, 205–208, 211, 226, 230, 237, 253, 259, 275, 279, 285
Clustering quality, 232, 255
CLVD. *See* Compensated linear vector dipole (CLVD)
Color code, 117, 119, 206–207, 211–212, 229, 284–285, 299
Compensated linear vector dipole (CLVD), 226
Complete link, 105–108, 280

Comprehensive Nuclear Test-Ban Treaty Organization (CTBTO), 139
Compressional wave, 7, 36, 222
Confusion matrix, 136, 137t, 144t, 150t, 239–240, 244–245
Convolution matrix, 9
Co-occurrence matrix, 8
Cosine similarity measure, 89
Cost function, 22, 42, 45, 55, 78–79, 101–102
Covariance matrix, 13, 28, 37, 43, 44f, 84, 96–97, 122, 193t, 265, 266f, 289, 297
Cover's theorem, 56–57
Cross validation, 26, 34, 137–139, 167, 239
CTBTO. *See* Comprehensive Nuclear Test-Ban Treaty Organization (CTBTO)
Cumulant, 12–13
Curse of dimensionality, 23–24, 254–255
Cybenko's theorem, 51, 151

D

DAG. *See* Directed acyclic graph (DAG)
Daubechies wavelet, 19, 19f
Davies Bouldin index. *See* DB index
DB clustering. *See* Density based clustering
DB index, 193–194, 194f, 197, 205, 230, 255
DBN. *See* Dynamic Bayesian neworks (DBN)
DBSCAN, 110–112, 113f
DBSCAN-STRATA, 112, 113f, 114, 200–203, 201f, 255–256
DC. *See* Double couple (DC)
DENCLUE, 112
Dendrogram, 104–107, 104f, 107f–108f, 281f
Density based clustering, 109–114, 197–203, 255–256, 280–282
Determinant criterion, 120–121, 196, 255
Determinant metric, 96
Dip, 156, 222, 311
Directed acyclic graph (DAG), 69–70, 73
Directional features, 222–230
Discriminant analysis, 35–39, 243f
Dispersion matrix, 91, 196, 299
Divisive scheme, 108–109
DMGA, 153, 307–311
Double couple (DC), 226, 256–257
Double stochastic process, 60–62, 168
Dunn index, 205, 231–232
Dynamic Bayesian neworks (DBN), 73, 179

E

Earthquake, 4, 6, 33–34, 36, 36f, 39, 44–45, 45f, 92–95, 97, 127–129, 151, 166f, 180–181, 191, 198, 201f, 226–227, 261, 263f
Eigenvalues, 11, 39, 91–92, 123, 195t, 261, 289, 298–299
Eigenvectors, 11, 28, 118–119, 195, 261, 266f, 289, 298
Ekström ice shelf, 174
Electric conductivity, 158
Electric resistivity, 128, 158–160
Electro-kinetic (EM) effect, 153
EM. *See* Expectation maximization (EM)
Emission matrix, 62, 68, 81, 83–84, 273–274
Emission probability, 62, 83–84, 168, 172
Empirical prediction of ground motion, 161–162
Energy, 9, 35–36, 131, 139, 213, 222–223
Entropy, 9, 233
Epsilon-SVR, 305
Euclidean distance, 25, 29f, 88, 92–95, 104–105, 121, 211, 226, 231–232
Exclusive OR (XOR), 40–41, 46, 46f, 48, 151
Expectation maximization (EM), 22, 73, 83, 168, 278–279
Explosion quake, 127, 131–133, 133f, 251

F

False alert, 252
False Negatives, 240
False positive rate, 240
False Positives, 240
Fast Fourier Transform, 215
F-distribution, 25–26, 31
Feature, 3–27, 33, 57, 59, 77, 87, 114–115, 134, 171–173, 178f, 189–191, 211, 222, 226, 237, 239, 253, 256–257
Feature vector, 3–5, 23, 33, 59, 77, 88, 92–95, 109, 121, 189–190, 192f, 215, 222, 226, 239, 254–257, 283–284, 300–301
Fisher's linear discriminant analysis (FLDA), 73–75
Flank eruption, 213–215
FLDA. *See* Fisher's linear discriminant analysis (FLDA)
Fourier transform, 10, 13–15, 250, 255
Fractal geometry, 21
Frequency bin, 191–192, 219f, 255
F-test, 119t
Fuzzy clustering, 98–104, 208, 210f, 217–218, 248f, 254f, 279–280
Fuzzy C-means, 102–103, 208, 279, 290–294, 299

G

Gap index, 232–233
Gap statistics, 232–233
Gaussian, 20, 27, 30–31, 48, 109–110, 169–170, 275
Gaussian mixture, 84, 109–110, 170–171

General Mapping Tools, 312
Genetic algorithm, 27
Geodesic domes, 115
Geology, 21, 160–161, 197–198
Geophysics, 4, 59, 127–128, 237
Glacial event, 131
GMDAS, 98, 101f
GPS, 152–153
Gravity field, 127–128, 153, 249
Great circles, 223
Ground deformation, 5, 127–128, 153–155, 157t

H

Haar function, 16
Haar transform, 16–18
Haar wavelets, 17f
Hidden layer, 46, 47f, 48, 51, 76, 134, 154–155, 162, 250, 300
Hidden Markov models (HMM), 34, 59–69, 128, 246–247
Hierarchical clustering, 104–109, 190, 280
Hist, 22
HMM. *See* Hidden Markov models (HMM)
Hotelling's *T*, 25
HTK, 275
Hybrid event, 130f, 131, 170f, 173, 246–247

I

ICA. *See* Independent component analysis (ICA)
Icequakes, 174–180, 175f
Icosahedron, 115–116, 115f
Image processing, 5–6, 8
Independent component analysis (ICA), 10, 12–13
Information entropy, 13
Infrasound, 5, 139–144, 245
Inversion, 127–128, 152f, 156, 160, 249–251, 258–259, 300
ISO, 226

J

Jaccard index, 89
Jack-knife, 27, 138–139

K

Kagan angle, 225–226
Kappa statistics, 237–238, 245
Karhunen Loève transformation, 173
Karush-Kuhn-Tucker (KKT), 185
Kernel, 56–59, 149t, 267–268
KKAnalysis, 117–118, 194, 212f, 231, 275, 282–299
KKT. *See* Karush-Kuhn-Tucker (KKT)
K-means, 49, 91–92, 95f–96f, 193–195, 195f, 208, 255, 275
Kohonen maps, 114
Köppen-Geiger classification. *See* Köppen-Geiger scheme
Köppen-Geiger scheme, 204, 130f
Köppen zones, 190–191, 206–207
Kullback-Leibler distance, 13
Kurtosis, 12–13, 30, 162

L

Lagrange function, 78–79
Lagrange multipliers, 54, 78–79, 120–121, 184
Lahar, 131
Landslide, 131–132
Laplace distribution, 148–151, 150f
Lava fountain, 213–215, 217–218, 221f
Learning rate, 117–118, 267, 301–302
Leave-one-out, 138–139, 215, 239
LIBSVM, 148, 150f
Likelihood, 48, 65, 75, 83, 174–176
Linear dichotomies, 56–57
Linear perceptron, 40
Listric geometry, 198–199
Logarithm, 21, 159–160, 297
Logistic, 21, 183–184, 297
Long period event, 129–131, 177, 246–247
Long time average (LTA), 6
LTA. *See* Long time average (LTA)
Lunar seismic network, 176–177

M

M0, 137t
Mahalanobis distance, 25, 29, 29f, 31, 75, 96–97, 196
Manhattan distance, 222, 258
Marginal distribution, 11, 37, 92, 261–265
Maximum flatness, 163
m_b, 36–37, 43, 92–96, 127, 240–241, 275, 276f
Mean, 6, 22, 27, 29, 31, 34, 83–85, 148–150, 172, 204, 232
Mercalli scale, 4
Metric, 13–14, 25, 88, 96–97, 114–115, 189–190, 197, 226, 231–232, 255
Mexican hat, 20
Meyer wavelet, 20, 20f
Minimum distance property, 120–121
Minkowski exponent, 92–95
Minkowski metric, 88
MLP. *See* Multilayer perceptron (MLP)
Moment magnitude, 224
Montserrat volcano, 169, 246–247
Moonquakes, 176–179
Morlet wavelet, 20
MRT. *See* Multivariate regression tree (MRT)
M_S, 35–36, 36f, 44–45, 45f, 92–95, 148, 264f
Mt Etna, 127, 129, 131, 142f, 152, 166f, 191, 200, 214–215
Mt S. Helens, 129, 180
Multiclass, 45, 146, 242–243, 257–258
Multidisciplinary analysis, 241–242, 253
Multilayer perceptron (MLP), 26, 46–52, 132–133, 257–258

Multivariate regression tree (MRT), 204
MultSeqTrainDoHMMBWsc, 274

N

Neumayer seismic network, 174
Neuron, 40, 41f, 46, 115, 154–155, 250
N-fold cross validation, 138, 239
Nodal plane, 223, 224f, 229, 256–257
Non-linear perceptron, 48–52
Nonnegativity, 25, 88
Non-uniqueness, 250
Normalization, 21–23, 92, 155–156, 183–184, 297
Nuclear explosion, 35–37, 43, 44f
Nuclear test, 35, 36f, 37, 44–45, 45f, 132–133, 240–241, 242f, 261, 263f, 276f
Numerical data, 189–190, 204
nu-SVC, 305
nu-SVR, 305

O

Object, 3, 33, 59, 90–91, 111–112, 114–115, 251–253
OPTICS, 111–112
Ordinal data, 3–4
Outlier, 22, 164f
Overfitting, 87, 134–135, 238, 268

P

Partial Nuclear Test-Ban Treaty (PTBT), 35
Partitioning clustering, 91–104, 190, 275
Pattern, 3–31, 33–34, 59, 62–63, 77, 87, 97–98, 105–106, 109–110, 114–115, 122, 123f, 134–135, 137, 146, 162, 168, 189–190, 205, 211, 215, 218, 227, 283–285, 289, 305
PCA. *See* Principal component analysis (PCA)
Percentile, 14–15, 22, 27, 218
Perceptron, 34, 40, 41f, 46, 52f, 56, 76, 267–268
Piezo-magnetic effect
Pinatubo, 129
Platonic polyhedra, 115
Poisson ratio, 157t
Pooled covariance matrix, 25, 39, 44–45, 74–75, 97, 265
Pooled variance, 31, 74
Precision, 7, 240
Principal component analysis (PCA), 10–11, 195, 261
Principal stress, 189–190, 223
Prototype, 63–65, 87–88, 114, 174–176, 190, 211–212
psmeca, 312–313
PTBT. *See* Partial Nuclear Test-Ban Treaty (PTBT)
P-wave, 7, 36, 159–160

Q

Quantization error (QE), 121, 123–124

R

Radial Basis Function (RBF), 57
Radiation pattern, 35–36, 222
Radon, 252
Rake, 311
Random confusion matrix, 244
Random fields, 68–69
Range, 9, 21, 28–29, 44–45, 92, 118–119, 167, 215
RBF. *See* Radial Basis Function (RBF)
Receiver operation curve (ROC), 237–238, 241f
Reflexivity, 25, 88
Regression, 28, 51, 128, 158–168, 249, 251, 302
Representation space, 114, 190, 228, 289
RGB, 208, 211–212, 228, 284–285
RMS. *See* Root mean square (RMS)
ROC. *See* Receiver operation curve (ROC)
Rockfall, 130f, 131, 169, 173–174, 246–247
Root mean square (RMS), 27, 191

S

SA. *See* Simulated annealing (SA)
Score of success, 239
Seismic moment, 129, 190–191
Seismic swarm, 129, 200
Seismic velocities, 128, 158–161, 161t
Self Organizing Maps (SOM), 87–88, 114–119, 189–190
Self-similar, 21
Semblance, 141
Shear wave, 7, 36, 160
Short time average (STA), 6
Short time Fourier transform (STFT), 10, 14–15, 218
Sigmoid function, 48, 51f
Silhouette index, 232
Similarity, 63–65, 189, 208, 228, 253
Simulated annealing (SA), 51–52, 153, 249
Single link, 105–107, 107f
Skewness, 12–13, 30
Slip vector, 222
SOM. *See* Self Organizing Maps (SOM)
Soufrière Hills, 169, 246
Source plane, 223
Spectrograms, 10, 14, 15f, 140, 213, 215
STA. *See* Short time average (STA)
Standard deviation, 14–15, 21, 25, 29f, 30, 274
Standardization, 21–23
Stereographic net, 223
STFT. *See* Short time Fourier transform (STFT)
Streckeisen diagram, 144, 146, 251, 302
Strike, 222, 223f, 311

Stromboli, 127, 131–132, 153, 190–191, 238
Student-t, 25, 31
Supervised classification, 26, 33, 132–133, 214–215
Supervised learning, 33–85, 87, 127–187, 189, 237
Support vector, 53–54, 78, 127, 166f
Support vector machines (SVM), 26, 34, 53–59, 258, 268
Surface wave magnitude. *See* M_S
SVM. *See* Support vector machines (SVM)
S-wave, 159–160, 169
Symmetry, 25, 88

T

Tanimoto distance, 4, 88–89, 90f
TAS. *See* Total alkali silica (TAS)
Test Ban Treaty, 35
Test set, 26, 44f, 134–135, 148, 156, 166, 268
Texture, 8–9
Tolbachik volcano, 180
Topological distance (TD), 121, 122f
Topological error, 121–122
Topological fidelity, 117–118, 190
Torus geometry, 228
Total alkali silica (TAS), 145–146
Training set, 26, 136f, 141, 168, 268
Transition matrix, 61, 83–84
Transition probability, 60–61, 168
Triangle inequality, 25, 88
True Negatives, 240
True positive rate, 240
True Positives, 240
Tsunami, 132
Tsunami early warning, 181–183

U

U-matrix, 287, 287f
Unrest, 129, 141, 196–197, 252
Unsupervised learning, 49, 87–124, 237, 275–299
Unzen, 129

V

Validation set, 138
Variance trade-off, 127–187
Variation of information (VI), 205
Vesuvius, 129
VI distance. *See* VI index
VI index, 205–206
VitDoHMMsc, 273
VitDoHMMst, 273
Viterbi algorithm, 65–66, 273
Volcanic tremor, 130f, 131, 191–197
Volcano-tectonic (VT) earthquake, 129, 169f, 246–247

W

Wald-Wolfowitz runs test, 254–255
Wavelet transforms, 10
Weight vector, 117–118
Winsorizing, 22
World Poverty Map, 116–117, 116f, 211–212
World Wide Standard Seismic Network (WWSSN), 35
WWSSN. *See* World Wide Standard Seismic Network (WWSSN)

X

XOR. *See* Exclusive OR (XOR)

9780128118429